Current Topics in Behavioral Neurosciences

Volume 45

Current Topics in Behavioral Neurosciences provides critical and comprehensive discussions of the most significant areas of behavioral neuroscience research, written by leading international authorities. Each volume in the series represents the most informative and contemporary account of its subject available, making it an unrivalled reference source. Each volume will be made available in both print and electronic form.

With the development of new methodologies for brain imaging, genetic and genomic analyses, molecular engineering of mutant animals, novel routes for drug delivery, and sophisticated cross-species behavioral assessments, it is now possible to study behavior relevant to psychiatric and neurological diseases and disorders on the physiological level. The Behavioral Neurosciences series focuses on translational medicine and cutting-edge technologies. Preclinical and clinical trials for the development of new diagnostics and therapeutics as well as prevention efforts are covered whenever possible. Special attention is also drawn on epigenetical aspects, especially in psychiatric disorders.

CTBN series is indexed in PubMed and Scopus.

More information about this series at http://www.springer.com/series/7854

Mohammed Shoaib • Tanya L. Wallace
Editors

Behavioral Pharmacology of the Cholinergic System

Springer

Editors
Mohammed Shoaib
Institute of Neuroscience
Newcastle University
Newcastle upon Tyne, UK

Tanya L. Wallace
BlackThorn Therapeutics
San Francisco, CA, USA

ISSN 1866-3370 ISSN 1866-3389 (electronic)
Current Topics in Behavioral Neurosciences
ISBN 978-3-030-56012-6 ISBN 978-3-030-56013-3 (eBook)
https://doi.org/10.1007/978-3-030-56013-3

This Springer imprint is published by the registered company Springer Nature Switzerland AG
The registered company address is: Gewerbestrasse 11, 6330 Cham, Switzerland

Preface

The focus of this CTBN volume aims to provide an overview of cholinergic systems in relation to their effects on behavior and disease. Our intent was to organize a broad series of reviews that summarizes the past and current knowledge of cholinergic systems with in-depth discussions on the potential opportunities for harnessing this information for therapeutic development across central nervous system (CNS) disorders. This volume contains a collection of multidisciplinary chapters authored by leading experts who have made significant contributions in the field.

Research on cholinergic neurotransmission has revealed the diversity of behaviors this system modulates, including cognitive processes, pain, dependence, and addiction, among others. Dysfunction of the cholinergic system underlies nicotine addiction and has been implicated in various neurological and psychiatric disorders including dementia, pain, and Alzheimer's disease. The molecular genetics of the cholinergic system including both muscarinic and nicotinic acetylcholine receptors, cholinesterases, acetylcholine synthesis and release have provided significant insights into potential targets for pharmacological intervention. Cholinergic drugs are being used and are being evaluated for the treatment of many CNS diseases. Thus, this volume aims to broaden our understanding of the current state of cholinergic mechanisms to enable the implementation of novel approaches for the development of more effective treatments.

Acetylcholine (ACh), identified as a key neurotransmitter, has been observed to exert multiple effects on physiological functions which has paved the way toward the identification and characterization of two distinct receptor subtypes: nicotinic (nAChR) and muscarinic receptors (mAChR). In both cases, the effects of naturally occurring drugs acting on nAChRs and mAChRs have been exploited for both therapeutic and nontherapeutic purposes for hundreds of years, well before these receptors were identified. The first review presented by Bertrand and Wallace in the chapter "A Review of the Cholinergic System and Therapeutic Approaches to Treat Brain Disorders" will act as a foundational chapter to introduce cholinergic neurobiology and provides an in-depth discussion on the nAChR and mAChR receptors systems, their function and the opportunity that targeting these systems provide for novel medications across disorders. The reader will note that several authors introduce the cholinergic receptor systems throughout the book for the purposes of their individual discussions, and this is the strength of the book in that each chapter can stand on its own, as well as contributes to the broader vision of the book.

Following the introduction, we evaluate cholinergic contributions to cognitive function including sensory perception and attention, memory and working memory processes from ex vivo preparations through rodents, nonhuman primates, and human behaviors. In the chapter "Acetylcholine and Spontaneous Recognition Memory in Rodents and Primates," Easton, Barros, and Lever assess the impact of cholinergic manipulations on memory and highlight important considerations of what is meant by memory and how it is being modeled. In particular, they focus on the issue of translation from animal research into the clinic which is a key concern and argue that it is essential to move away from considering the debate to be one of ACh involvement in broadly described terms of "memory" or "attention," but instead argue that one needs to look carefully at the precise nature of the behavioral task being used and the nature of the behavioral impairment. Specific hypotheses about the nature of acetylcholine's function can be best arrived at through the careful consideration of specific elements of behavior and its relation to the manipulation at hand.

In the chapter "Endogenous Acetylcholine and Its Modulation of Cortical Microcircuits to Enhance Cognition," Venkatesan, Jeoung, Chen, Power, Liu, and Lambe discuss the role of ACh in shaping sensory perception and attentional processes using electrophysiological and optogenetic methods to elucidate how and where ACh acts within the cortex to help shape cognitive processing and open the door to new approaches for identifying novel treatments for the perceptual and attention deficits found in multiple psychiatric and neurological disorders. Parikh and Bangasser expand on our understanding of the cholinergic mechanisms of attention highlighting the recent developments in the chapter "Cholinergic Signaling Dynamics and Cognitive Control of Attention." Specifically, the evidence that phasic cholinergic signaling in the prefrontal cortex (PFC) is a causal mediator of signal detection is discussed. Moreover, studies that support the tonic neuromodulatory role of cholinergic inputs in top-down attentional control, and those that provide insights into the potential cellular substrates that integrate the phasic and neuromodulatory cholinergic signaling modes, are reviewed. The authors also incorporate considerations of sex differences that exist in the central cholinergic-attention system.

In the chapter "Involvement of Nicotinic Receptors in Working Memory Function," Galvin, Arnsten, and Wang provide an in-depth discussion of the cholinergic system in the prefrontal cortex on working memory functions in nonhuman primates. The authors highlight how the dorsolateral prefrontal cortex is particularly critical for rule representation and working memory, or the ability to hold information "in mind" in the absence of sensory input, and present the emerging evidence that supports a prominent and permissive role for ACh in these excitatory circuits, through actions at cholinergic nAChRs.

In the chapter "Nicotinic Receptors Underlying Nicotine Dependence: Evidence from Transgenic Mouse Models," we move away from the involvement of the cholinergic system in cognitive processes and begin the discussion on the role of this system in dependence and addiction. Gipson and Fowler provide an elegant discussion on the contributions from transgenic mouse models that have added to our understanding of nicotine's effects on the reward-related mesolimbic pathway and the aversion-related habenulo-interpeduncular pathway. Following in the chapter "Cholinergic Receptors and Addiction," Papke, Brunzell, and De Biasi review the structure and diversity of nAChR subunits and how different nAChR subtypes play specific roles in the phenomenon of nicotine addiction. The authors also provide a compelling discussion on how brain cholinergic receptors are involved with areca addiction and the unique challenges for dealing with addiction to this substance, which is a major public health issue across most of South Asia.

In the chapter "Behavioral and Molecular Basis of Cholinergic Modulation of Pain: Focus on Nicotinic Acetylcholine Receptors," Damaj and colleagues focus their review on the recent progress in our understanding of the cholinergic system as a target for pain modulation. In particular, they focus on nAChR subunit biology and balance the data supporting the therapeutic potential for this system in chronic pain with the inherent complexities associated with drug development in this space.

Chapter "An Evolving Therapeutic Rationale for Targeting the α7 Nicotinic Acetylcholine Receptor in Autism Spectrum Disorder" provides a comprehensive evaluation of the involvement of the cholinergic system in autism spectrum disorder (ASD). Deutsch and Burket review historic and recent literature supporting selective therapeutic targeting of the α7 nAChR in persons affected with ASD and summarize evolving literature that support the therapeutic exploration of selectively targeted cholinergic interventions for the treatment of ASD.

The discussion of therapeutic development for the cholinergic system continues in the chapter "Activators of α7 nAChR as Potential Therapeutics for Cognitive Impairment" by Wang, Bell, and Uslaner, who present rationale to support the α7 nAChR as a promising target for the treatment of cognitive deficits associated with psychiatric and neurological disorders, including schizophrenia and Alzheimer's disease (AD). The authors highlight the progression of several compounds targeting the α7 nAChR, including agonists and positive allosteric modulators (PAMs) into early clinical trials and highlight none of the α7 nAChR ligands has been approved for clinical use. The chapter focuses on ligands that have advanced to clinical studies and provides an excellent overview that allows us to explore the reasons why these agents have not met with unequivocal clinical success.

The intent of this book is to maximize the reader's insight into the translational perspective on targeting nicotinic and muscarinic acetylcholine receptors and to provide insight into the diverse array of disease states affected by dysfunction of cholinergic systems. In summary, we aim for this volume to be thought-provoking and to continue the discussions and important research and development efforts into the cholinergic system for treating diseases of the central nervous system.

Newcastle upon Tyne, UK
San Francisco, CA, USA

Mohammed Shoaib
Tanya L. Wallace

Contents

A Review of the Cholinergic System and Therapeutic Approaches to Treat Brain Disorders

Daniel Bertrand and Tanya L. Wallace

Contents

Abstract Since its identification over a hundred years ago, the neurotransmitter acetylcholine (ACh) has proven to play an essential role in supporting many diverse functions. Some well-characterized functions include: chemical transmission at the neuromuscular junction; autonomic function in the peripheral nervous system; and,

---

D. Bertrand
HiQScreen Sàrl, Geneva, Switzerland
e-mail: daniel.bertrand@hiqscreen.com

T. L. Wallace (✉)
BlackThorn Therapeutics, San Francisco, CA, USA
e-mail: tanya@blackthornrx.com

Curr Topics Behav Neurosci (2020) 45: 1–28
https://doi.org/10.1007/7854_2020_141
Published Online: 26 May 2020

sustained attention, sleep/wake regulation, and learning and memory within the central nervous system. Within the brain, major cholinergic projection pathways from the basal forebrain and the brainstem support these centrally mediated processes, and dysregulation of the cholinergic system is implicated in cognitive decline associated with aging and dementias including Alzheimer's disease. ACh exerts its effects by binding to two different membrane-bound receptor classes: (1) G-protein coupled muscarinic acetylcholine receptors (mAChRs), and (2) ligand-gated nicotinic acetylcholine receptors (nAChRs). These receptor systems are described in detail within this chapter along with discussion on the successes and failures of synthetic ligands designed to selectively target receptor subtypes for treating brain disorders. New molecular approaches and advances in our understanding of the target biology combined with opportunities to re-purpose existing cholinergic drugs for new indications continue to highlight the exciting opportunities for modulating this system for therapeutic purposes.

Keywords Acetylcholine · mAChR · Muscarinic · nAChR · Nicotinic

1 Cholinergic System Overview

1.1 Acetylcholine

The activity of a chemical substance that reduced heart rate frequency was first observed by the pharmacologist Otto Loewi in 1921 and was later identified as acetylcholine (ACh) by Henry Dale. Subsequently, ACh was shown to play a role in many essential functions including (1) chemical transmission at the neuromuscular junction, (2) autonomic function in the peripheral nervous system, and (3) centrally mediated cognitive processes such as attention, learning, and memory. ACh is synthesized from choline and acetyl-CoA through the enzyme choline acetyltransferase that occurs in different neurons as well as non-neuronal cells and is released locally (Wessler and Kirkpatrick 2008; Schubert et al. 2012; Beckmann and Lips 2013). ACh is hydrolyzed by acetylcholinesterase enzymes, which are abundant in the synaptic cleft, after its release from presynaptic neurons.

Two major cholinergic projection pathways occur in the brain (Fig. 1): (1) The magnocellular basal forebrain cholinergic system, which includes the nucleus basalis of Meynert, the medial septal nucleus, and the vertical and horizontal limbs of the diagonal band of Broca. The basal forebrain cholinergic system has extensive projections to neocortical regions, as well as to basolateral amygdala and olfactory bulb, hippocampus, and entorhinal cortices. (2) The brainstem cholinergic system which includes the pedunculopontine nucleus and the laterodorsal pontine tegmental nucleus and projects primarily to thalamic structures and to basal forebrain regions.

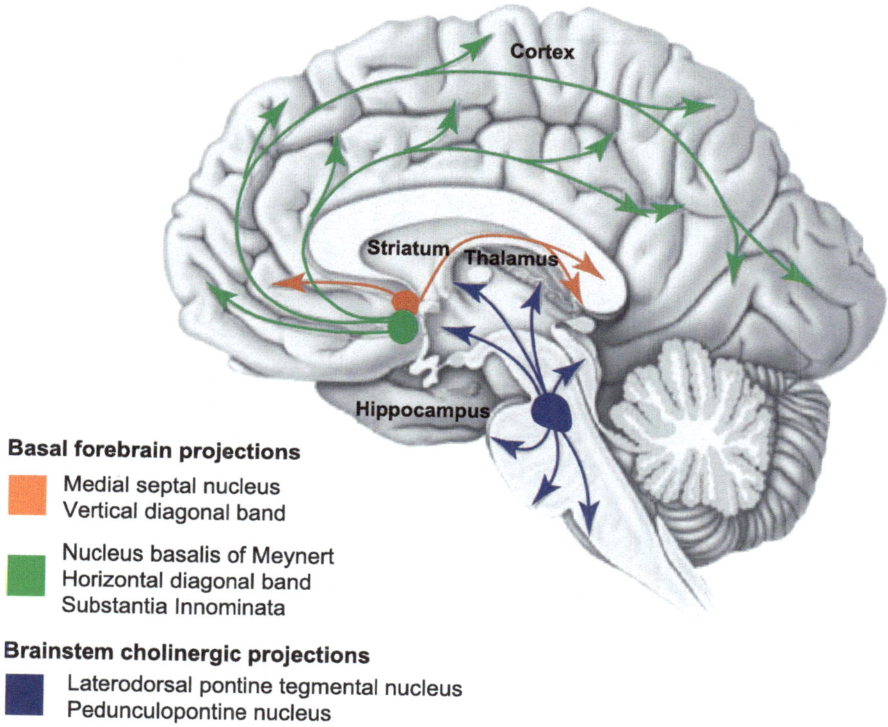

Fig. 1 Schematic of major cholinergic projections in the human brain

ACh exerts its effects by binding to two different membrane-bound receptor classes: (1) G protein-coupled muscarinic acetylcholine receptors (mAChRs), present in both the peripheral and central nervous systems, and (2) ligand-gated nicotinic acetylcholine receptors (nAChRs), which function in the peripheral and central nervous systems, in the neurons from the parasympathetic ganglia, at the neuromuscular junction, as well as in non-neuronal cells (Fig. 2). These receptor systems will be described in more detail in the following text.

1.2 Muscarinic ACh Receptors (mAChRs)

Muscarinic acetylcholine receptors (mAChRs) are class A G protein-coupled receptors (GPCRs) and exist as five distinct subtypes (M1–M5) expressed in different brain regions and the periphery (Kruse et al. 2014). M1, M2, M3, M4, and M5 mAChR subtypes are encoded by separate genes (*CHRM1–CHRM5)* and are classified based on their tissue localization, molecular conformation, and activation of different intracellular signaling pathways. The M1, M3, and M5 mAChRs are

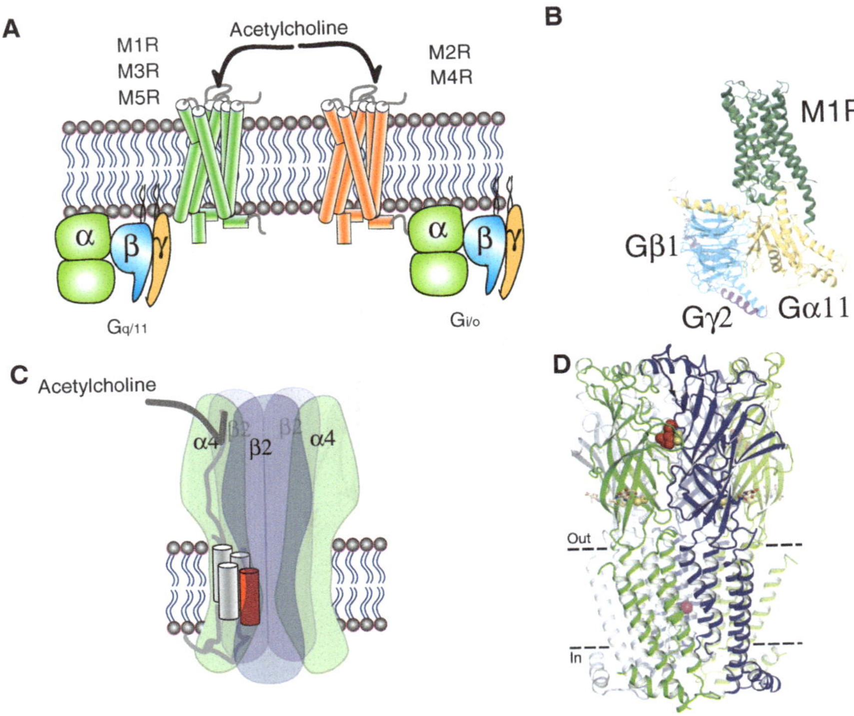

Fig. 2 Illustration of mAChR subtypes and a typical structure of a heteromeric nAChR. Schematic representation of the muscarinic and nicotinic receptors inserted in the membrane are represented in panels (**a**) and (**c**). Crystal structures of the muscarinic and nicotinic receptors in a side view are shown for comparison in panels (**b**) and (**d**). Structures correspond to publications Maeda et al. (2019) and Morales-Perez et al. (2016)

excitatory and couple to Gq proteins, activating phospholipase C and subsequently mobilizing intracellular calcium, whereas M2 and M4 mAChRs act in an inhibitory manner by coupling to Gi/o proteins and inhibiting adenylate cyclase. M1–M4 mAChR subtypes have all been crystallized in an inactive state (Haga et al. 2012; Kruse et al. 2012; Thal et al. 2016). The M5 mAChR subtype is the most recent mAChR to be cloned (Bender et al. 2019).

M1–M5 mAChRs are integral membrane proteins with seven transmembrane segments that form a pocket in which ACh can penetrate from the extracellular space, bind with at high affinity, and activate intracellular GTP binding regulatory proteins (G proteins). Heteromeric G proteins consist of an α subunit, a β subunit, and a γ subunit, and when ACh or other agonists bind to the extracellular mAChR binding site, it causes a conformational change in the receptor that promotes the α subunit to separate from the βγ moiety which then binds to effector proteins engaging multiple signaling cascades that amplify the initial ligand-receptor interaction. Additionally, following the ligand-receptor interaction, internalization of the receptor can occur through phosphorylation of the G proteins by intracellular kinases that lead to their uncoupling from the receptor and aim to regulate the ensuing biological response.

Additionally, elegant cryo-electron microscopy of the M1 and M2 mAChRs by Maeda and colleagues has demonstrated unique features of the receptors and their interactions with G proteins that provide a provocative basis for how they may impact their signaling properties (Maeda et al. 2019). Whereas it was originally thoughts that GPCRs are formed by a single protein which spans the membrane seven times, advances in the field of $GABA_B$ receptors have highlighted that functional receptors can result from the dimerization of two subunits (Kuner et al. 1999). Although evidence of heterodimers between the M2 and M3 proteins were identified, several outstanding questions about the quaternary structure of the mAChRs still remain (Marsango et al. 2018).

1.3 Nicotinic ACh Receptors (nAChRs)

Neuronal nicotinic acetylcholine receptors (nAChRs) belong to the superfamily of ligand-gated ion channels (LGIC). These highly specialized membrane proteins consist of a high-affinity binding site for a given ligand (e.g., ACh) and a pore-forming domain which is normally closed and opens upon binding of the ligand. Multiple forms of LGICs can be identified and classified according to their structural determinant, pharmacology, or ionic selectivity. Work conducted initially at the neuromuscular junction revealed that nAChRs are formed by the assembly of five subunits around a central ionic pore, and each subunit spans the membrane four times with N- and C-terminal ends lying in the extracellular space (Bertrand et al. 2015). Maintained throughout evolution, these structural features can already be identified in LGICs expressed by bacteria which present a striking similarity with mammalian nAChRs (Hilf and Dutzler 2009; Nemecz et al. 2016). Today, 16 genes (α1–α10, β1–β4, γ, δ, ε; note that α8 was identified only in chicken) encoding nAChR subunits have been identified in the mammalian genome (Schaaf 2014; Bertrand et al. 2015). Although the nAChR genes show a high degree of conservation, variations exist among species, and functional differences have already been identified between rodents and humans, for example (Paradiso et al. 2001; Curtis et al. 2002; Shorey-Kendrick et al. 2015).

While crystallography and high-resolution electron microscopy brought our understanding of the structural features of LGIC to an entirely new level, these studies have also highlighted the underlying complexity of nAChRs (Morales-Perez et al. 2016; Walsh et al. 2018). One complexity is that a single receptor results from the assembly of 5 subunits, and given the possibility offered by the combinations of the 16 subunits, multiple forms of receptors have been identified (Bertrand et al. 2015). In their simplest form, nAChRs are homomeric or comprised of five identical subunits, such as the α7 nAChRs; however, most generally, nAChRs are composed of subunits of at least two or more forms such as the α4β2 which is the major brain nAChR subtype (Fig. 2). Moreover, even in a receptor containing two types of subunits (e.g., α and β), the α versus β ratio has been shown to yield structural differences in the protein interfaces and also to modify the receptor properties (Zwart and Vijverberg 1998; Nelson et al. 2003; Zhou et al. 2003; Tapia et al. 2007; Walsh et al. 2018). The α4β2 nAChR demonstrates high (nM) binding affinity to nicotine and ACh, whereas the α7 nAChR shows lower (μM) sensitivity to these same ligands. The α4β2nAChR can occur with two distinct stoichiometries: (α4)2 and (β2)3 subunits (high sensitivity state) or (α4)3 and (β2)2 subunits (low sensitivity state) that are characterized by their binding affinity for ACh (Zwart and Vijverberg 1998; Zhou et al. 2003). Interestingly, functional properties associated with differences in stoichiometric ratio are not just an in vitro phenomenon, but also have been observed in vivo (Lamotte d'Incamps et al. 2018).

To understand the role and contribution of nAChRs in brain function, it is therefore indispensable to know their precise brain localization as well as their structural arrangement. For example, while some receptors will be composed of α4 and β2, the introduction of an additional subunit in the receptor complex, such as α5 or other, will be accompanied by modifications in the functional and pharmacological properties of the receptors (Brown et al. 2007; Kuryatov et al. 2008; Grady et al. 2010; Besson et al. 2016). Moreover, as multiple studies have already highlighted, a single cell can express more than one receptor subtype, and therefore it is necessary to understand the precise receptor distribution to be able to evaluate their functional outcome (Klink et al. 2001).

2 Localization of AChRs in the Central Nervous System

2.1 mAChRs

The mAChR system is widespread throughout the brain and periphery; however, for the purposes of this chapter, we will focus on the mAChR brain expression and its related functions (Table 1). mAChRs show the densest expression within the caudate nucleus and putamen regions. The M1 and M2 subtypes are the most abundant mAChR subtypes in the brain; however, the M1, M2, and M4 subtypes have all received a great deal of attention as drug targets for neuropsychiatric and neurological disorders.

The M1 mAChR represents approximately 35–60% of the total mAChRs in the human brain (Volpicelli and Levey 2004). It is localized postsynaptically and is prominently expressed in the cerebral cortex, including frontal, temporal, parietal, and occipital cortices, and also is abundant in the hippocampus, striatum, amygdala, and thalamic brain regions (Levey et al. 1991; Crook et al. 2001). The M1 mAChR subtype is involved in learning and memory functions, and selective activation of the M1 mAChR has been investigated for its therapeutic potential as a cognitive-enhancing agent in disorders such as Alzheimer's disease in which there is associated decline of these processes (Wess et al. 2007; Scarr 2012). Initial attempts to

Table 1 Localization of nAChRs and mAChRs in the central nervous system

Region	Cholinergic receptor subtype	
Olfactory bulb	α4β2	M3
	α7	
Prefrontal cortex		M1
		M4
Cerebral cortex	α4β2	M1
	α7	M3
	α4α5β2	M4
Occipital cortex		M2
Hippocampus	α4β2	M1
	α7	M2
	α4α5β2	M3
	α3β4	M4
		M5
Striatum	α4β2	M1
	α4α5β2	
	α6β2β3	
	α6α4β2β3	
Amygdala	α4β2	M1
	α7	M2
		M3
Thalamus	α4β2	M1
		M2
		M3
Medial habenula	α4β2	
	α7	
	α3β3β4	
	α3β4	
Hypothalamus	α4β2	M5
	α7	
Substantia nigra	α4β2	
	α7	
	α3β4	
	α6β2β3	
	α4α5β2	
Ventral tegmental area	α4β2	M5
	α7	
	α3β4	
	α6β2β3	
	α4α5β2	
Interpeduncular nucleus	α4β2	
	α7	
	α3β3β4	
	α2β2	

(continued)

Table 1 (continued)

Region	Cholinergic receptor subtype	
Raphe nucleus	$\alpha4\beta2$	
Locus coeruleus	$\alpha3\beta4$	
	$\alpha6\beta2\beta3$	
Pons		M3
Pineal gland	$\alpha7$	
	$\alpha3\beta4$	
Cerebellum	$\alpha4\beta2$	M2
	$\alpha7$	
	$\alpha3\beta4$	
	$\alpha3\beta2$	
Spinal cord	$\alpha4\beta2$	
	$\alpha7$	
	$\alpha3\beta2$	

Summarized from Gotti et al. (2007) and Ahmed et al. (2017)

develop selective agonists to the M1 mAChR highlighted the challenge with the highly conserved homology that exists within the orthosteric binding site of mAChRs making it difficult to design selective subtype-specific ligands. Compounds such as xanomeline with M1- and M4-preferring agonist activity were tested for cognitive-enhancing potential, but dose-limiting side effects attributed to "off-target" parasympathomimetic activity at peripheral M2 and M3 mAChR subtypes were considered a limitation to their investigation (Bymaster et al. 2003; Wess et al. 2007).

The M2 mAChR is a cholinergic inhibitory autoreceptor localized on presynaptic terminals in many regions throughout the brain. M2 mAChRs are present on large cholinergic interneurons in the striatum and have high expression in cerebellum, thalamus, and nucleus basalis of Meynert along with some limbic structures, e.g., amygdala and hippocampus. Stimulating M2 mAChRs decreases cholinergic neurotransmission and impairs memory. This has led to investigation of M2 mAChR antagonists as an approach to restore ACh release and improve learning and memory for cognitive-impairing diseases in which the cholinergic system is compromised (Billard et al. 1995; Langmead et al. 2008). The limitations to this approach thus far have included the challenges associated with developing M2 mAChR selective antagonists that do not have off-target activity at other mAChRs (as described above for M1) but, also importantly, that do not engender liabilities in peripheral organs in which high levels of expression of M2 mAChRs have been shown (e.g., from M2 mAChR expression in the heart and consequently cardiovascular complications).

Relative to M1 and M2, both M3 and M4 mAChR subtypes show much lower expression in the brain. For example, the M3 subtype is estimated to constitute only 5–10% of all mAChRs (Levey et al. 1994). Both M3 and M4 mAChRs are involved in neurotransmitter regulation and are prominent in hippocampal subregions, cerebral cortex, and striatum. In the brain, the M3 subtype has a functional role in regulating insulin secretion making it an interesting target to investigate for its role in type 2 diabetes mellitus (Gautam et al. 2006). However, the primary functions attributed to the M3 subtype are peripheral, e.g., smooth muscle contraction, exocrine secretion (e.g., saliva), and endocrine function (Matsui et al. 2000).

The distribution of the M4 subtype largely overlaps with that of the M1 and M3 subtypes, and it functions primarily as an inhibitory autoreceptor that decreases ACh release. Additionally, with its expression in striatum, the M4 subtype has demonstrated a regulatory control on dopamine-mediated functions in this region. Genetic deletion of the M4 subtype receptor in mice results in increased locomotor stimulation in response to dopamine agonists (e.g., amphetamine, cocaine) (Wess et al. 2007). An exciting area of ongoing research is to target the M4 subunit as a therapeutic approach for movement disorders such as Parkinson's disease (Langmead et al. 2008).

The M5 mAChR shows low expression in the brain, but is localized in dopamine-rich areas such as the ventral tegmental area and the substantia nigra suggesting that it could play a role in reward processing and movement, respectively. It also has been identified in the hippocampus, outermost layers of the cerebral cortex, and caudate putamen areas.

Developing a better understanding of the receptor distribution with the highest possible granularity including the homomer and/or heterodimer expression is expected to offer new alternatives to develop molecules displaying enough specificity to target a precise and well-localized subtype.

2.2 *nAChRs*

The homomeric α7 and heteromeric α4β2 nAChRs are the most prominent nAChRs in the mammalian brain, and both are involved in diverse functions. The α4β2 nAChRs have been identified in all layers of the cerebral cortex, hippocampal subregions, substantia nigra, and ventral tegmental area (Gotti et al. 2007). With high expression in dopamine-rich regions, the α4β2 nAChR has been studied for its involvement in hedonic processes and addiction particularly with regard to tobacco smoking. To this end, the α4β2 nAChR partial agonist, varenicline, has been brought to the market as a smoking cessation product.

High expression of the α7 nAChR in hippocampal subregions (CA1, CA3, dentate gyrus), the prefrontal cortex (layers I–VI), and also subcortical structures has directed attention to its role in cognitive processes (e.g., long-term memory, working memory, sensory gating). The α7 nAChR is localized presynaptically, postsynaptically, and perisynaptically contributing to its wide-ranging effects on neurotransmission (Jones and Wonnacott 2004). In the hippocampus, it has been identified postsynaptically on GABAergic neurons, whereas in brainstem nuclei such as the ventral tegmental area and the substantia nigra, the α7 nAChR localized presynaptically and is important in regulating excitatory neurotransmission (McGehee et al. 1995; Cheng and Yakel 2014).

3 nAChR Neuronal Expression and Neurotransmission

3.1 *Ionic Selectivity and Ca2+ Permeability*

Agonist stimulation of neuronal nAChRs causes the rapid opening of the ion channel and, in most cases, a depolarization of the cell in which these receptors are expressed. Structurally similar to the nAChRs expressed at the neuromuscular junction, the neuronal nAChRs are permeable to cations (Ballivet et al. 1988). Determination of the ionic selectivity using electrophysiological methods and ionic substitutions revealed that these receptors are permeable to sodium, potassium, and calcium (Vernino et al. 1992; Bertrand et al. 1993a, 1993b; Castro and Albuquerque 1995; Fucile et al. 2004). The calcium permeability was shown to vary as a function of the nAChR composition and is the highest for α7 and α9 nAChRs which are comparable or superior to the ionic permeability of the N-methyl-D-aspartic-acid (NMDA) receptor (Bertrand et al. 1993a, b; Séguéla et al. 1993; Elgoyhen et al. 1994; Castro and Albuquerque 1995; Sgard et al. 2002; Katz et al. 2000; Fucile et al. 2005; Uteshev 2012). Opening of divalent permeable channels can increase the intracellular calcium concentration in the proximity of the membrane and trigger the opening of the other ionic pore. For example, in the case of the α9 α10 receptors expressed in the outer hair cells of the inner ear, it was shown that activation of these receptors can activate potassium channels and indirectly cause the hyperpolarization of the cell (Janssen et al. 2004; Nie et al. 2004; Dani and Bertrand 2007; Roux et al. 2011).

Experiments conducted using site-directed mutagenesis at the homomeric α7 nAChRs revealed the presence of two binding sites, located at the inner mouth and in the upper part of the pore, that determine the level of calcium permeability (Bertrand et al. 1993a, b). The high degree of conservation of the second transmembrane segment (TM2) and, consequently, the cationic permeability in addition to the calcium permeability of these homomeric receptors suggest an important overall physiological role (Devillers-Thiery et al. 1992). Intracellular calcium homeostasis is an important physiological mechanism regulated, for example, by the calcium influx through receptors such as the N-methyl-D-aspartic acid (NMDA) or by the α7 nAChRs and by its sequestration in intracellular compartments such as the endoplasmic reticulum (ER). The intracellular calcium concentration is a ubiquitous second messenger regulating the activity of several membrane proteins including the calcium-activated potassium channels up to the control of many enzymatic pathways. Disruption of the intracellular calcium concentration is thought to be associated with several neurological disorders (Glaser et al. 2018).

Taking advantage of the homomeric nature of the α7 nAChRs, site-directed mutagenesis further allowed a more detailed characterization of the ionic selectivity filter of these cationic channels. It was shown that amino acids at the inner mouth of the channel determine the cation or anion selectivity, and interestingly, the introduction of a single proline amino acid is sufficient to switch the ionic selectivity (Galzi et al. 1992; Corringer et al. 1999). These observations were confirmed subsequently in invertebrates which have nicotinic-like channels that are natively permeable to anions and possess the critical amino acids earlier identified by mutagenesis (van Nierop et al. 2006; Juneja et al. 2014). Moreover, the same biophysical rules were found to apply for naturally occurring anion-permeable channels that were successfully switched to cation permeable by the introduction of the corresponding amino acids (Keramidas et al. 2000).

3.2 Voltage Dependence

Early characterization of the functional properties of nAChRs revealed that these receptors display a peculiar voltage sensitivity. While the nAChRs at the neuromuscular junction present an essentially Ohmic behavior, neuronal nAChRs display a strong inward rectification (Bertrand et al. 1991). Activation of the neuronal nAChRs causes the opening of the channels only when the cell membrane potential is negatively charged but the ionic pore becomes shut for resting potential smaller than −40 mV. As single channel conductance measurements revealed a resistive Ohmic behavior, it was concluded that rectification probably occurs by intracellular blockade (Bertrand et al. 1991; Haghighi and Cooper 2000). The mechanism responsible for this channel closure is thought to be caused by the blockade at the intracellular mouth by magnesium ions and/or intracellular polyamines (Forster and Bertrand 1995; Bonfante-Cabarcas et al. 1996; Stauderman et al. 1998). Inward rectification was reported for the different nAChR subtypes including the homomeric α7 and the heteromeric α4β2, α4α∗β∗ receptors and was observed for both recombinant and native nAChRs (Buisson et al. 1996; Gerzanich et al. 1997; Stauderman et al. 1998; Zaninetti et al. 1999; Alkondon et al. 2000; Nelson et al. 2001). Highly conserved across species, rectification is another hallmark of neuronal nAChRs. To understand the relevance of this mechanism, consider that release of ACh in the vicinity of the receptors will cause a physiological effect only when the cell is near its resting potential. On the contrary, if the cell is depolarized by any other activity, the ACh release will not provoke further signaling. This can also be summarized as a mechanism of coincidence detection in which the sequence of events is determinant for the functional outcome.

4 Synaptic Plasticity

A prominent role of the cholinergic system is its involvement in attention, learning, and memory processes, which has been established over decades of research using animal models (cholinergic lesions, receptor pharmacology, genetic manipulations), as well as in humans with clinically effective therapies that are prescribed for patients with cognitive dysfunction (e.g., cholinesterase inhibitors for Alzheimer's disease). Underlying these in vivo studies and contributing to our mechanistic understanding of the cognitive involvement of the cholinergic system are in vitro models of synaptic plasticity (long-term potentiation and long-term depression).

The mAChR-dependent modulation of synaptic plasticity has been demonstrated within hippocampal neurocircuits, and this is perhaps best represented by studies investigating the M1 mAChR. M1 mAChRs localized on glutamatergic pyramidal neurons of the CA1 subregion of the hippocampus provide a direct excitatory outlet for cholinergic basal forebrain afferents and are believed to underlie the cholinergic potentiation of glutamate-mediated neurotransmission that results in a robust strengthening of glutamatergic synapses in pyramidal neurons in this region (Dennis et al. 2016). Studies conducted in the CA1 region of the mouse hippocampus have shown that mAChR agonists applied at low concentrations modulate plasticity of glutamatergic synapses in this region (Shinoe et al. 2005). It remains to be determined if the enhanced synaptic plasticity with M1 mAChR agonists extends to other brain regions such as the neocortex.

The characteristic high Ca2+permeability of the α7 nAChR combined with its localization on glutamatergic axon terminals can lead to enhanced synaptic plasticity following stimulation (e.g., with nicotine or with selective α7 nAChR agonists) as has been shown using the long-term potentiation (LTP), an in vitro model of learning and memory in the dentate gyrus region of the hippocampus. In addition, genetic deletion of the α7 nAChR in mice produces deficits in LTP following nicotine administration (Criscuolo et al. 2015).

5 Intracellular Signaling

mAChR and nAChR subtypes are all activated by acetylcholine, but as will be discussed within the current section, each receptor system is coupled to different second messenger pathways that can yield divergent signaling effects across cell types. Interestingly, activation of different classes of mAChRs and nAChRs, distinguished both by their location on the neurons and by their subunit composition within a single cell, can regulate differences in the sources of calcium mobilized (e.g., extracellular, intracellular stores) and result in altered physiological and dynamic intracellular responses that ultimately control cellular function at an individual neuron level (Rathouz et al. 1995), as will be highlighted more below.

5.1 mAChRs

As introduced earlier in the chapter, the M1, M3, and M5 mAChRs couple with Gq-type G proteins and mediate activation of phospholipase C (PLC) and inositol triphosphate/calcium signaling via pertussis toxin-insensitive G proteins of the Gq family. M1 receptors are the most abundant mAChR in the brain and have the richest biology at this point with which to focus our attention.

M1 mAChRs localized postsynaptically on pyramidal neurons in the cerebral cortex and hippocampus receive innervation from cholinergic projections from basal forebrain. Activation of the M1 receptor causes calcium release from intracellular stores via IP_3-dependent calcium release that subsequently causes a transient inhibition driven by calcium-dependent small conductance potassium channels. In addition to the Gq-coupled activity, activation of M1 mAChRs in cortical pyramidal neurons produces a longer-lasting and voltage-dependent excitation that involves additional cation channels that have not been fully characterized to date (Dasari et al. 2017). Together, this signaling is hypothesized to underlie some of the functional effects of M1 activation on attention, learning, and memory.

M2 and M4 receptors signal through the pertussis-sensitive Gi/Go subfamily of G proteins and mediate inhibition of cAMP production. Targeting the M2 autoreceptor with selective antagonist ligands has been hypothesized to be a therapeutic strategy for treating Alzheimer's disease through increasing ACh neurotransmission by blocking its Gi-coupled inhibitory actions on cholinergic neurons. This has been supported by animal studies demonstrating that antagonism of the M2 mAChR elevates extracellular ACh concentrations (Billard et al. 1995). However, the limitations with this approach have proved to be several-fold and include that M2 receptors are localized on both cholinergic and non-cholinergic terminals in the cortex and hippocampus; therefore, increasing cholinergic tone with an M2 antagonist yields a more complicated pharmacology that is not necessarily therapeutic (Levey 1996). In addition and perhaps a more prominent restraint is that activation of the M2 receptor plays an important role in the physiological regulation of cardiac function through its inhibitory actions on cAMP, as well as through its modulation of muscarinic potassium channels. Upon stimulation of the M2 receptor by ACh or other agonists, the α subunit separates from the $\beta\gamma$ moiety, causing a decrease in adenylate cyclase, cAMP activity, and deceasing downstream signaling cascades. Activation of M2 also causes the $\beta\gamma$ moiety to act on potassium channels. In the heart, M2 activates potassium channel and decreases heart rate (Krejci et al. 2004). Due to its prominent effects on cardiac function, it remains challenging to target the M2 receptor due to dose-limiting side effects that could prevent the therapeutic benefit from being achieved.

Much of the work investigating M4 signaling and function has been based on its abundant expression in the striatum, specifically in the dopamine D1 receptor containing spiny projections neurons of the direct output pathway. In this region where cholinergic projections densely innervate the dopamine pathways, the M4 subunit has been shown to suppress dopamine D1 receptor signaling through its inhibitory actions on the adenylate cyclase and cAMP pathway and diminish regulator of G protein signaling type 4 (RGS4) at corticostriatal glutamatergic neuronal synapses (Shen et al. 2015). This combination of activities is hypothesized to have therapeutic potential as it may be a mechanism to regulate aberrant corticostriatal synaptic plasticity that is involved in symptoms such as L-Dopa therapy-induced dyskinesias observed in patients with Parkinson's disease who have been on chronic dopamine agonist-based therapies, and early evidence in animal models has supported this approach (Shen et al. 2015).

5.2 *nAChRs*

Signal transduction does not occur exclusively at the cell membrane, but also intracellularly. The high calcium permeability of the α7 or α9 containing nAChRs induces changes in the concentration of these divalent ions in the close vicinity of the membrane which, in turn, can trigger downstream signaling cascades and alteration in gene transcription. For the highly permeable α9 containing receptor, this can be best exemplified in the outer hair cells from the inner ear, wherein release of ACh triggers a hyperpolarization of the cells. Detailed analysis of these mechanisms revealed that the primary mechanism is the activation of α9α10 nAChRs which increases the intracellular calcium and, indirectly, triggers the opening of small conductance (SK) potassium channels that are at the origin of the cell hyperpolarization (Nie et al. 2004; Roux et al. 2011; Katz et al. 2000).

Similarly, activation of α7 containing nAChRs is known to increase the cytosolic calcium concentration, the outcome of which directly depends on other proteins through which this intracellular signaling will manifest. The physiological relevance of such mechanisms is readily understood in light of examples such as data from cortical pyramidal neurons. Accordingly, increased intracellular calcium in this cell segment activates potassium currents yielding modification of the signal processing (Berger and Lüscher 2003). An increase in the intracellular calcium concentration can yield multiple downstream cellular effects ranging from changes in the microtubules to alteration of the growth cone (King and Kabbani 2018).

Moreover, stimulation of α7 nAChRs triggers multiple intracellular cascades such as the neuroprotective Janus Kinase 2 (JAK2/STAT3/NF-kB (Marrero and Bencherif 2009)), extracellular related kinase 1 (ERK1) (Bencherif and Lippiello 2009), or mitogen-activated protein (MAP) kinase pathways (Gubbins et al. 2010). Activation of α7 nAChRs and subsequent engagement of intracellular signaling pathways can exert neurotrophic effects and has been suggested as a potential mechanism of neuroprotection in degenerative diseases such as Alzheimer's disease (Ma and Qian 2019). Additionally, stimulation with the α7 nAChR agonist PNU-282987 and activation of key intracellular pathways exert a protective effect against myocardial reperfusion injury (Hou et al. 2018).

6 Receptor Desensitization

The response of receptors to a sustained exposure of agonist is generally characterized by first a rapid reaction which progressively declines over time. Known as desensitization, it is an active physiological mechanism aimed to return the system to a homeostatic state following continued exposure to an external stimulus. A simple example of desensitization occurs with the application of perfume; while it is easy to recognize its scent when first applied, after a few minutes, it is difficult to recognize whether it is on or not. While both mAChRs and nAChRs show desensitization properties, the magnitude of their effects and their mechanisms are clearly distinct, as it will be described below.

6.1 mAChRs

The desensitization mechanisms of G-coupled proteins are mainly due to internalization of these integral membrane proteins. Namely, stimulation of the receptors by the agonist causes changes in the receptor conformation and its interaction with the G proteins and, indirectly, triggers its internalization (van Koppen and Kaiser 2003). The progressive decline of the receptors at the cell surface causes a reduction in sensitivity to the agonist. The mAChR desensitization is a function of the receptor subtype and conditions. Recovery requires the incorporation of new receptors which are translocated from the intracellular pool into the plasma membrane. Given the complex cellular mechanisms involve in desensitization and recovery, the timing for these processes is rather slow.

6.2 nAChRs

Nicotinic acetylcholine receptors display desensitization and recovery mechanisms that are quite distinct from the mAChRs. In contrast to the G-coupled proteins, ligand-gated channels are fast responders and do not internalize during desensitization. Exposing nAChRs to brief pulses of agonist causes a brisk activation of the receptors, but in contrast, prolonged agonist application causes a desensitization of the receptors leading to a response with a peak and a plateau. The ratio between peak and plateau markedly differs with the receptor subtype with the fastest desensitization being observed with the $\alpha 7$ nAChRs and the slowest desensitization being observed at the nAChRs from the neuromuscular junction receptors. Major brain $\alpha 4\beta 2$ nAChRs display a clear peak and plateau response during agonist exposure indicative of multiple phases of desensitization (Hogg and Bertrand 2007). The ability of the receptors to maintain a response during agonist exposure is thought to relate to their physiological function. Insertion in the plasma membrane of fast desensitizing receptors, such as the α7 nAChRs, is expected to cause only a transient activation of the cell, whereas a more sustained depolarization and, in consequence, physiological outcome will be observed in cells expressing more slowly desensitizing receptors such as the α4β2 nAChR or even further for the α3β4 nAChR. Given the specificity of cellular expression of some receptors in a particular neuronal pathway, such as in the interpeduncular nucleus and the fasciculus retroflexus (Perry et al. 2002), suggests that these brain areas will be more susceptible to prolonged stimulations by nicotinic agonists. The transition from peak to plateau characterizes the desensitization occurring during agonist exposure ranging from milliseconds to a few seconds, as its recovery is rather fast.

Desensitization is often subdivided into short- and long-term depending upon the exposure time of the receptor to the agonist and the persistence of the effects. For compound exposures of several minutes or hours, such as that caused by nicotine intake during smoking, a sustained inhibition of the receptors is observed. Various sets of experiments have shown that sustained exposure to 100 nM nicotine, which corresponds to the brain concentration observed after smoking a cigarette, is sufficient to cause a long-lasting receptor desensitization at the α4β2 nAChR (Ochoa et al. 1989; Lester and Dani 1995; Fenster et al. 1997; Paradiso and Brehm 1998; Dani et al. 2000; Besson et al. 2007). Long-term desensitization is closely related to the receptor composition (Vibat et al. 1995; Gerzanich et al. 1998; Dani and Bertrand 2007; Rollema et al. 2015). Desensitization to low concentrations of agonist can be observed by monitoring the amplitude of the response to a brief pulse of ACh, and it was shown that natural variants in the α4, β2, or α5 subunits influence the profile of desensitization and its recovery (Hoda et al. 2008; Improgo et al. 2010; Tammimäki et al. 2012).

7 Areas of Continued Investigation

A general division in neurotransmission is often made between ligand-gated ion channels which encompasses all the fast transmission mediated by ionotropic receptors and the opening of ion channels in the cell membrane, versus metabotropic receptors which are coupled to intracellular mechanisms such as the G proteins. This division was initially defined in terms of the pharmacology with nAChRs and their high sensitivity to the alkaloid nicotine found in the tobacco plant *Nicotiana* and for mAChRs and their sensitivity to alkaloids found in certain mushrooms such as the *Amanita muscaria* or the deadly poisoning *Clitocybe dealbata*. Based on the selectivity of these two substances, it was subsequently found that nicotine activates receptors that are ionotropic, which acts by the opening of the ionic pore (nAChRs), whereas muscarine indirectly modifies level of second messengers by its interaction with receptors (mAChRs) belonging to the family of GPCRs or seven transmembrane proteins.

Although these definitions largely hold true for most of the nAChR subtypes, their limit became obvious with the cloning and characterization of the pharmacological properties of the α9 nAChRs which displays a mixed nicotinic/muscarinic profile (Elgoyhen et al. 1994; Verbitsky et al. 2000). Nonetheless, the ionotropic properties of these receptors continued to fulfill the original distinction between fast-acting channels and receptors acting on a slower and prolonged time scale.

As a number of studies expanded our understanding of nAChR biology, especially within the homomeric α7 or heteromeric α9α10 receptors, several peculiarities also emerged, for example, recognition that the receptor expression occurred in many areas throughout the brain and periphery, as well as in the immune system (Wang et al. 2003; Peng et al. 2004; Razani-Boroujerdi et al. 2007). These results cast doubt about the sole ionotropic activity of these receptors with effects that could be associated only with ion fluxes. Follow-up investigation by different groups led to the proposition of a direct interaction between α7 nAChRs and metabotropic receptors that is reviewed in Kabbani and Nichols (2018). Evidence for the interaction between the α7 and metabotropic receptors has pointed to the specific intracellular sequence of the α7 protein which presents a unique amino acid sequence "LRMKRP" that is highly conserved throughout species (King et al. 2015). This observation might reconcile previous observations suggesting a mixed effect of α7 nAChRs on phosphorylation of multiple intracellular pathways (de Jonge and Ulloa 2007; Bencherif and Lippiello 2009; Maanen et al. 2009; Gubbins et al. 2010; Dhawan et al. 2012).

Genomic analysis of *CHRNA7* in multiple species revealed a further complexity specific to this particular nAChR gene with a duplication of exons 6–10 that is observed only in human. Leading to a form that was subsequently termed *CHRFAM7A*, this genomic duplication encodes for a protein that closely resembles α7, but which is missing the N-terminal domain (Gault et al. 1998; Sinkus et al. 2015). Several studies have replicated this initial observation, and the key questions that were opened by this initial discovery included (a) "Is the dupα7 able to form functional receptors?" and (b) "Does the dupα7 assemble with α7 itself to form a hetero-oligomer?". Experiments conducted by many different laboratories have concluded so far that while widely expressed, the dupα7 does not yield functional receptors. However, it was shown that α7 and the dupα7 can assemble in heteromers to form functional channels (Wang et al. 2010, Araud et al. 2011, de Lucas-Cerrillo et al. 2011,Lasala et al. 2019). Given the homologies between *CHRNA7* and *CHRFAM7A*, this indicates that heteromeric receptors containing α7 and the dupα7 can equally interact with GPCRs and could participate to modulation of cellular functions.

Genetic studies conducted in humans have correlated variations in *CHRFAM7A* and neurological pathologies (Flomen et al. 2012; Williams et al. 2012; Rozycka et al. 2013; Kunii et al. 2015; Sinkus et al. 2015). Moreover, the potential role of the dupα7 was underlined by its high degree of expression in the immune system (Villiger et al. 2002; de Lucas-Cerrillo et al. 2011; Costantini et al. 2015; Dang et al. 2015; Baird et al. 2016). Studies conducted in human-induced pluripotent stem cells (iPSCs) might provide a model to get a better understanding of the physiological role of *CHRFAM7A* (Ihnatovych et al. 2019).

8 Cholinergic Receptors as Therapeutic Targets

8.1 *mAChRs*

Of the mAChR subtypes that have been investigated clinically for brain disorders, the focus largely has been on M1, M2, and M4 selective molecules. The non-selective M1 agonist xanomeline was taken into patients with Alzheimer's disease, but was stopped due to an undesirable safety profile combined with minimal therapeutic benefit. More recently this molecule has been brought back into clinical development combined with the peripherally active non-selective mAChR antagonist, trospium chloride, that is approved for patients with overactive bladder. The hypothesis is that the combined therapy should allow engagement of centrally active mAChRs and antagonize peripheral mAChRs associated with adverse effects and medication discontinuation. The xanomeline/trospium chloride combined therapy is being investigated in patients with schizophrenia (ClinicalTrials.gov Identifier: NCT03697252).

An alternative approach to targeting mAChR subtype-specific molecules that bind to the orthosteric site is to identify molecules that can selectively modulate the mAChR subtype by targeting the allosteric site. By binding at a site distinct from the agonist (orthosteric site), these molecules bind at a distinct location on the receptor and are therefore more specific for a given amino acid sequence and tri-dimensional organization. Most recent advances in the optimization of allosteric modulators, which alter the affinity or efficacy of orthosteric ligands and offer selective and localized receptor modulation, have invigorated drug discovery efforts and resulted in several highly selective tool compounds that have demonstrated potential in preclinical studies and progressed to early clinical development (Felder et al. 2018). All of the mAChRs have at least one allosteric site that has been identified and provide an approach to selectively modulate mAChR subtypes independent of one another (Bock et al. 2018). In particular, positive allosteric modulators of the M1 and M4 mAChRs have been developed and are an exciting area of research that is enabling the biology of these receptors to be selectively interrogated (Kruse et al. 2012).

8.2 *nAChRs*

Development of nAChR subtype-specific compounds led to the discovery of several molecules which showed sufficient selectivity and robust pharmacological characteristics to be supported for testing in human clinical trials. The best examples can be shown for the α7 nAChRs, with the discovery of quinuclidine-based molecules such as the PNU-282987, TC-5619, MEM-3454/RG3487, or encenicline. Today, no less than 12 molecules showing specificity for the α7 nAChRs were brought into clinical trials (reviewed in Rollema et al. 2014)). However, despite these efforts, either beneficial effects were insufficient or side effects were not tolerable. Similar results were, unfortunately, observed with α4β2 nAChR-specific agonists such as ABT-418, sofinicline, pozanicline, ispronicline, or rivanicline (reviewed in Rollema et al. 2014).

In view of these efforts and failure to achieve the desired therapeutic effects, research of agonists specific to a given nAChR subtype and subsequent investments were slowed bringing this field almost to a complete halt in recent years. Nonetheless, it is important to underline that the development of an α4β2 nAChR partial agonist, varenicline, made a significant impact in the field of smoking cessation and was successfully introduced more than a decade ago (Rollema and Hurst 2018).

Although multiple brain diseases, such as Alzheimer's, schizophrenia, or Parkinson's, were investigated in clinical trials with selective nAChR ligands and highlight the diverse potential of these molecules, no nAChR drugs, except for varenicline, have made it as to the market as therapeutics (Terry 2008; Wallace and Bertrand 2013a, b; Perez-Lloret and Barrantes 2016; Hampel et al. 2019). Information about failures of clinical trials are limited, and it is often not possible to know if a compound was abandoned because of its side effects or lack of efficacy; however, it is possible to speculate about the encountered difficulties.

A first and an obvious difficulty are concerns around insufficient selectivity across receptor systems. For example, as in the case of encenicline, it was shown that this compound was active at the α7 nAChRs and was equipotent at the $5HT_3$ receptors. This cross interaction might be at the origin of the gastrointestinal difficulties that were reported in the clinical Phase 3 trial for this compound (Barbier et al. 2015; Keefe et al. 2015; Hayward et al. 2017; Godyń et al. 2016). A second and more difficult point concerns the desired mechanism of action in vivo. While it is easy to assess the effects of an agonist in vitro under controlled experimental conditions, the beneficial outcomes of such molecules in vivo are more difficult to comprehend. Especially when considering the desensitization properties of the nAChRs, it has remained a challenge how to effectively anticipate the outcome of sustained exposure to a low concentration of agonist under various disease conditions in which cholinergic tone and/or receptor expression may be different than in the healthy condition. Experiments conducted in vitro and in vivo with the α7 nAChRs and agonists selective for this subtype have highlighted that low concentrations of agonists can enhance the response to ACh by a mechanism described as priming (Prickaerts et al. 2012; Stoiljkovic et al. 2015). Similarly, priming was observed with the nicotine metabolite cotinine or the $5HT_3$ receptor antagonist tropisetron (Terry et al. 2015; Callahan et al. 2017). This mechanism would elegantly reconcile the enhancement of the cognitive performances observed with nicotinic derived compounds as well as the inverted U-shape observed between the compound concentration and performance increases (Wallace and Bertrand 2013a, b).

In view of a large body of evidence correlating the cholinergic system with cognitive performances, it is tempting to speculate that development of nAChR-specific ligands still has a promising future. In addition, such compounds might find additional and unexpected benefits by acting outside the CNS, such as the current repurposing of varenicline to treat dry eye afflictions.

References

Ahmed T, Zahid S, Mahboob A, Farhat SM (2017) Cholinergic system and post-translational modifications: an insight on the role in Alzheimer's disease. Curr Neuropharmacol 15 (4):480–494

Alkondon M, Braga MF, Pereira EF, Maelicke A, Albuquerque EX (2000) alpha7 nicotinic acetylcholine receptors and modulation of gabaergic synaptic transmission in the hippocampus. Eur J Pharmacol 393(1–3):59–67

Araud T, Graw S, Berger R, Lee M, Neveu E, Bertrand D, Leonard S (2011) The chimeric gene CHRFAM7A, a partial duplication of the CHRNA7 gene, is a dominant negative regulator of alpha7*nAChR function. Biochem Pharmacol 82(8):904–914

Baird A, Coimbra R, Dang X, Eliceiri BP, Costantini TW (2016) Up-regulation of the human-specific CHRFAM7A gene in inflammatory bowel disease. BBA Clin 5:66–71

Ballivet M, Nef P, Couturier S, Rungger D, Bader CR, Bertrand D, Cooper E (1988) Electrophysiology of a chick neuronal nicotinic acetylcholine receptor expressed in Xenopus oocytes after cDNA injection. Neuron 1(9):847–852

Barbier AJ, Hilhorst M, Van Vliet A, Snyder P, Palfreyman MG, Gawryl M, Dgetluck N, Massaro M, Tiessen R, Timmerman W, Hilt DC (2015) Pharmacodynamics, pharmacokinetics, safety, and tolerability of encenicline, a selective α7 nicotinic receptor partial agonist, in single ascending-dose and bioavailability studies. Clin Ther 37(2):311–324

Beckmann J, Lips KS (2013) The non-neuronal cholinergic system in health and disease. Pharmacology 92(5–6):286–302

Bencherif M, Lippiello PM (2009) Alpha7 neuronal nicotinic receptors: the missing link to understanding Alzheimer's etiopathology? Med Hypotheses 74(2):281–285

Bender AM, Garrison AT, Lindsley CW (2019) The muscarinic acetylcholine receptor M5: therapeutic implications and allosteric modulation. ACS Chem Neurosci 10(3):1025–1034

Berger T, Lüscher H-R (2003) Timing and precision of spike initiation in layer V pyramidal cells of the rat somatosensory cortex. Cereb Cortex 13(3):274–281

Bertrand D, Cooper E, Valera S, Rungger D, Ballivet M (1991) Electrophysiology of neuronal nicotinic acetylcholine receptors expressed in Xenopus oocytes following nuclear injection of genes or cDNA. In: Conn M (ed) Methods in neuroscience, vol 4. Academic Press, New York, pp 174–193

Bertrand D, Galzi JL, Devillers-Thiery A, Bertrand S, Changeux JP (1993a) Mutations at two distinct sites within the channel domain M2 alter calcium permeability of neuronal alpha 7 nicotinic receptor. Proc Natl Acad Sci U S A 90(15):6971–6975

Bertrand D, Galzi JL, Devillers-Thiery A, Bertrand S, Changeux JP (1993b) Stratification of the channel domain in neurotransmitter receptors. Curr Opin Cell Biol 5(4):688–693

Bertrand D, Lee CH, Flood D, Marger F, Donnelly-Roberts D (2015) Therapeutic potential of alpha7 nicotinic acetylcholine receptors. Pharmacol Rev 67(4):1025–1073

Besson M, Granon S, Mameli-Engvall M, Cloëz-Tayarani I, Maubourguet N, Cormier A, Cazala P, David V, Changeux J-P, Faure P (2007) Long-term effects of chronic nicotine exposure on brain nicotinic receptors. Proc Natl Acad Sci U S A 104(19):8155–8160

Besson M, Guiducci S, Granon S, Guilloux J-P, Guiard B, Repérant C, Faure P, Pons S, Cannazza G, Zoli M, Gardier AM, Maskos U (2016) Alterations in alpha5* nicotinic acetylcholine receptors result in midbrain- and hippocampus-dependent behavioural and neural impairments. Psychopharmacology (Berl) 233(18):3297–3314

Billard W, Binch H 3rd, Crosby G, McQuade RD (1995) Identification of the primary muscarinic autoreceptor subtype in rat striatum as m2 through a correlation of in vivo microdialysis and in vitro receptor binding data. J Pharmacol Exp Ther 273(1):273–279
Bock A, Schrage R, Mohr K (2018) Allosteric modulators targeting CNS muscarinic receptors. Neuropharmacology 136(Pt C):427–437
Bonfante-Cabarcas R, Swanson KL, Alkondon M, Albuquerque EX (1996) Diversity of nicotinic acetylcholine receptors in rat hippocampal neurons. IV. Regulation by external Ca++ of alpha-bungarotoxin-sensitive receptor function and of rectification induced by internal Mg++. J Pharmacol Exp Ther 277(1):432–444
Brown RWB, Collins AC, Lindstrom JM, Whiteaker P (2007) Nicotinic alpha5 subunit deletion locally reduces high-affinity agonist activation without altering nicotinic receptor numbers. J Neurochem 103(1):204–215
Buisson B, Gopalakrishnan M, Arneric SP, Sullivan JP, Bertrand D (1996) Human alpha4beta2 neuronal nicotinic acetylcholine receptor in HEK 293 cells: a patch-clamp study. J Neurosci 16 (24):7880–7891
Bymaster FP, Carter PA, Yamada M, Gomeza J, Wess J, Hamilton SE, Nathanson NM, McKinzie DL, Felder CC (2003) Role of specific muscarinic receptor subtypes in cholinergic parasympathomimetic responses, in vivo phosphoinositide hydrolysis, and pilocarpine-induced seizure activity. Eur J Neurosci 17(7):1403–1410
Callahan PM, Bertrand D, Bertrand S, Plagenhoef MR, Terry AV Jr (2017) Tropisetron sensitizes alpha7 containing nicotinic receptors to low levels of acetylcholine in vitro and improves memory-related task performance in young and aged animals. Neuropharmacology 117:422–433
Castro NG, Albuquerque EX (1995) Alpha-Bungarotoxin-sensitive hippocampal nicotinic receptor channel has a high calcium permeability. Biophys J 68(2):516–524
Cheng Q, Yakel JL (2014) Presynaptic α7 nicotinic acetylcholine receptors enhance hippocampal mossy Fiber glutamatergic transmission via PKA activation. J Neurosci 34(1):124–133
Corringer PJ, Bertrand S, Galzi JL, Devillers-Thiery A, Changeux JP, Bertrand D (1999) Mutational analysis of the charge selectivity filter of the alpha 7 nicotinic acetylcholine receptor. Neuron 22(4):831–843
Costantini TW, Dang X, Yurchyshyna MV, Coimbra R, Eliceiri BP, Baird A (2015) A human-specific α7-nicotinic acetylcholine receptor gene in human leukocytes: identification, regulation and the consequences of CHRFAM7A expression. Mol Med 3(21):323–336
Criscuolo C, Accorroni A, Domenici L, Origlia N (2015) Impaired synaptic plasticity in the visual cortex of mice lacking alpha7-nicotinic receptor subunit. Neuroscience 294:166–171
Crook JM, Tomaskovic-Crook E, Copolov DL, Dean B (2001) Low muscarinic receptor binding in prefrontal cortex from subjects with schizophrenia: a study of Brodmann's areas 8, 9, 10, and 46 and the effects of neuroleptic drug treatment. Am J Psychiatry 158(6):918–925
Curtis L, Buisson B, Bertrand S, Bertrand D (2002) Potentiation of human alpha4beta2 neuronal nicotinic acetylcholine receptor by estradiol. Mol Pharmacol 61(1):127–135
Dang X, Eliceiri BP, Baird A, Costantini TW (2015) CHRFAM7A: a human-specific α7-nicotinic acetylcholine receptor gene shows differential responsiveness of human intestinal epithelial cells to LPS. FASEB J 29(6):2292–2302
Dani JA, Bertrand D (2007) Nicotinic acetylcholine receptors and nicotinic cholinergic mechanisms of the central nervous system. Annu Rev Pharmacol Toxicol 47:699–729
Dani JA, Radcliffe KA, Pidoplichko VI (2000) Variations in desensitization of nicotinic acetylcholine receptors from hippocampus and midbrain dopamine areas. Eur J Pharmacol 393 (1–3):31–38
Dasari S, Hill C, Gulledge AT (2017) A unifying hypothesis for M1 muscarinic receptor signalling in pyramidal neurons. J Physiol 595(5):1711–1723
de Jonge WJ, Ulloa L (2007) The alpha7 nicotinic acetylcholine receptor as a pharmacological target for inflammation. Br J Pharmacol 151(7):915–929

de Lucas-Cerrillo AM, Maldifassi MC, Arnalich F, Renart J, Atienza G, Serantes R, Cruces J, Sánchez-Pacheco A, Andrés-Mateos E, Montiel C (2011) Function of partially duplicated human α77 nicotinic receptor subunit CHRFAM7A gene: potential implications for the cholinergic anti-inflammatory response. J Biol Chem 286(1):594–606

Dennis SH, Pasqui F, Colvin EM, Sanger H, Mogg AJ, Felder CC, Broad LM, Fitzjohn SM, Isaac JT, Mellor JR (2016) Activation of muscarinic M1 acetylcholine receptors induces long-term potentiation in the Hippocampus. Cereb Cortex 26(1):414–426

Devillers-Thiery A, Galzi JL, Bertrand S, Changeux JP, Bertrand D (1992) Stratified organization of the nicotinic acetylcholine receptor channel. Neuroreport 3(11):1001–1004

Dhawan S, Cailotto C, Harthoorn LF, de Jonge WJ (2012) Cholinergic signalling in gut immunity. Life Sci 91(21–22):1038–1042

Elgoyhen AB, Johnson DS, Boulter J, Vetter DE, Heinemann S (1994) Alpha 9: an acetylcholine receptor with novel pharmacological properties expressed in rat cochlear hair cells. Cell 79 (4):705–715

Felder CC, Goldsmith PJ, Jackson K, Sanger HE, Evans DA, Mogg AJ, Broad LM (2018) Current status of muscarinic M1 and M4 receptors as drug targets for neurodegenerative diseases. Neuropharmacology 136(Pt C):449–458

Fenster CP, Rains MF, Noerager B, Quick MW, Lester RA (1997) Influence of subunit composition on desensitization of neuronal acetylcholine receptors at low concentrations of nicotine. J Neurosci 17(15):5747–5759

Flomen RH, Shaikh M, Walshe M, Schulze K, Hall M-H, Picchioni M, Rijsdijk F, Toulopoulou T, Kravariti E, Murray RM, Asherson P, Makoff AJ, Bramon E (2012) Association between the 2-bp deletion polymorphism in the duplicated version of the alpha7 nicotinic receptor gene and P 50 sensory gating. Eur J Hum Genet 21(1):76–81

Forster I, Bertrand D (1995) Inward rectification of neuronal nicotinic acetylcholine receptors investigated by using the homomeric alpha 7 receptor. Proc Biol Sci 260(1358):139–148

Fucile S, Renzi M, Lauro C, Limatola C, Ciotti T, Eusebi F (2004) Nicotinic cholinergic stimulation promotes survival and reduces motility of cultured rat cerebellar granule cells. Neuroscience 127 (1):53–61

Fucile S, Sucapane A, Eusebi F (2005) Ca2+ permeability of nicotinic acetylcholine receptors from rat dorsal root ganglion neurones. J Physiol 565(Pt 1):219–228

Galzi JL, Devillers-Thiery A, Hussy N, Bertrand S, Changeux JP, Bertrand D (1992) Mutations in the channel domain of a neuronal nicotinic receptor convert ion selectivity from cationic to anionic. Nature 359(6395):500–505

Gault J, Robinson M, Berger R, Drebing C, Logel J, Hopkins J, Moore T, Jacobs S, Meriwether J, Choi MJ, Kim EJ, Walton K, Buiting K, Davis A, Breese C, Freedman R, Leonard S (1998) Genomic organization and partial duplication of the human alpha7 neuronal nicotinic acetylcholine receptor gene (CHRNA7). Genomics 52(2):173–185

Gautam D, Han SJ, Hamdan FF, Jeon J, Li B, Li JH, Cui Y, Mears D, Lu H, Deng C, Heard T, Wess J (2006) A critical role for beta cell M3 muscarinic acetylcholine receptors in regulating insulin release and blood glucose homeostasis in vivo. Cell Metab 3(6):449–461

Gerzanich V, Kuryatov A, Anand R, Lindstrom J (1997) "Orphan" alpha 6 nicotinic AChR subunit can form a functional heteromeric acetylcholine receptor. Mol Pharmacol 51(2):320–327

Gerzanich V, Wang F, Kuryatov A, Lindstrom J (1998) Alpha 5 subunit alters desensitization, pharmacology, Ca++ permeability and Ca++ modulation of human neuronal alpha 3 nicotinic receptors. J Pharmacol Exp Ther 286(1):311–320

Glaser T, Arnaud Sampaio VF, Lameu C, Ulrich H (2018) Calcium signalling: a common target in neurological disorders and neurogenesis. Semin Cell Dev Biol S1084-9521(18):30068–30065

Godyń J, Jończyk J, Panek D, Malawska B (2016) Therapeutic strategies for Alzheimer's disease in clinical trials. Pharmacol Rep 68(1):127–138

Gotti C, Moretti M, Gaimarri A, Zanardi A, Clementi F, Zoli M (2007) Heterogeneity and complexity of native brain nicotinic receptors. Biochem Pharmacol 74(8):1102–1111

Grady SR, Salminen O, McIntosh JM, Marks MJ, Collins AC (2010) Mouse striatal dopamine nerve terminals express alpha4alpha5beta2 and two stoichiometric forms of alpha4beta2-*-nicotinic acetylcholine receptors. J Mol Neurosci 40(1–2):91–95

Gubbins EJ, Gopalakrishnan M, Li J (2010) Alpha7 nAChR-mediated activation of MAP kinase pathways in PC12 cells. Brain Res 1328:1–11

Haga K, Kruse AC, Asada H, Yurugi-Kobayashi T, Shiroishi M, Zhang C, Weis WI, Okada T, Kobilka BK, Haga T, Kobayashi T (2012) Structure of the human M2 muscarinic acetylcholine receptor bound to an antagonist. Nature 482(7386):547–551

Haghighi AP, Cooper E (2000) A molecular link between inward rectification and calcium permeability of neuronal nicotinic acetylcholine alpha3beta4 and alpha4beta2 receptors. J Neurosci 20(2):529–541

Hampel H, Mesulam MM, Cuello AC, Khachaturian AS, Vergallo A, Farlow MR, Snyder PJ, Giacobini E, Khachaturian ZS (2019) Revisiting the cholinergic hypothesis in Alzheimer's disease: emerging evidence from translational and clinical research. J Prev Alzheimers Dis 6 (1):2–15

Hayward A, Adamson L, Neill JC (2017) Partial agonism at the α7 nicotinic acetylcholine receptor improves attention, impulsive action and vigilance in low attentive rats. Eur Neuropsychopharmacol 7(4):325–335

Hilf RJC, Dutzler R (2009) Structure of a potentially open state of a proton-activated pentameric ligand-gated ion channel. Nature 457(7225):115–118

Hoda JC, Gu W, Friedli M, Phillips HA, Bertrand S, Antonarakis SE, Goudie D, Roberts R, Scheffer IE, Marini C, Patel J, Berkovic SF, Mulley JC, Steinlein OK, Bertrand D (2008) Human nocturnal frontal lobe epilepsy: pharmocogenomic profiles of pathogenic nicotinic acetylcholine receptor beta-subunit mutations outside the ion channel pore. Mol Pharmacol 74 (2):379–391

Hogg RC, Bertrand D (2007) Partial agonists as therapeutic agents at neuronal nicotinic acetylcholine receptors. Biochem Pharmacol 73(4):459–468

Hou Z, Zhou Y, Yang H, Liu Y, Mao X, Qin X, Li X, Zhang X, Hu Y (2018) Alpha7 nicotinic acetylcholine receptor activation protects against myocardial reperfusion injury through modulation of autophagy. Biochem Biophys Res Commun 500(2):357–364

Ihnatovych I, Nayak TK, Ouf A, Sule N, Birkaya B, Chaves L, Auerbach A, Szigeti K (2019) iPSC model of CHRFAM7A effect on α7 nicotinic acetylcholine receptor function in the human context. Transl Psychiatry 9(1):59

Improgo MRD, Scofield MD, Tapper AR, Gardner PD (2010) The nicotinic acetylcholine receptor CHRNA5/A3/B4 gene cluster: dual role in nicotine addiction and lung cancer. Prog Neurobiol 92:212–226

Janssen I, Shepard DS, Katzmarzyk PT, Roubenoff R (2004) The healthcare costs of sarcopenia in the United States. J Am Geriatr Soc 52(1):80–85

Jones IW, Wonnacott S (2004) Precise localization of alpha7 nicotinic acetylcholine receptors on glutamatergic axon terminals in the rat ventral tegmental area. J Neurosci 24(50):11244–11252

Juneja P, Horlacher R, Bertrand D, Krause R, Marger F, Welte W (2014) An internally modulated, thermostable, pH-sensitive Cys loop receptor from the hydrothermal vent worm Alvinella pompejana. J Biol Chem 289(21):15130–15140

Kabbani N, Nichols RA (2018) Beyond the channel: metabotropic signaling by nicotinic receptors. Trends Pharmacol Sci 39(4):354–366

Katz E, Verbitsky M, Rothlin CV, Vetter DE, Heinemann SF, Elgoyhen AB (2000) High calcium permeability and calcium block of the alpha 9 nicotinic acetylcholine receptor. Hear Res 141 (1–2):117–128

Keefe RSE, Meltzer HA, Dgetluck N, Gawryl M, Koenig G, Moebius HJ, Lombardo I, Hilt DC (2015) Randomized, double-blind, placebo-controlled study of Encenicline, an Alpha-7 nicotinic acetylcholine receptor agonist as a treatment for cognitive impairment in schizophrenia. Neuropsychopharmacology 40(13):3053–3060

Keramidas A, Moorhouse AJ, French CR, Schofield PR, Barry PH (2000) M2 pore mutations convert the glycine receptor channel from being anion- to cation-selective. Biophys J 79 (1):247–259
King JR, Kabbani N (2018) Alpha 7 nicotinic receptors attenuate neurite development through calcium activation of calpain at the growth cone. PLoS One 13(5):e0197247
King JR, Nordman JC, Bridges SP, Lin M-K, Kabbani N (2015) Identification and characterization of a G protein-binding cluster in α7 nicotinic acetylcholine receptors. J Biol Chem 290 (33):20060–20070
Klink R, de Kerchove d'Exaerde A, Zoli M, Changeux JP (2001) Molecular and physiological diversity of nicotinic acetylcholine receptors in the midbrain dopaminergic nuclei. J Neurosci 21 (5):1452–1463
Krejci A, Michal P, Jakubik J, Ricny J, Dolezal V (2004) Regulation of signal transduction at M2 muscarinic receptor. Physiol Res 53(Suppl 1):S131–S140
Kruse AC, Hu J, Pan AC, Arlow DH, Rosenbaum DM, Rosemond E, Green HF, Liu T, Chae PS, Dror RO, Shaw DE, Weis WI, Wess J, Kobilka BK (2012) Structure and dynamics of the M3 muscarinic acetylcholine receptor. Nature 482(7386):552–556
Kruse AC, Hu J, Kobilka BK, Wess J (2014) Muscarinic acetylcholine receptor X-ray structures: potential implications for drug development. Curr Opin Pharmacol 16:24–30
Kuner R, Köhr G, Grünewald S, Eisenhardt G, Bach A, Kornau H-C (1999) Role of Heteromer formation in GABAB receptor function. Science 283:74–77
Kunii Y, Zhang W, Xu Q, Hyde TM, McFadden W, Shin JH, Deep-Soboslay A, Ye T, Li C, Kleinman JE, Wang KH, Lipska BK (2015) CHRNA7 and CHRFAM7A mRNAs: co-localized and their expression levels altered in the postmortem dorsolateral prefrontal cortex in major psychiatric disorders. Am J Psychiatry 172(11):1122–1130
Kuryatov A, Onksen J, Lindstrom J (2008) Roles of accessory subunits in alpha4beta2(*) nicotinic receptors. Mol Pharmacol 74(1):132–143
Lamotte d'Incamps B, Zorbaz T, Dingova D, Krejci E, Ascher P (2018) Stoichiometry of the Heteromeric nicotinic receptors of the Renshaw cell. J Neurosci 38(21):4943–4956
Langmead CJ, Watson J, Reavill C (2008) Muscarinic acetylcholine receptors as CNS drug targets. Pharmacol Ther 117(2):232–243
Lasala M, Fabiani C, Corradi J, Antollini S, Bouzat C (2019) Molecular modulation of human alpha7 nicotinic receptor by amyloid-beta peptides. Front Cell Neurosci 13:37
Lester RA, Dani JA (1995) Acetylcholine receptor desensitization induced by nicotine in rat medial habenula neurons. J Neurophysiol 74(1):195–206
Levey AI (1996) Muscarinic acetylcholine receptor expression in memory circuits: implications for treatment of Alzheimer disease. Proc Natl Acad Sci U S A 93(24):13541–13546
Levey AI, Kitt CA, Simonds WF, Price DL, Brann MR (1991) Identification and localization of muscarinic acetylcholine receptor proteins in brain with subtype-specific antibodies. J Neuroscience 11:3218–3226
Levey AI, Edmunds SM, Heilman CJ, Desmond TJ, Frey KA (1994) Localization of muscarinic m3 receptor protein and M3 receptor binding in rat brain. Neuroscience 63(1):207–221
Ma K-G, Qian Y-H (2019) Alpha 7 nicotinic acetylcholine receptor and its effects on Alzheimer's disease. Neuropeptides 73:96–106
Maanen MAV, Stoof SP, Zanden EPVD, Jonge WJD, Janssen RA, Fischer DF, Vandeghinste N, Brys R, Vervoordeldonk MJ, Tak PP (2009) The alpha7 nicotinic acetylcholine receptor on fibroblast-like synoviocytes and in synovial tissue from rheumatoid arthritis patients: a possible role for a key neurotransmitter in synovial inflammation. Arthritis Rheum 60(5):1272–1281
Maeda S, Qu Q, Robertson MJ, Skiniotis G, Kobilka BK (2019) Structures of the M1 and M2 muscarinic acetylcholine receptor/G-protein complexes. Science 364(6440):552–557
Marrero MB, Bencherif M (2009) Convergence of alpha 7 nicotinic acetylcholine receptor-activated pathways for anti-apoptosis and anti-inflammation: central role for JAK2 activation of STAT3 and NF-kappa B. Brain Res 1256:1–7

Marsango S, Ward R, Alvarez-Curto E, Milligan G (2018) Muscarinic receptor oligomerization. Neuropharmacology 136:401–410

Matsui M, Motomura D, Karasawa H, Fujikawa T, Jiang J, Komiya Y, Takahashi S, Taketo MM (2000) Multiple functional defects in peripheral autonomic organs in mice lacking muscarinic acetylcholine receptor gene for the M3 subtype. Proc Natl Acad Sci U S A 97(17):9579–9584

McGehee DS, Heath MJ, Gelber S, Devay P, Role LW (1995) Nicotine enhancement of fast excitatory synaptic transmission in CNS by presynaptic receptors. Science 269 (5231):1692–1696

Morales-Perez CL, Noviello CM, Hibbs RE (2016) X-ray structure of the human α4β2 nicotinic receptor. Nature 538(7625):411–415

Nelson ME, Wang F, Kuryatov A, Choi CH, Gerzanich V, Lindstrom J (2001) Functional properties of human nicotinic AChRs expressed by IMR-32 neuroblastoma cells resemble those of alpha3beta4 AChRs expressed in permanently transfected HEK cells. J Gen Physiol 118(5):563–582

Nelson ME, Kuryatov A, Choi CH, Zhou Y, Lindstrom J (2003) Alternate stoichiometries of alpha4beta2 nicotinic acetylcholine receptors. Mol Pharmacol 63(2):332–341

Nemecz Á, Prevost MS, Menny A, Corringer P-J (2016) Emerging molecular mechanisms of signal transduction in Pentameric ligand-gated ion channels. Neuron 90(3):452–470

Nie L, Song H, Chen M-F, Chiamvimonvat N, Beisel KW, Yamoah EN, Vázquez AE (2004) Cloning and expression of a small-conductance Ca (2+)-activated K+ channel from the mouse cochlea: coexpression with alpha9/alpha10 acetylcholine receptors. J Neurophysiol 91 (4):1536–1544

Ochoa EL, Chattopadhyay A, McNamee MG (1989) Desensitization of the nicotinic acetylcholine receptor: molecular mechanisms and effect of modulators. Cell Mol Neurobiol 9(2):141–178

Paradiso K, Brehm P (1998) Long-term desensitization of nicotinic acetylcholine receptors is regulated via protein kinase A-mediated phosphorylation. J Neurosci 18(22):9227–9237

Paradiso K, Zhang J, Steinbach JH (2001) The C terminus of the human nicotinic alpha4beta2 receptor forms a binding site required for potentiation by an estrogenic steroid. J Neurosci 21 (17):6561–6568

Peng H, Ferris RL, Matthews T, Hiel H, Lopez-Albaitero A, Lustig LR (2004) Characterization of the human nicotinic acetylcholine receptor subunit alpha (alpha) 9 (CHRNA9) and alpha (alpha) 10 (CHRNA10) in lymphocytes. Life Sci 76(3):263–280

Perez-Lloret S, Barrantes FJ (2016) Deficits in cholinergic neurotransmission and their clinical correlates in Parkinson's disease. NPJ Parkinsons Dis 2:16001

Perry DC, Xiao Y, Nguyen HN, Musachio JL, Dávila-García MI, Kellar KJ (2002) Measuring nicotinic receptors with characteristics of alpha4beta2, alpha3beta2 and alpha3beta4 subtypes in rat tissues by autoradiography. J Neurochem 82(3):468–481

Prickaerts J, van Goethem NP, Chesworth R, Shapiro G, Boess FG, Methfessel C, Reneerkens OA, Flood DG, Hilt D, Gawryl M, Bertrand S, Bertrand D, Konig G (2012) EVP-6124, a novel and selective alpha7 nicotinic acetylcholine receptor partial agonist, improves memory performance by potentiating the acetylcholine response of alpha7 nicotinic acetylcholine receptors. Neuropharmacology 62(2):1099–1110

Rathouz MM, Vijayaraghavan S, Berg DK (1995) Acetylcholine differentially affects intracellular calcium via nicotinic and muscarinic receptors on the same population of neurons. J Biol Chem 270(24):14366–14375

Razani-Boroujerdi S, Boyd RT, Dávila-García MI, Nandi JS, Mishra NC, Singh SP, Pena-Philippides JC, Langley R, Sopori ML (2007) T cells express alpha7-nicotinic acetylcholine receptor subunits that require a functional TCR and leukocyte-specific protein tyrosine kinase for nicotine-induced Ca2+ response. J Immunol 179(5):2889–2898

Rollema H, Hurst RS (2018) The contribution of agonist and antagonist activities of α4β2∗ nAChR ligands to smoking cessation efficacy: a quantitative analysis of literature data. Psychopharmacology (Berl) 235(9):2479–2505

Rollema H, Bertrand D Hurst R (2014) Nicotinic agonists and antagonists. Encyclopedia of psychopharmacology, vol 16(6), Springer, Heidelberg, pp 733–742
Rollema H, Bertrand D, Hurst R (2015) Nicotinic agonists and antagonists. In: Stolerman IP, Price LH (eds) Encyclopedia of psychopharmacology. Springer, Berlin
Roux I, Wersinger E, McIntosh JM, Fuchs PA, Glowatzki E (2011) Onset of cholinergic efferent synaptic function in sensory hair cells of the rat cochlea. J Neurosci 31(42):15092–15101
Rozycka A, Dorszewska J, Steinborn B, Kempisty B, Lianeri M, Wisniewska K, Jagodzinski PP (2013) A transcript coding for a partially duplicated form of α7 nicotinic acetylcholine receptor is absent from the CD4+ T-lymphocytes of patients with autosomal dominant nocturnal frontal lobe epilepsy (ADNFLE). Folia Neuropathol 51(1):65–75
Scarr E (2012) Muscarinic receptors: their roles in disorders of the central nervous system and potential as therapeutic targets. CNS Neurosci Ther 18(5):369–379
Schaaf CP (2014) Nicotinic acetylcholine receptors in human genetic disease. Genet Med 16 (9):649–656
Schubert J, Beckmann J, Hartmann S, Morhenn H-G, Szalay G, Heiss C, Schnettler R, Lips KS (2012) Expression of the non-neuronal cholinergic system in human knee synovial tissue from patients with rheumatoid arthritis and osteoarthritis. Life Sci 91(21–22):1048–1052
Séguéla P, Wadiche J, Dineley-Miller K, Dani JA, Patrick JW (1993) Molecular cloning, functional properties, and distribution of rat brain alpha 7: a nicotinic cation channel highly permeable to calcium. J Neurosci 13(2):596–604
Sgard F, Charpantier E, Bertrand S, Walker N, Caput D, Graham D, Bertrand D, Besnard F (2002) A novel human nicotinic receptor subunit, alpha10, that confers functionality to the alpha9-subunit. Mol Pharmacol 61(1):150–159
Shen W, Plotkin JL, Francardo V, Ko WK, Xie Z, Li Q, Fieblinger T, Wess J, Neubig RR, Lindsley CW, Conn PJ, Greengard P, Bezard E, Cenci MA, Surmeier DJ (2015) M4 muscarinic receptor signaling ameliorates striatal plasticity deficits in models of L-DOPA-induced dyskinesia. Neuron 88(4):762–773
Shinoe T, Matsui M, Taketo MM, Manabe T (2005) Modulation of synaptic plasticity by physiological activation of M1 muscarinic acetylcholine receptors in the mouse hippocampus. J Neurosci 25(48):11194–11200
Shorey-Kendrick LE, Ford MM, Allen DC, Kuryatov A, Lindstrom J, Wilhelm L, Grant KA, Spindel ER (2015) Nicotinic receptors in non-human primates: analysis of genetic and functional conservation with humans. Neuropharmacology 96(Pt B):263–273
Sinkus ML, Graw S, Freedman R, Ross RG, Lester HA, Leonard S (2015) The human CHRNA7 and CHRFAM7A genes: a review of the genetics, regulation, and function. Neuropharmacology 96(Pt B):274–288
Stauderman KA, Mahaffy LS, Akong M, Veliçelebi G, Chavez-Noriega LE, Crona JH, Johnson EC, Elliott KJ, Gillespie A, Reid RT, Adams P, Harpold MM, Corey-Naeve J (1998) Characterization of human recombinant neuronal nicotinic acetylcholine receptor subunit combinations alpha2beta4, alpha3beta4 and alpha4beta4 stably expressed in HEK293 cells. J Pharmacol Exp Ther 284(2):777–789
Stoiljkovic M, Leventhal L, Chen A, Chen T, Driscoll R, Flood D, Hodgdon H, Hurst R, Nagy D, Piser T, Tang C, Townsend M, Tu Z, Bertrand D, Koenig G, Hajos M (2015) Concentration-response relationship of the alpha7 nicotinic acetylcholine receptor agonist FRM-17874 across multiple in vitro and in vivo assays. Biochem Pharmacol 97(4):576–589
Tammimäki A, Herder P, Li P, Esch C, Laughlin JR, Akk G, Stitzel JA (2012) Impact of human D398N single nucleotide polymorphism on intracellular calcium response mediated by α3β4α5 nicotinic acetylcholine receptors. Neuropharmacology 63(6):1002–1011
Tapia L, Kuryatov A, Lindstrom J (2007) Ca2+ permeability of the (alpha4)3(beta2)2 stoichiometry greatly exceeds that of (alpha4)2(beta2)3 human acetylcholine receptors. Mol Pharmacol 71 (3):769–776
Terry AV (2008) Role of the central cholinergic system in the therapeutics of schizophrenia. Curr Neuropharmacol 6(3):286–292

Terry AV Jr, Callahan PM, Bertrand D (2015) R-(+) and S-(−) isomers of cotinine augment cholinergic responses in vitro and in vivo. J Pharmacol Exp Ther 352(2):405–418
Thal DM, Sun B, Feng D, Nawaratne V, Leach K, Felder CC, Bures MG, Evans DA, Weis WI, Bachhawat P, Kobilka TS, Sexton PM, Kobilka BK, Christopoulos A (2016) Crystal structures of the M1 and M4 muscarinic acetylcholine receptors. Nature 531(7594):335–340
Uteshev VV (2012) α7 nicotinic ACh receptors as a ligand-gated source of Ca(2+) ions: the search for a Ca(2+) optimum. Adv Exp Med Biol 740:603–638
van Koppen CJ, Kaiser B (2003) Regulation of muscarinic acetylcholine receptor signaling. Pharmacol Ther 98(2):197–220
van Nierop P, Bertrand S, Munno DW, Gouwenberg Y, van Minnen J, Spafford JD, Syed NI, Bertrand D, Smit AB (2006) Identification and functional expression of a family of nicotinic acetylcholine receptor subunits in the central nervous system of the mollusc Lymnaea stagnalis. J Biol Chem 281(3):1680–1691
Verbitsky M, Rothlin CV, Katz E, Elgoyhen AB (2000) Mixed nicotinic-muscarinic properties of the alpha9 nicotinic cholinergic receptor. Neuropharmacology 39(13):2515–2524
Vernino S, Amador M, Luetje CW, Patrick J, Dani JA (1992) Calcium modulation and high calcium permeability of neuronal nicotinic acetylcholine receptors. Neuron 8(1):127–134
Vibat CR, Lasalde JA, McNamee MG, Ochoa EL (1995) Differential desensitization properties of rat neuronal nicotinic acetylcholine receptor subunit combinations expressed in Xenopus laevis oocytes. Cell Mol Neurobiol 15(4):411–425
Villiger Y, Szanto I, Jaconi S, Blanchet C, Buisson B, Krause KH, Bertrand D, Romand JA (2002) Expression of an alpha7 duplicate nicotinic acetylcholine receptor-related protein in human leukocytes. J Neuroimmunol 126(1–2):86–98
Volpicelli LA, Levey AI (2004) Muscarinic acetylcholine receptor subtypes in cerebral cortex and hippocampus. In: Acetylcholine in the cerebral cortex. Elsevier, Amsterdam, pp 59–66
Wallace TL, Bertrand D (2013a) Alpha7 neuronal nicotinic receptors as a drug target in schizophrenia. Expert Opin Ther Targets 17(2):139–155
Wallace TL, Bertrand D (2013b) Importance of the nicotinic acetylcholine receptor system in the prefrontal cortex. Biochem Pharmacol 85(12):1713–1720
Walsh RM Jr, Roh SH, Gharpure A, Morales-Perez CL, Teng J, Hibbs RE (2018) Structural principles of distinct assemblies of the human alpha4beta2 nicotinic receptor. Nature 557 (7704):261–265
Wang H, Yu M, Ochani M, Amella CA, Tanovic M, Susarla S, Li JH, Wang H, Yang H, Ulloa L, Al-Abed Y, Czura CJ, Tracey KJ (2003) Nicotinic acetylcholine receptor alpha7 subunit is an essential regulator of inflammation. Nature 421(6921):384–388
Wang X-J, Liu Y-F, Wang Q-Y, Tsuruoka M, Ohta K, Wu S-X, Yakushiji M, Inoue T (2010) Functional expression of alpha 7 nicotinic acetylcholine receptors in human periodontal ligament fibroblasts and rat periodontal tissues. Cell Tissue Res 340(2):347–355
Wess J, Eglen RM, Gautam D (2007) Muscarinic acetylcholine receptors: mutant mice provide new insights for drug development. Nat Rev Drug Discov 6(9):721–733
Wessler I, Kirkpatrick CJ (2008) Acetylcholine beyond neurons: the non-neuronal cholinergic system in humans. Br J Pharmacol 154(8):1558–1571
Williams NM, Franke B, Mick E, Anney RJL, Freitag CM, Gill M, Thapar A, O'Donovan MC, Owen MJ, Holmans P, Kent L, Middleton F, Zhang-James Y, Liu L, Meyer J, Nguyen TT, Romanos J, Romanos M, Seitz C, Renner TJ, Walitza S, Warnke A, Palmason H, Buitelaar J, Rommelse N, Vasquez AA, Hawi Z, Langley K, Sergeant J, Steinhausen H-C, Roeyers H, Biederman J, Zaharieva I, Hakonarson H, Elia J, Lionel AC, Crosbie J, Marshall CR, Schachar R, Scherer SW, Todorov A, Smalley SL, Loo S, Nelson S, Shtir C, Asherson P, Reif A, Lesch K-P, Faraone SV (2012) Genome-wide analysis of copy number variants in attention deficit hyperactivity disorder: the role of rare variants and duplications at 15q13.3. Am J Psychiatry 169(2):195–204

Zaninetti M, Tribollet E, Bertrand D, Raggenbass M (1999) Presence of functional neuronal nicotinic acetylcholine receptors in brainstem motoneurons of the rat. Eur J Neurosci 11 (8):2737–2748
Zhou Y, Nelson ME, Kuryatov A, Choi C, Cooper J, Lindstrom J (2003) Human alpha4beta2 acetylcholine receptors formed from linked subunits. J Neurosci 23(27):9004–9015
Zwart R, Vijverberg HP (1998) Four pharmacologically distinct subtypes of alpha4beta2 nicotinic acetylcholine receptor expressed in Xenopus laevis oocytes. Mol Pharmacol 54(6):1124–1131

Acetylcholine and Spontaneous Recognition Memory in Rodents and Primates

Alexander Easton, Marilia Barros, and Colin Lever

Contents

Abstract Whilst acetylcholine has long been linked to memory, there have been significant questions about its specific role. In particular, the effects of cholinergic manipulations in primates and rodents has often been at odds. Here, we review the work in primates and rodents on the specific function of acetylcholine in memory, and episodic memory in particular. We propose that patterns of impairment can best be understood in terms of a role for hippocampal acetylcholine in resolving spatial interference and we discuss the benefits of new tasks of episodic memory in animals allowing clearer translation of findings to the clinic.

Keywords Acetylcholine · Episodic memory · Interference

Acetylcholine has long been linked to a role in memory (Drachman 1977; Hasselmo 2006; Micheau and Marighetto 2011), with loss of the transmitter evident in the early stages of Alzheimer's disease associated specifically with memory loss (Bierer et al. 1995). In particular, the projections of cholinergic cells from the basal forebrain to

A. Easton (✉) and C. Lever
Department of Psychology, Durham University, Durham, UK

Learning and Memory Processes Centre, Durham University, Durham, UK
e-mail: alexander.easton@durham.ac.uk

M. Barros
Department of Pharmacy, University of Brasilia, Brasilia, Brazil

Curr Topics Behav Neurosci (2020) 45: 29–46
https://doi.org/10.1007/7854_2020_132
Published Online: 28 May 2020

the cerebral cortex and hippocampus rising within the medial septum (MS) and vertical limb of the diagonal band (vDB) have been linked to memory in monkeys (e.g. Easton et al. 2002; Fine et al. 1997; Ridley et al. 1999), and lesions of the basal forebrain produce a profound amnesia in humans (e.g. Deluca and Diamond 1995; Norlen and Olivecrona 1953). However, cholinergic cells are not the only cells present within this region of the basal forebrain, and damage in animals and humans has rarely been restricted to the MS/VDB and even more rarely to only the cholinergic projections from this region. As a result, the necessary involvement of these cholinergic cells in memory has been much debated (e.g. Baxter and Chiba 1999; Easton et al. 2012a; Hasselmo 2006; Parent and Baxter 2004), with evidence that the role of acetylcholine may be more specific to attentional mechanisms than memory per se (e.g. Baxter and Chiba 1999).

It is therefore very important when assessing the impact of cholinergic manipulations on memory to consider what is meant by memory and what is being modelled. As in much of neuroscience, the issue of translation from animal research into the clinic is a key concern. It is essential that we move away, therefore, from considering the debate to be one of acetylcholine's involvement in broadly described terms of 'memory' or 'attention'. Rather, we would argue that one needs to look carefully at the precise nature of the behavioural task being used (Ameen-Ali et al. 2015) and the nature of the behavioural impairment. Specific hypotheses about the nature of acetylcholine's function can be best arrived at through the careful consideration of specific elements of behaviour and its relation to the manipulation at hand (Easton et al. 2012a).

If we intend to model the clinical concerns of memory loss in ageing and Alzheimer's, then we need to consider specific aspects of memory. In particular, early stages of Alzheimer's (those which are associated primarily with a specific decline in cholinergic markers; Bierer et al. 1995) are associated with loss of episodic memory (Collie and Maruff 2000). Episodic memory is the memory for specific, personally experienced events in one's life (Tulving 1983). In humans it usually comes with the explicit conscious experience of recollecting and reliving the experience as it was originally experienced (so called mental time travel; Suddendorf and Corballis 2007). The element of conscious experience in episodic memory has led some to argue that it is a uniquely human form of memory (Suddendorf and Corballis 2007; Tulving 2002). However, many researchers have now presented behavioural models of episodic memory in a wide range of non-human species (e.g. Babb and Crystal 2006; Clayton and Dickinson 1998; Eacott and Norman 2004; Eacott et al. 2005; Ferkin et al. 2008; Kart-Teke et al. 2006; Singer and Zentall 2007) meaning that we can now explore the role of acetylcholine in this specific form of memory.

1 A Content-Based Approach to Episodic Memory

In humans, episodic memory is primarily associated with the conscious experience of recollection. When one remembers what one ate for breakfast, you remember it not in isolation, but as a relived experience, remembering who was there, what time it was, the taste and smells, the emotions of being rushed getting ready for work, etc. In Tulving's original description of episodic memory, he described it as memory which 'receives and stores information about temporally dated episodes or events, and tempero-spatial relations between them' (Tulving 1983). The conscious reliving of this experience has been termed 'mental time travel', and it is this critical inclusion of conscious re-experience of the memory which pushes some towards the view that only humans are capable of episodic memories (Suddendorf and Corballis 2007). Without being able to conclusively be persuaded that non-human animals have a conscious experience, it is impossible to conclude that they have a form of memory so intrinsically tied to it.

However, conscious experience occurs for all sorts of cognitive phenomena, yet it does not impact on the description of the cognitive process to the same degree as it does in episodic memory. When we see an object, we have a clear conscious experience of perception, and yet this has not prevented us from using animal models to carefully explore the neural basis of such perception, even when the conscious experience of the animal model cannot be understood. In short, one would not deny that a monkey or a rat can see, just because they may not have the conscious experience afforded to humans when they see. Therefore the question remains why animals are not easily afforded the concept of episodic memory purely on the basis of the potential absence of a conscious experience.

As a result of this limitation imposed by consciousness, Clayton and Dickinson (1998) proposed an alternative approach to defining episodic memory in a way that could be modelled outside of humans. Their demonstration of what-where-when (WWW) memory in scrub jays showed that these birds could adapt their behavioural response to a particular food (what; worms or peanuts) they had cached in a particular location (where) and at a particular time (recently or several days previously). They argued that this memory of what happened, where and when met Tulving's description of episodic memory. Also, there are still additional criteria which need to be met to ensure such a content-based description truly captures the essence of this clinically important form of memory. For example, one might remember what (you were born), where and when, and yet this would not be an episodic memory as it is not the recollection of a personally experienced event; rather it stems from semantic memory. Therefore Clayton et al. (Clayton et al. 2003) set out a series of other criteria that were required for episodic memory to be demonstrated in animals, including its structure (which must be an integrated single memory, not the combination of multiple memories) and flexibility to remember things with no explicit reason for knowing that they needed to be remembered.

This content-based description of episodic memory has become more widely accepted in recent years. However, what-where-when memory has not been consistently demonstrated across species, including non-human primates (Hampton et al. 2005). One critical limit has been the importance of time to WWW memory. Although Tulving spoke of 'temporally dated' events, actual dating of memories in humans is very difficult and often relies on non-episodic information (Friedman 1993, 2007). Alternate versions of WWW were subsequently developed with the aim of maintaining the content of the memory (what happened, where and the occasion it happened on) but allowing the occasion to be defined in ways other than purely information about when it happened. In particular, context has been used to define one event as being separate from another (Eacott and Norman 2004; Robertson et al. 2015).

In the what-where-which occasion task, Eacott and Norman (2004) used a spontaneous recognition task in rodents, where animals demonstrate their memory through preferential exploration of novel items. In exploring a novel item (or combination of features), they demonstrate their memory for having seen the more familiar item before (Ennaceur and Delacour 1988). In the first sample event, rats were exposed to two objects (e.g. A and B) in left and right positions within an open field with a particular visuo-tactile context present in the arena (e.g. context X; a metal mesh on the walls and floors of the arena). After a short delay, a second sample event was presented in the same arena but with a new context present (e.g. context Y; a patterned and ridged plastic floor). The same objects (A and B) were presented again but now in reversed positions (i.e. in context X, object A would be on the right of the arena, but in context Y it would be on the left of the arena). After a delay period, the animal would be returned to the arena with one of the previous contexts present and two new copies of one of the previously seen objects. For example, the rat may have been returned to context X and seen two new copies of object A in the left and right positions. In this case the context is familiar, as is the object, the combination of object A on the left and on the right, and in context X. However, there is novelty in this test stage as in this particular example when object A was seen in context X, it was on the right-hand side of the arena. Therefore the presence of object A on the left side of the arena in context X is novel (A has only been seen on the left in context Y previously). Therefore novelty in this task is not defined by an individual feature (object, location or context) but rather as a coherent single memory of what has been seen, where it has been seen and which occasion (context X or Y) it was seen there.

2 The Role of Acetylcholine in What-Where-Which Occasion Memory

As discussed earlier, the importance of an animal model of episodic memory is that it provides a close match to the clinically relevant form of memory that is impaired early in diseases such as Alzheimer's (Collie and Maruff 2000). In the case of the proposed relationship between acetylcholine and memory loss in Alzheimer's, the ability to explore this relationship in the type of memory loss seen in the clinic improves the ability to translate findings from animal to human studies.

Using an immunotoxic lesion (IgG-saporin) designed to specifically target cells that express acetylcholine as a transmitter (Wiley et al. 1991), Easton and colleagues investigated the role of specific cholinergic input to the hippocampus (Easton et al. 2010). Within the basal forebrain where cholinergic projections arise, the MS/vDB project directly to the hippocampus (Mesulam et al. 1983). Targeting these structures with the immunologic lesion therefore aimed to reduce cholinergic input specifically to the hippocampus. The hippocampus itself is known to be critical for both episodic memory in humans (e.g. Aggleton and Brown 1999; Bayley and Squire 2003; Scoville and Milner 1957) and for what-where-which occasion memory in rodents (Eacott and Norman 2004; Langston and Wood 2010). However, cholinergic depletion of the hippocampus had no effect on this episodic memory task (Easton et al. 2010). The lesion was selective for acetylcholine (GABAergic cells were reliably intact following the lesion) and effective as another behavioural task was found to be impaired in these same animals (a where-which task; see below). The lack of impairment in the episodic task was also not simply a result of the lesion being slow to develop as returning to the episodic task after seeing an impairment in the where-which task still showed no impairment in what-where-which occasion memory (Easton et al. 2010).

The where-which task impaired in these animals was another spontaneous recognition memory task, but this time novelty was defined by the combination of location and context (i.e. at test one location was filled with an object which had not been previously occupied in that context at sample but had been occupied in a sample with another context). The task is based on spatial-context conditional discriminations in reward-based tasks which had previously been shown to be impaired following cholinergic lesions to the hippocampus in both marmosets (Ridley et al. 1999) and rats (Janisiewicz et al. 2004).

As a result, the pattern of results from this study leaves us with two unusual observations. First, the hippocampus is necessary for both the episodic memory task and the where-which task, and yet cholinergic inputs to the hippocampus are necessary only for the where-which task. This means that there must be dissociation of function within the hippocampus on the basis of cholinergic input. Some hippocampal tasks rely on acetylcholine, and some don't, and these tasks can be manipulated independently of each other. However, with only a single dissociation in evidence, it remains possible that some simpler explanation remains for this pattern of results, such as task difficulty, with only more difficult tasks being sensitive to the removal of acetylcholine. However, in the case of this current set of data, this would require the two-component where-which task to be *more* difficult than the three-component what-where-which occasion task, even though there are overlapping features between the tasks. Indeed, the discrimination ratio in both tasks is very similar in these animals (Easton et al. 2009) implying that the episodic task is not obviously more difficult as control animals are equally able to show memory ability in both tasks. Such a task difficulty explanation therefore remains unlikely, leaving us to conclude that there is dissociation within the hippocampus based on the necessity for acetylcholine in performing memory tasks. The hippocampus is not the only site where such dissociation on the basis of cholinergic involvement is seen. Only a portion of tasks dependent on the prefrontal cortex in macaques depend on the cholinergic projections to prefrontal regions (Croxson et al. 2011).

The second unusual observation from this data is that procedures similar to the what-where-which occasion task have been impaired following cholinergic lesions in non-human primates (Easton et al. 2002). In a scene learning task, monkeys were taught a visual discrimination task (to simply learn which one of two objects presented was correct by trial and error), and these discriminations took place against a background scene. Each time the visual discrimination problem was presented, the same objects were in the same locations against the same trial unique background. New problems were presented against a different background and with the objects in different spatial locations. This scene task has also been argued to model episodic memory in monkeys (Gaffan 1994), is reliant on the hippocampus (Gaffan 1994) and requires the animals to solve problems with the content of object (what), location (where) and background scene (which occasion). It seems, then, that the results of immunotoxic lesions in primates impairing this episodic task (Easton et al. 2002) but the same lesions showing no effect on a similar episodic task in rats (Easton et al. 2010) cause some problems in interpretation.

These tasks of episodic memory in primates and rats are somewhat different, despite the apparent similarity in their content. In rodents, the what-where-which occasion task is one of spontaneous recognition. Animals require no training to perform the task and are not rewarded for their behavioural choices. In contrast, the scene learning task in monkeys is one of visual discrimination and therefore requires that animals learn to choose one object over another in order to achieve maximum food reward. In humans, episodic memory is spontaneous and requires no explicit effort to encode information. It is possible, then, that this difference in reward motivation and learning between the rodent and primate task is sufficient to explain the difference in outcome following cholinergic lesions. However, both the primate scene learning task and the spontaneous rodent episodic memory task have been adapted for use in humans, and they show either phenomenological similarity to episodic memory (Easton et al. 2012b) or impairment in amnesic patients (Aggleton et al. 2000). Subtle differences between the tasks then seem unlikely to cause such significant differences in the effect of cholinergic lesions.

A more likely cause of the difference between the results in rodents and primates is the scale of the cholinergic lesion used. In the rodent task, the cholinergic lesion was targeted at the hippocampus, with the lesion made in the MS/vDB that projects directly to the hippocampus. However, in the primate studies, lesions extended beyond the MS/vDB and into the nucleus basalis of Meynert (NBM), meaning cortical regions including the perirhinal cortex were also depleted of their cholinergic input. Although the primate scene learning task is dependent upon the hippocampus (Gaffan 1994), structures in the temporal and medial temporal cortices are also necessary (Easton and Gaffan 2000; Murray et al. 1998). It is unclear, then, whether in primates cholinergic lesions of the hippocampus alone may have impaired the scene learning task, or whether a lack of impairment would have been seen, as in the rodents' episodic memory task. Similarly, it remains unclear whether a more widespread lesion of the cholinergic system in rodents might have produced an impairment in the episodic memory task.

3 Differential Roles of Acetylcholine in the Hippocampus and Perirhinal Cortex

Whether cholinergic input to both the hippocampus and perirhinal cortex is necessary for episodic memory in animals is an important question, as cholinergic inputs to these two regions are known to have very different patterns of impairment. In rats, intraperitoneal (i.p.) injection of scopolamine (which will have central effects across both regions) impairs spontaneous recognition memory, although only at higher doses than those which impaired the same animals on a radial maze spatial learning task (Ennaceur and Meliani 1992). Barros and colleagues have recently shown object-location memory deficit in marmosets as well after scopolamine was given i.p. but at doses that also impaired their contextual fear-conditioning (Melamed et al. 2017). Closer investigation shows that whilst i.p.-administered scopolamine impairs spontaneous recognition, it only does so when delivered during the encoding (sample) phase, but not when administered during the retrieval (test) phase (Melamed et al. 2017; Warburton et al. 2003). Scopolamine-induced impairment in rat spontaneous object recognition and marmoset object discrimination can also be reversed by a number of nootropic drugs (e.g. rat, Milić et al. 2013; Rutten et al. 2006; Woolley et al. 2009; marmoset, Carey et al. 1992).

Direct infusion of scopolamine into the perirhinal cortex mirrors the effects of systemic administration in impairing object recognition (rat, Warburton et al. 2003; macaque, Tang et al. 1997), and so the impairments from systemic administration cannot be ascribed simply to peripheral effects which might serve to cause particular confounds in a task of spontaneous exploratory behaviour. In contrast, scopolamine infusions into the hippocampus produce impairments in spatial memory (e.g. Blokland et al. 1992; Givens and Olton 1995).

Immunotoxic lesions of the cholinergic projections to either the perirhinal cortex or hippocampus mirror the effects of direct scopolamine administration. Cholinergic lesions of the perirhinal cortex through direct injections into the cortex impair object recognition memory in rats (Winters and Bussey 2005). Perirhinal lesions in macaques (Turchi et al. 2005) and NBM lesions in marmosets (Ridley et al. 1999) lead to similar object discrimination impairments. In contrast, specific cholinergic lesions of the hippocampus produce a reliable impairment in spontaneous recognition of spatial locations in rats (Cai et al. 2012) and visuospatial discriminations in marmosets (Ridley et al. 1999). Cholinergic agents can minimize these immunotoxin-induced performance impairments (Ridley et al. 1999).

4 Does the What-Where-Which Task Measure Episodic Memory?

Given the impact of cholinergic manipulations on spontaneous recognition of objects, locations and object-locations, the lack of impairment in what-where-which stands out. The perirhinal cortex supports object recognition (Eacott and Gaffan 2005; Murray et al. 1998), and cholinergic inputs to perirhinal cortex are required for this, whether in spontaneous recognition tasks in rodents (Winters and Bussey 2005) or in rewarded object recognition tasks such as delayed match to sample in primates (Turchi et al. 2005). Similarly the hippocampus is required for many spatial learning tasks, and at least some of these tasks are also dependent upon the cholinergic inputs to the hippocampus (e.g. Cai et al. 2012). However, the episodic memory task (what-where-which) appears to be different in that it requires the hippocampus (Eacott and Norman 2004; Langston and Wood 2010) but not the cholinergic inputs to the hippocampus (Easton et al. 2010).

In humans, there are data that implicate the cholinergic system in episodic memory particularly. Lesions within the basal forebrain give rise to significant amnesia (e.g. Deluca and Diamond 1995), although cholinergic cells will not be the only types affected by such lesions. However, in the early stages of Alzheimer's disease, where episodic memory is primarily affected (Collie and Maruff 2000), it is biomarkers of cholinergic activity that best predict memory performance (Bierer et al. 1995). One possibility, then, is that the what-where-which task in rodents simply does not measure episodic memory and that the content-based approach to modelling episodic memory may not be sufficient. However, there are several reasons to believe this is not the case.

As discussed above, any task of episodic memory should meet a number of criteria, not just show what-where-which occasion content (Clayton et al. 2003). One of those criteria is that the memory should be a single coherent memory, and not the result of summation of independent memories for different components of the event (such as what or where) on their own. The very lack of impairment in the what-where-which task strongly suggests that the memory is not the result of simple summations of component memories. Although manipulations of the cholinergic system in the hippocampus do not impair the episodic memory task, they do impair where-which memory (Easton et al. 2010). If the episodic memory task were merely a summation of smaller component tasks, then the failure to be able to process some of these components (e.g. where-which memory) should prevent the overall completion of a task requiring those components. That where-which recognition and what-where-which recognition are dissociable in terms of their requirement for acetylcholine in the hippocampus shows us that the episodic memory task uses a single coherent memory for the entire event rather than just combining components together.

Further evidence that the what-where-which task measures episodic memory comes from human data. Human participants run on versions of the rodent what-where-which task have shown that these memories require recollection and cannot be solved using familiarity alone, even though a what-where-when version of the same task can be, implying a clear link to episodic memory processes (Easton et al. 2012b; Persson et al. 2016). In addition, we have run human participants on an object recognition task in which participants only have to make old or new judgements about individual objects. However, without it being necessary for solving the object recognition task aspects of the location, the background context or both were altered between encoding and retrieval of the object memory. In this case the degree to which recollection was used (compared to familiarity) increased markedly when the spatial location of the object and the background context were identical at encoding and retrieval, whilst matching either location or background context on its own did not lead to the same increase in recollection (Ameen-Ali et al. 2017). Together these studies show that in humans, what-where-which memories rely on recollection and phenomenologically appear very similar to episodic memory as defined through non-content-based descriptions.

Together, then, it appears very unlikely that the lack of impairment in the what-where-which task in rodents with lesions of the cholinergic input to the hippocampus is a result simply of a mismatch between the task and the cognitive process it aims to model. Rather, we may be able to explain the lack of impairment in the episodic memory task by looking to the role of acetylcholine in encoding and retrieval.

5 The Role of Acetylcholine in Encoding and Retrieval

Acetylcholine is released during exposure to novelty, and higher levels of acetylcholine boost a wide array of novelty-oriented processes, such as exploration (e.g. rearing on hind legs) and synaptic plasticity, reviewed in (Easton et al. 2012a; Hasselmo 2012; Lever et al. 2006; Poulter et al. 2018). One of acetylcholine's effects is to reduce proactive interference, i.e. interference from previously encoded associations. Scopolamine administration to the perirhinal cortex impairs object recognition memory when given at encoding, but not at retrieval (Warburton et al. 2003). This impairment of encoding but not retrieval is also a common outcome of scopolamine administration in other domains such as hippocampal-dependent spatial memory (Deiana et al. 2011; Easton et al. 2012a) and sits alongside observations of the role of acetylcholine in interference (Winters et al. 2007). When interfering stimuli are presented (i.e. stimuli similar to those used in the experiment irrelevant to the experiment) in the presence of scopolamine administration, there is a surprising improvement in object recognition memory (Winters et al. 2006). This effect has been attributed to acetylcholine's involvement in encoding all object information. If information about irrelevant objects is encoded after the experimental encoding stage, then this can interfere with the experimentally relevant memories. In contrast, if scopolamine is administered when these interfering stimuli are presented, then they will fail to be encoded well and therefore will have a lesser interfering effect on the experimental stimuli meaning those experimental stimuli will be better remembered as a result.

These interference-related problems have been seen in computational modelling of encoding and retrieval and have led to a set of high-profile models of the way in which acetylcholine allows the separation of encoding and retrieval states (Douchamps et al. 2013; Hasselmo 1999, 2006, 2012; Meeter et al. 2004). By suppressing recurrent inputs within the hippocampus, notably those mediated by region CA3, interference can be reduced by preventing the retrieval of previously learned associations from pattern completion. Instead, pattern separation is encouraged allowing distinct items to be encoded separately from one another with reduced interference (Duncan et al. 2012; Hasselmo 1999, 2006; Meeter et al. 2004). In contrast, low levels of acetylcholine would then improve retrieval and consolidation of information, and such low levels can be seen in states such as slow wave sleep in which memory consolidation is thought to occur (Gais and Born 2004).

Such models explain that in any task in which proactive interference is likely to occur, acetylcholine is important in order to help encode novel information in spite of interfering information. How might these models explain the role of acetylcholine in the hippocampus in a where-which but not a what-where-which task in rats (Easton et al. 2011) where levels of interference might be expected to be very similar? Indeed, interference could be even higher in the *what-where-which* task as the same objects are experienced by the animal at each phase of the trial. However, in these animals only cholinergic inputs to the hippocampus are lesioned, and we know the hippocampus has a high level of involvement in spatial memory. In the *what-where-which* task, the location of a particular object changes across the trial and across contexts, but every time the animal goes into the arena, objects are always to the left and right of the animal. As a result the purely spatial component of this memory does not change. In contrast, locations of objects in the *where-which* task constantly shift within the trial. On no two entries into the arena are objects in the same two locations. As a result there is more potential for *spatial* interference in this task as a series of entries into the arena have to be separated in memory by distinguishing between highly similar but not identical locations within that arena. If acetylcholine in the hippocampus was particularly important for reducing the impact of potential interference in spatial memory, then we might expect it to be of more importance in the where-which task than the what-where-which task because of the instability of spatial locations over trials in the where-which task (Easton et al. 2012a) (Fig. 1).

To explicitly test this hypothesis, we recently investigated both the where-which and what-where-which tasks in rats with lesions of the cholinergic projections to the hippocampus but in versions of those tasks where many trials were run consecutively rather than in a one trial a day manner (Seel et al. 2018). By running the tasks using this continual trials approach (Ameen-Ali et al. 2012; Chan et al. 2018), we were able to raise the levels of proactive interference in both tasks. The nature of running many trials consecutively means that each trial is highly similar (with overlap of object features, contexts and spatial locations) and therefore proactive interference is a feature of the design and can be seen in performance of normal animals on some tasks (Chan et al. 2018). However, although overall interference levels will have gone up in both task versions, because there are more trials than in standard versions

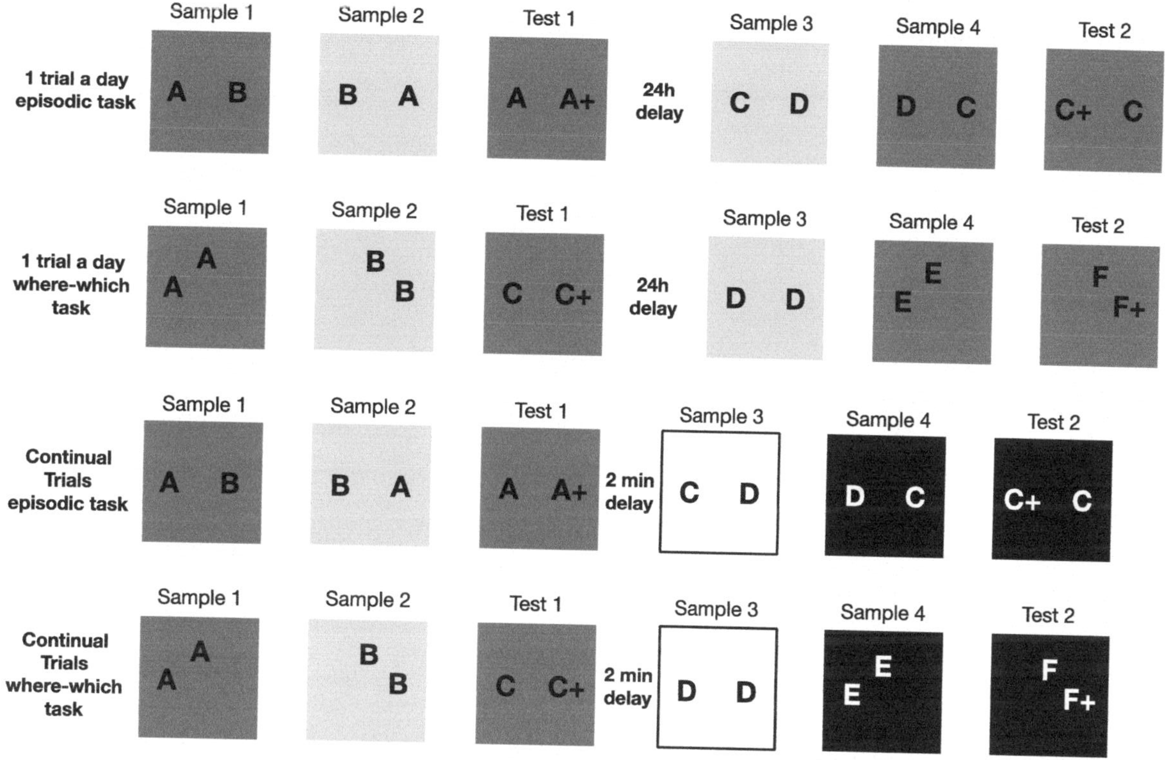

Fig. 1 Schematic representations of the episodic (what-where-which) and where-which tasks in rats. For comparison one trial a day versions (top two panels) are presented where there is a 24 h delay between trials, and 16 trials are run over 16 days of testing. The continual trials version (bottom two panels) have only a 2 min delay between trials with all trials (16 in total) happening in a single session. To allow animals to distinguish the separate trials run within a continual trials session, more contexts are used in the continual trials versions of the task than the one trial a day versions. Objects are real-world junk objects, with unique

Fig. 1 (continued) objects being identified by different letters in the figure above. In all cases + indicates the novel feature at test for the trial shown. Differences in spatial interference can be seen between the tasks with the location of objects fixed in left/right positions every time the animal enters the apparatus for the episodic memory task. In contrast the positions of objects vary not just on a trial by trial basis but also across samples and tests for the where-which task, leading to an increased need to maintain accurate representations of object-locations in this task (Easton et al. 2010; Seel et al. 2018)

of the tasks, it remains the case that spatial variability remains high only in the where-which task. As a result, if acetylcholine in the hippocampus was required to resolve all interference, then we would expect the loss of acetylcholine in the hippocampus to impact on both tasks using continual trials. If on the other hand acetylcholine in the hippocampus is only necessary to resolve spatial interference, then we would expect still to see a role for hippocampal acetylcholine in where-which memory but not in what-where-which memory, as before. We found that IgG-saporin lesions of the MS/vDB in rats continued to only impair where-which memory and not what-where-which memory, even when run with this high level of interference (Seel et al. 2018).

These findings support the idea that whilst the hippocampus is necessary for what-where-which memory, acetylcholine in the hippocampus is only required for identification of spatial novelty. This may also, then, explain the difference between the what-where-which task in rodents and the scene learning task in primates. The what-where-which task involves objects being presented in stable spatial locations within and across trials, meaning there is limited opportunity for interference in the spatial component of this memory (Easton et al. 2012a; Easton et al. 2011). In contrast, the scene learning task in monkeys more closely resembles the where-which task in rats in that the locations of objects are trial unique and therefore these highly similar spatial locations need to be separated in memory. This separation requires the cholinergic system to promote encoding of separate locations despite high levels of spatial interference. With the use of spontaneous recognition tasks to explore cholinergic function across rodents (Easton et al. 2011; Seel et al. 2018; Winters and Bussey 2005) and primates (Melamed et al. 2017), we will be able to make more reliable comparisons across species. In addition, evidence that spontaneous recognition tasks of episodic memory can be translated to episodic memory in humans (Ameen-Ali et al. 2017; Easton et al. 2012b) takes us to a position where we are now able to improve translation from animal studies to the clinic.

References

Aggleton JP, Brown MW (1999) Episodic memory, amnesia and the hippocampal-anterior thalamic axis. Behav Brain Sci 22:425–444

Aggleton JP, McMakin D, Carpenter K, Hornak J, Kapur N, Halpin S et al (2000) Differential cognitive effects of colloid cysts in the third ventricle that spare or compromise the fornix. Brain 123:800–815

Ameen-Ali KE, Eacott MJ, Easton A (2012) A new behavioral apparatus to reduce animal numbers in multiple types of spontaneous object recognition paradigms in rats. J Neurosci Methods 211 (1):66. https://doi.org/10.1016/j.jneumeth.2012.08.006

Ameen-Ali KE, Easton A, Eacott MJ (2015) Moving beyond standard procedures to assess spontaneous recognition memory. Neurosci Biobehav Rev 53:37. https://doi.org/10.1016/j.neubiorev.2015.03.013

Ameen-Ali KE, Norman LJ, Eacott MJ, Easton A (2017) Incidental context information increases recollection. Learn Mem 24(3):136. https://doi.org/10.1101/lm.042622.116

Babb SJ, Crystal JD (2006) Episodic-like memory in the rat. Curr Biol 16(13):1317–1321

Baxter MG, Chiba AA (1999) Cognitive functions of the basal forebrain. Curr Opin Neurobiol 9:178–183
Bayley PJ, Squire LR (2003) The medial temporal lobe and declarative memory. Int Congr Ser 1250:245–259. http://www.sciencedirect.com/science/article/B7581-49N7DHR-R/2/0390d73876c3ab1dda8cf1533659f027
Bierer LM, Haroutunian V, Gabriel S, Knott PJ, Carlin LS, Purohit DP et al (1995) Neurochemical correlates of dementia severity in Alzheimer's disease: relative importance of the cholinergic deficits. J Neurochem 64:749–760
Blokland A, Honig W, Raaijmakers WGM (1992) Effects of intra-hippocampal scopolamine injections in a repeated spatial acquisition task in the rat. Psychopharmacology 109 (3):373–376. https://doi.org/10.1007/BF02245886
Cai L, Gibbs RB, Johnson DA (2012) Recognition of novel objects and their location in rats with selective cholinergic lesion of the medial septum. Neurosci Lett 506(2):261–265. https://doi.org/10.1016/j.neulet.2011.11.019
Carey GJ, Costall B, Domeney AM, Gerrard PA, Jones DNC, Naylor RJ, Tyers MB (1992) Ondansetron and arecoline prevent scopolamine-induced cognitive deficits in the marmoset. Pharmacol Biochem Behav 42(1):75–83. https://doi.org/10.1016/0091-3057(92)90449-P
Chan M, Eacott MJ, Sanderson DJ, Wang J, Sun M, Easton A (2018) Continual trials spontaneous recognition tasks in mice: reducing animal numbers and improving our understanding of the mechanisms underlying memory. Front Behav Neurosci 12:214. https://doi.org/10.3389/fnbeh.2018.00214
Clayton NS, Dickinson A (1998) Episodic-like memory during cache recovery by scrub jays. Nature 395:272–274
Clayton NS, Bussey TJ, Dickinson A (2003) Can animals recall the past and plan for the future? Nat Rev Neurosci 4(8):685–691. https://doi.org/10.1038/nrn1180
Collie A, Maruff P (2000) The neuropsychology of preclinical Alzheimer's disease and mild cognitive impairment. Neurosci Biobehav Rev 24:365–374
Croxson PL, Kyriazis DA, Baxter MG (2011) Cholinergic modulation of a specific memory function of prefrontal cortex. Nat Neurosci 14(12):1510–1512. https://doi.org/10.1038/nn.2971
Deiana S, Platt B, Riedel G (2011) The cholinergic system and spatial learning. Behav Brain Res 221(2):389–411. https://doi.org/10.1016/j.bbr.2010.11.036
Deluca J, Diamond BJ (1995) Aneurysm of the anterior communicating artery - a review of neuroanatomical and neuropsychological sequelae. J Clin Exp Neuropsychol 17(1):100–121. https://doi.org/10.1080/13803399508406586
Douchamps V, Jeewajee A, Blundell P, Burgess N, Lever C (2013) Evidence for encoding versus retrieval scheduling in the Hippocampus by Theta phase and acetylcholine. J Neurosci 33 (20):8689–8704. https://doi.org/10.1523/JNEUROSCI.4483-12.2013
Drachman DA (1977) Memory and cognitive function in man - does cholinergic system have a specific role. Neurology 27(8):783–790
Duncan K, Sadanand A, Davachi L (2012) Memory's penumbra: episodic memory decisions induce lingering mnemonic biases. Science 337(6093):485–487. https://doi.org/10.1126/science.1221936
Eacott MJ, Gaffan EA (2005) The roles of perirhinal cortex, postrhinal cortex, and the fornix in memory for objects, contexts, and events in the rat. Q J Exp Psychol B 58:202–217
Eacott MJ, Norman G (2004) Integrated memory for object, place, and context in rats: a possible model of episodic-like memory? J Neurosci 24(8):1948–1953
Eacott MJ, Easton A, Zinkivskay A (2005) Recollection in an episodic-like memory task in the rat. Learn Mem 12(3):221–223. https://doi.org/10.1101/lm.92505
Easton A, Gaffan D (2000) Comparison of perirhinal cortex ablation and crossed unilateral lesions of the medial forebrain bundle from the inferior temporal cortex in the rhesus monkey: effects on learning and retrieval. Behav Neurosci 114(6):1041. https://doi.org/10.1037/0735-7044.114.6.1041

Easton A, Ridley RM, Baker HF, Gaffan D (2002) Unilateral lesions of the cholinergic basal forebrain and fornix in one hemisphere and inferior temporal cortex in the opposite hemisphere produce severe learning impairments in rhesus monkeys. Cerebral Cortex 12(7):729–736. https://doi.org/10.1093/cercor/12.7.729

Easton A, Fitchett A, Baxter MG, Eacott MJ (2009) Cholinergic lesions of the medial septum impair where-which memory but not episodic memory in the rat. European Brain and Behavior Society: EBBS

Easton A, Fitchett AE, Eacott MJ, Baxter MG (2010) Medial septal cholinergic neurons are necessary for context-place memory but not episodic-like memory. Hippocampus 21:1–7. https://doi.org/10.1002/hipo.20814

Easton A, Fitchett A, Eacott MJ, Baxter MG (2011) Medial septal cholinergic neurons are necessary for context-place memory but not episodic-like memory. Hippocampus 21:1021–1027

Easton A, Douchamps V, Eacott M, Lever C (2012a) A specific role for septohippocampal acetylcholine in memory? Neuropsychologia 50(13):3156. https://doi.org/10.1016/j.neuropsychologia.2012.07.022

Easton A, Webster LAD, Eacott MJ (2012b) The episodic nature of episodic-like memories. Learn Mem 19(4):146. https://doi.org/10.1101/lm.025676.112

Ennaceur A, Delacour J (1988) A new one-trial test for neurobiological studies of memory in rats. 1: behavioral data. Behav Brain Res 31:47–59

Ennaceur A, Meliani K (1992) Effects of physostigmine and scopolamine on rats' performances in object-recognition and radial-maze tests. Psychopharmacology 109(3):321–330. https://doi.org/10.1007/BF02245880

Ferkin MH, Combs A, delBarco-Trillo J, Pierce AA, Franklin S (2008) Meadow voles, Microtus pennsylvanicus, have the capacity to recall the "what", "where", and "when" of a single past event. Anim Cogn 11(1):147–159. https://doi.org/10.1007/s10071-007-0101-8

Fine A, Hoyle C, MacLean CJ, Levatte TL, Baker HF, Ridley RM (1997) Learning impairments following injection of a selective cholinergic immunotoxin, ME20.4 IgG-saporin, into the basal nucleus of Meynert in monkeys. Neuroscience 81(2):331–343. https://doi.org/10.1016/s0306-4522(97)00208-x

Friedman WJ (1993) Memory for the time of past events. Psychol Bull 113:44–66

Friedman WJ (2007) The meaning of "time" in episodic memory and mental time travel. Behav Brain Sci 30:323

Gaffan D (1994) Scene-specific memory for objects: a model of episodic memory impairment in monkeys with fornix transection. J Cogn Neurosci 6:305–320

Gais S, Born J (2004) Low acetylcholine during slow-wave sleep is critical for declarative memory consolidation. Proc Natl Acad Sci U S A 101(7):2140–2144. https://doi.org/10.1073/pnas.0305404101

Givens B, Olton DS (1995) Bidirectional modulation of scopolamine-induced working memory impairments by muscarinic activation of the medial septal area. Neurobiol Learn Mem 63(3):269–276. https://doi.org/10.1006/NLME.1995.1031

Hampton RR, Hampstead BM, Murray EA (2005) Rhesus monkeys (Macaca mulatta) demonstrate robust memory for what and where, but not when, in an open-field test of memory. Learn Motiv 36:245–259

Hasselmo ME (1999) Neuromodulation: acetylcholine and memory consolidation. Trends Cogn Sci 3(9):351–359. https://doi.org/10.1016/s1364-6613(99)01365-0

Hasselmo ME (2006) The role of acetylcholine in learning and memory. Curr Opin Neurobiol 16(6):710–715. https://doi.org/10.1016/j.conb.2006.09.002

Hasselmo ME (2012) How we remember: brain mechanisms of episodic memory. The M.I.T. Press, Cambridge

Janisiewicz AM, Jackson O, Firoz EF, Baxter MG (2004) Environment-spatial conditional learning in rats with selective lesions of medial septal cholinergic neurons. Hippocampus 14(2):265–273. https://doi.org/10.1002/hipo.10175

Kart-Teke E, De Souza Silva MA, Huston JP, Dere E (2006) Wistar rats show episodic-like memory for unique experiences. Neurobiol Learn Mem 85(2):173–182

Langston RF, Wood ER (2010) Associative recognition and the hippocampus: differential effects of hippocampal lesions on object-place, object-context and object-place-context memory. Hippocampus 20(10):1139–1153. http://www.ncbi.nlm.nih.gov/pubmed/19847786

Lever C, Burton S, O'Keefe J (2006) Rearing on hind legs, environmental novelty, and the hippocampal formation. Rev Neurosci 17(1–2):111–133. http://www.ncbi.nlm.nih.gov/pubmed/16703946

Meeter M, Murre JMJ, Talamini LM (2004) Mode shifting between storage and recall based on novelty detection in oscillating hippocampal circuits. Hippocampus 14(6):722–741. https://doi.org/10.1002/hipo.10214

Melamed JL, de Jesus FM, Maior RS, Barros M (2017) Scopolamine induces deficits in spontaneous object-location recognition and fear-learning in marmoset monkeys. Front Pharmacol 8:395. https://doi.org/10.3389/fphar.2017.00395

Mesulam MM, Mufson EJ, Wainer BH, Levey AI (1983) Central cholinergic pathways in the rat - an overview based on an alternative nomenclature (Ch1-Ch6). Neuroscience 10(4):1185–1201. https://doi.org/10.1016/0306-4522(83)90108-2

Micheau J, Marighetto A (2011) Acetylcholine and memory: a long, complex and chaotic but still living relationship. Behav Brain Res 221(2):424–429. https://doi.org/10.1016/j.bbr.2010.11.052

Milić M, Timić T, Joksimović S, Biawat P, Rallapalli S, Divljaković J et al (2013) PWZ-029, an inverse agonist selective for α5 GABAA receptors, improves object recognition, but not water-maze memory in normal and scopolamine-treated rats. Behav Brain Res 241:206–213. https://doi.org/10.1016/J.BBR.2012.12.016

Murray EA, Baxter MG, Gaffan D (1998) Monkeys with rhinal cortex damage or neurotoxic hippocampal lesions are impaired on spatial scene learning and object reversals. Behav Neurosci 112(6):1291–1303. https://doi.org/10.1037/0735-7044.112.6.1291

Norlen G, Olivecrona H (1953) The treatment of aneurysms of the circle of Willis. J Neurosurg 10 (4):404–415. https://doi.org/10.3171/jns.1953.10.4.0404

Parent MB, Baxter MG (2004) Septohippocampal acetylcholine: involved in but not necessary for learning and memory? Learn Mem 11(1):9–20. https://doi.org/10.1101/lm.69104

Persson BM, Ainge JA, O'Connor AR (2016) Disambiguating past events: accurate source memory for time and context depends on different retrieval processes. Neurobiol Learn Mem 132:40–48. https://doi.org/10.1016/J.NLM.2016.05.002

Poulter S, Hartley T, Lever C (2018) The neurobiology of mammalian navigation. Curr Biol 28(17): R1023–R1042. https://doi.org/10.1016/J.CUB.2018.05.050

Ridley RM, Barefoot HC, Maclean CJ, Pugh P, Baker HF (1999) Different effects on learning ability after injection of the cholinergic immunotoxin ME20.4IgG-saporin into the diagonal band of Broca, basal nucleus of Meynert, or both in monkeys. Behav Neurosci 113(2):303–315

Robertson B-A, Eacott MJ, Easton A (2015) Putting memory in context: dissociating memories by distinguishing the nature of context. Behav Brain Res 285:99. https://doi.org/10.1016/j.bbr.2014.10.045

Rutten K, Prickaerts J, Blokland A (2006) Rolipram reverses scopolamine-induced and time-dependent memory deficits in object recognition by different mechanisms of action. Neurobiol Learn Mem 85(2):132–138. https://doi.org/10.1016/j.nlm.2005.09.002

Scoville WB, Milner B (1957) Loss of recent memory after bilateral hippocampal lesions. J Neurol Neurosurg Psychiatry 20:11–21

Seel SV, Eacott MJ, Langston RF, Easton A (2018) Cholinergic input to the hippocampus is not required for a model of episodic memory in the rat, even with multiple consecutive events. Behav Brain Res 354:48–54. https://doi.org/10.1016/J.BBR.2017.06.001

Singer RA, Zentall TR (2007) Pigeons learn to answer the question "where did you just peck?" and can report peck location when unexpectedly asked. Learn Behav 35(3):184–189

Suddendorf T, Corballis MC (2007) The evolution of foresight: what is mental time travel, and is it unique to humans? Behav Brain Sci 30(3):299. https://doi.org/10.1017/s0140525x07001975

Tang Y, Mishkin M, Aigner TG (1997) Effects of muscarinic blockade in perirhinal cortex during visual recognition. Proc Natl Acad Sci 94(23):12667–12669. https://doi.org/10.1073/PNAS.94.23.12667

Tulving E (1983) Elements of episodic memory. Oxford University Press, London

Tulving E (2002) Episodic memory: from mind to brain. Annu Rev Psychol 53:1–25

Turchi J, Saunders RC, Mishkin M (2005) Effects of cholinergic deafferentation of the rhinal cortex on visual recognition memory in monkeys. Proc Natl Acad Sci U S A 102(6):2158–2161. https://doi.org/10.1073/pnas.0409708102

Warburton EC, Koder T, Cho K, Massey PV, Duguid G, Barker GRI et al (2003) Cholinergic neurotransmission is essential for perirhinal cortical plasticity and recognition memory. Neuron 38(6):987–996. https://doi.org/10.1016/S0896-6273(03)00358-1

Wiley RG, Oeltmann TN, Lappi DA (1991) Immunolesioning - selective destruction of neurons using immunotoxin to rat NGF receptor. Brain Res 562(1):149–153. https://doi.org/10.1016/0006-8993(91)91199-b

Winters BD, Bussey TJ (2005) Removal of cholinergic input to perirhinal cortex disrupts object recognition but not spatial working memory in the rat. Eur J Neurosci 21(8):2263–2270. https://doi.org/10.1111/j.1460-9568.2005.04055.x

Winters BD, Saksida LM, Bussey TJ (2006) Paradoxical facilitation of object recognition memory after infusion of scopolamine into Perirhinal cortex: implications for cholinergic system function. J Neurosci 26:9520–9529

Winters BD, Bartko SJ, Saksida LM, Bussey TJ (2007) Scopolamine infused into perirhinal cortex improves object recognition memory by blocking the acquisition of interfering object information. Learn Mem 14(9):590–596. https://doi.org/10.1101/lm.634607

Woolley ML, Waters KA, Gartlon JE, Lacroix LP, Jennings C, Shaughnessy F et al (2009) Evaluation of the pro-cognitive effects of the AMPA receptor positive modulator, 5-(1-piperidinylcarbonyl)-2,1,3-benzoxadiazole (CX691), in the rat. Psychopharmacology 202(1–3):343–354. https://doi.org/10.1007/s00213-008-1325-2

Endogenous Acetylcholine and Its Modulation of Cortical Microcircuits to Enhance Cognition

Sridevi Venkatesan, Ha-Seul Jeoung, Tianhui Chen, Saige K. Power, Yupeng Liu, and Evelyn K. Lambe

Contents

Abstract Acetylcholine regulates the cerebral cortex to sharpen sensory perception and enhance attentional focus. The cellular and circuit mechanisms of this cholinergic modulation are under active investigation in sensory and prefrontal cortex, but the universality of these mechanisms across the cerebral cortex is not clear. Anatomical maps suggest that the sensory and prefrontal cortices receive distinct cholinergic projections and have subtle differences in the expression of cholinergic receptors and the metabolic enzyme acetylcholinesterase. First, we briefly review this anatomical literature and the recent progress in the field. Next, we discuss in detail the electrophysiological effects of cholinergic receptor subtypes and the cell and circuit consequences of their stimulation by endogenous acetylcholine as established by recent optogenetic work. Finally, we explore the behavioral ramifications of in vivo manipulations of endogenous acetylcholine. We find broader similarities than we

Sridevi Venkatesan and Ha-Seul Jeoung contributed equally to this work.

S. Venkatesan, H.-S. Jeoung, T. Chen, S. K. Power, and Y. Liu
Department of Physiology, University of Toronto, Toronto, ON, Canada

E. K. Lambe (✉)
Department of Physiology, University of Toronto, Toronto, ON, Canada

Department of Obstetrics and Gynaecology, University of Toronto, Toronto, ON, Canada

Department of Psychiatry, University of Toronto, Toronto, ON, Canada
e-mail: evelyn.lambe@utoronto.ca

Curr Topics Behav Neurosci (2020) 45: 47–70
https://doi.org/10.1007/7854_2020_138
Published Online: 30 June 2020

expected between the cholinergic regulation of sensory and prefrontal cortex, but there are some differences and some gaps in knowledge. In visual, auditory, and somatosensory cortex, the cell and circuit mechanisms of cholinergic sharpening of sensory perception have been probed in vivo with calcium imaging and optogenetic experiments to simultaneously test mechanism and measure the consequences of manipulation. By contrast, ascertaining the links between attentional performance and cholinergic modulation of specific prefrontal microcircuits is more complicated due to the nature of the required tasks. However, ex vivo optogenetic manipulations point to differences in the cholinergic modulation of sensory and prefrontal cortex. Understanding how and where acetylcholine acts within the cerebral cortex to shape cognition is essential to pinpoint novel treatment targets for the perceptual and attention deficits found in multiple psychiatric and neurological disorders.

Keywords Acetylcholine · Attention · Muscarinic receptors · Nicotinic receptors · Prefrontal cortex · Sensory cortex · Sensory perception · Visual cortex

1 Introduction

Cholinergic regulation of cerebral cortex is critical for attention and sensory processing (Ballinger et al. 2016; Sarter 2015; Metherate et al. 2012). Identifying the specific cellular and circuit mechanisms through which endogenous acetylcholine modulates cortical microcircuits is important to find new treatment targets for cognitive disruption. Research in sensory cortex can link mechanisms efficiently to behavioral consequences, but it remains uncertain whether cholinergic modulation recruits the same mechanisms across cortical regions. Here we examine recent research into the regional diversity of cortical cholinergic innervation, receptor expression, and functional and behavioral consequences of manipulation of endogenous acetylcholine release. We find that broad similarities exist across sensory and prefrontal domains, but differences emerge, particularly in the cholinergic regulation of output neurons in sensory and prefrontal cortex. Intriguingly, these differences between the mechanisms of cholinergic modulation in sensory and prefrontal cortex can be observed across species. The mechanisms of cholinergic modulation across cerebral cortex and their integration in cognitive performance present a number of critical points for future research.

2 Anatomy of the Cholinergic System in Prefrontal and Sensory Cortices

2.1 Cholinergic Innervation of Cerebral Cortex

Cortical acetylcholine is primarily synthesized by neurons of the basal forebrain in their axonal projections throughout the cortex. A growing body of evidence suggests that the cholinergic neurons of the basal forebrain are organized into distinct populations that project to different parts of the cortex. Individual cholinergic neurons have widely divergent projection patterns within the cerebral cortex (Li et al. 2018) and appear to manifest different electrophysiological phenotypes (Unal et al. 2012; Ahmed et al. 2019; Laszlovszky et al. 2019). There appears to be little overlap between the axons of specific cholinergic neuron populations that project to the prefrontal cortex and those that project to the sensory cortex (Pinto et al. 2013). The basal forebrain consists of the substantia innominata (SI), the horizontal and vertical limbs of the diagonal band of Broca (HDB/VDB), and the nucleus basalis (NB) (Jones 2004; Zaborszky et al. 2018; Bloem et al. 2014b; Huppe-Gourgues et al. 2018). The cholinergic neurons are just one population of neurons within the basal forebrain that project to the cerebral cortex, making modern tracing techniques essential for appreciating the complicated relationships between the basal forebrain and the cortex (Chandler and Waterhouse 2012; Chandler et al. 2013; Kim et al. 2016; Huppe-Gourgues et al. 2018; Bloem et al. 2014b).

Figure 1 illustrates cholinergic innervation of different cortical regions after viral transfection of a subset of ChAT-positive neurons in the basal forebrain. In rodent, the prefrontal cortex and primary visual cortex receive denser cholinergic innervation than other cortical regions in mice (Do et al. 2016; Mechawar et al. 2000). Regional differences are also seen in human and non-human primates (Lewis 1991; Mrzljak et al. 1995; Mesulam et al. 1992; Mrzljak and Goldman-Rakic 1993; Coppola and Disney 2018; Obermayer et al. 2017; Galvin et al. 2018). The laminar specificity of cholinergic innervation differs across association and sensory cortical regions. In the monkey and human brain, the cholinergic fibers are prominent in both supragranular and deeper layers in prefrontal cortex, and they strongly innervate layers 1 and 4 in V1 and S1 (Ghashghaei and Barbas 2001; Lewis 1991; Mrzljak

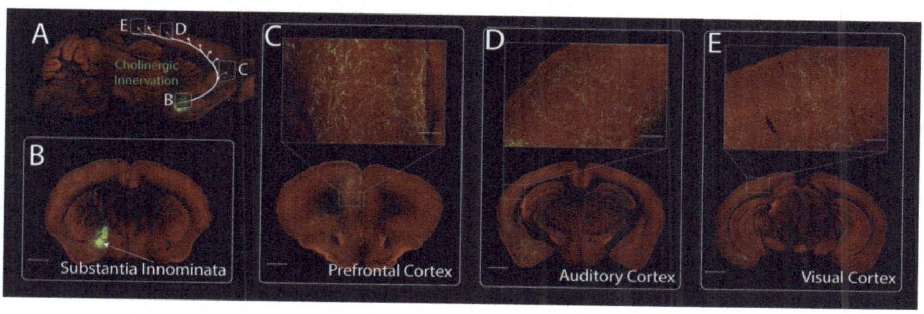

Fig. 1 Cholinergic projections from the basal forebrain to the cortex in the mouse brain. (**a**) Midsagittal view of a mouse brain with white arrows outlining the cholinergic projections from the basal forebrain to the cortices. (**b**) Coronal view of a ChAT-IRES-Cre mouse brain slice showing the EGFP anterograde tracer injection site (injected volume, 0.104 μL) in the substantia innominata. Lower panel scale bars: 1000 μm; inset scale bars: 150 μm. Coronal images of tracer label axons in the prefrontal cortex (**c**), the auditory cortex (**d**), and the visual cortex (**e**). Image credit: Allen Institute

et al. 1995; Mesulam et al. 1992; Mrzljak and Goldman-Rakic 1993). There appears to be dense innervation of rodent prefrontal cortex (Bloem et al. 2014a; Eckenstein et al. 1988; Li et al. 2018). Layers 1 and 4/5 receive preferential cholinergic innervation in the rodent visual and somatosensory cortices (Li et al. 2018; Eckenstein et al. 1988; Lysakowski et al. 1989; Mechawar et al. 2000). In addition to the cholinergic projections from the basal forebrain, a small minority of cortical interneurons are cholinergic (von Engelhardt et al. 2007; Obermayer et al. 2019; Granger et al. 2018), releasing acetylcholine as well as GABA (Obermayer et al. 2019). Furthermore, it has recently been shown that there are several populations of cortically projecting glutamatergic neurons that are transiently cholinergic in early postnatal development, including subsets of neurons in the ventromedial thalamus, lateral hypothalamus, and presubiculum (Nasirova et al. 2019).

2.2 Cortical Cholinergic Synapses

Although the precise mode of cholinergic transmission is still a matter of debate, the ability of acetylcholine to act in a localized manner has been well supported by recent functional studies (Sarter et al. 2009; Turrini et al. 2001; Smiley et al. 1997; Mechawar et al. 2000; Jing et al. 2018; Dasgupta et al. 2018). A typical cortical cholinergic "synapse" from a basal forebrain axonal projection is illustrated in Fig. 2. Such a synapse has a presynaptic terminal containing choline acetyltransferase (ChAT) enzymes to synthesize acetylcholine from choline and vesicular acetylcholine transporters (VAChT) that transport acetylcholine into vesicles for release. When a cholinergic neuron fires an action potential, acetylcholine is released and diffuses through the synaptic cleft to act on postsynaptic nicotinic and/or muscarinic receptors. Of note, the function of cholinergic synapses appears under relatively tight feedback regulation through autoinhibition and temporal control via enzymatic breakdown of acetylcholine. Presynaptic autoinhibitory M2/M4 receptors limit subsequent acetylcholine release from cortical cholinergic terminals (Levey et al. 1991; Zhang et al. 2002; Venkatesan and Lambe 2020). Acetylcholine is rapidly broken down into choline, by acetylcholinesterase (AChE) enzymes localized on both the presynaptic terminal and the postsynaptic neuron. This breakdown product is then transported back into the presynaptic terminal by choline transporters.

The cholinergic ionotropic receptors are nicotinic acetylcholine receptors, which are nonselective cation channels. These pentameric channels can be homomeric (e.g., $\alpha7$ in cortex) or heteromeric in nature, with nicotinic α and β subunits expressed in various combinations which contribute to distinct receptor properties (Wilking and Stitzel 2015). The nicotinic $\alpha7$ homomers have low acetylcholine sensitivity, high calcium permeability, faster kinetics, and rapid desensitization, whereas the nicotinic $\alpha4\beta2$ heteromers typically show slower kinetics and a higher acetylcholine sensitivity (Gotti et al. 2006; Dani and Bertrand 2007). However, the properties of heteromeric nicotinic receptors depend on the identity of the subunit filling the fifth or "accessory" position, which has been recently suggested to

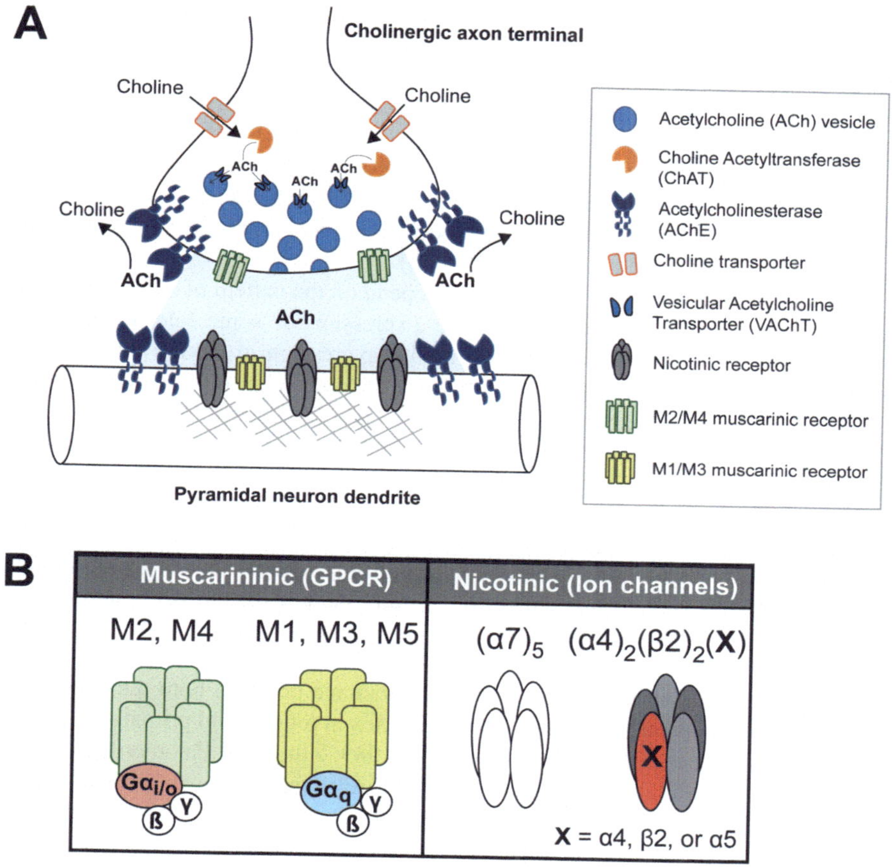

Fig. 2 (**a**) Schematic of a typical cholinergic "synapse" on a pyramidal neuron in the cortex (adapted from Venkatesan and Lambe 2020). (**b**) Depiction of major nicotinic and muscarinic receptors found in the cerebral cortex

participate in an unorthodox binding site that potentiates receptor conductance (Wang et al. 2015; Jain et al. 2016; Wang and Lindstrom 2018). An α4 accessory subunit decreases affinity but increases calcium permeability, while an accessory β2 increases affinity but decreases calcium permeability (Tapia et al. 2007; Kuryatov et al. 2008). Alternatively, an accessory α5 subunit can be included in a nicotinic α4β2 heteromer, increasing affinity to acetylcholine, calcium permeability, and resistance to desensitization (Tapia et al. 2007; Kuryatov et al. 2008).

The cholinergic metabotropic receptors are the family of G protein-coupled receptors called muscarinic acetylcholine receptors. These channels have electrophysiological consequences through direct channel effectors and can also change cellular properties through their second messenger cascades. The M1-like receptors (M1, M3, and M5 receptors) are Gαq-coupled excitatory receptors well known for their "M" current resulting from the inhibition of potassium ion channels active near the action potential threshold (Jentsch 2000). These typically excitatory muscarinic receptors also activate second messenger effects such as increasing phospholipase C levels and triggering release of calcium ions from intracellular stores. By contrast, the M2 and M4 muscarinic receptors are Gαi/o-coupled inhibitory receptors that can activate GIRK channels (Gerber et al. 1991; Fernandez-Fernandez et al. 1999; Seeger and Alzheimer 2001; Kohlmeier et al. 2012) and inhibit terminal voltage-gated calcium channels (Allen 1999; Shapiro et al. 1999; Kohlmeier et al. 2012), as well as trigger signaling pathways that decrease intracellular adenylyl cyclase and cAMP levels (Anderson and McKinney 1988).

The consequences of acetylcholine will depend on the pattern of expression of its receptors. Figure 3 illustrates the laminar expression of some relevant nicotinic subunits and muscarinic receptors with in situ expression in mouse and Fig. 4 in human *postmortem* tissue. RNAseq databases are increasing our awareness of the different patterns of co-expression in specific cell types (Keil et al. 2018; Hodge et al. 2019) as well as the conserved nature of cell and receptor subtypes from rodent to human (Hodge et al. 2019). Some broad patterns can be observed across species even at the level of in situ hybridization, such as the strong expression of the α4 nicotinic subunit in the deepest layer of rodent prefrontal and sensory cortex and the recapitulation of this pattern in the human prefrontal cortex. One receptor subunit with a clear expression difference between prefrontal and visual cortex in mouse is the α5 nicotinic subunit that is expressed in layer 6 and is particularly strong in the prefrontal cortex and weak in the visual cortex (Wada et al. 1990; Salas et al. 2003). By contrast, the excitatory muscarinic receptors have peaks in both the superficial and deep lamina in many regions of cerebral cortex in rodents and primates (Buckley et al. 1988; Rossner et al. 1993; Lidow et al. 1989; Zilles and Palomero-Gallagher 2017; Palomero-Gallagher and Zilles 2019). Of relevance, there are protein modulators of the nicotinic response that can complicate the interpretation of nicotinic receptor subunit expression and its linkage with a physiological impact. Lynx1, for example, belonging to the Ly6 superfamily of proteins, is able to suppress nicotinic receptor signaling in the adult mouse visual cortex (Morishita et al. 2010; Miwa et al. 2012). While this is just one example, it urges active examination of the physiological impact of cortical cholinergic receptors.

The enzyme AChE breaks down acetylcholine, strongly controlling its availability. It is therefore essential to understand regional and laminar differences in the expression of this enzyme with its repercussions for cholinergic receptor stimulation. The expression of AChE by cortical pyramidal neurons follows an interesting pattern: in humans, AChE-expressing pyramidal neurons in the cortex are absent at birth, only appearing in early childhood. They continue to mature and increase in numbers and AChE expression intensity into adulthood, with greatest numbers in layers 3 and 5 (Mesulam and Geula 1991; Janeczek et al. 2018). These neurons that strongly express AChE are the acetylcholine-sensitive or "cholinoceptive" neurons in the cortex likely to play key cognitive roles in which the timing of cholinergic stimulation must be tightly controlled. In the rodent cortex, there is evidence to support differential expression of AChE by neurons based on the cortical region (Anderson et al. 2009). An examination of in situ hybridization for AChE in the

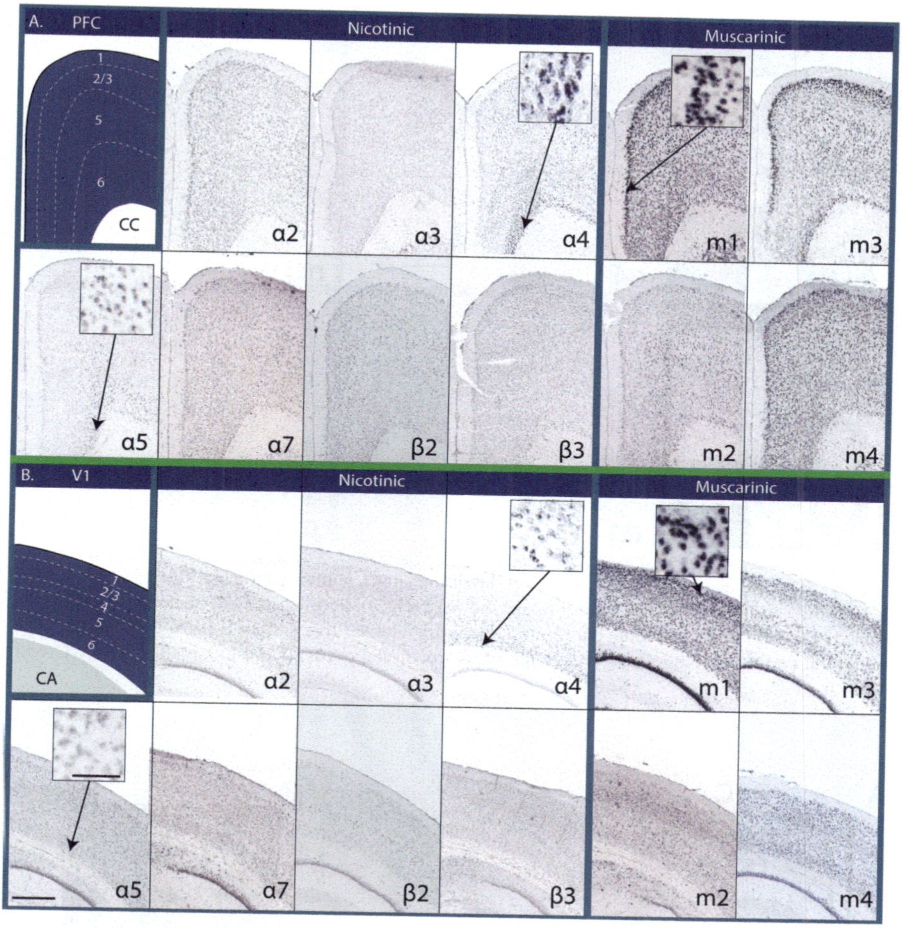

Fig. 3 Adult mouse brain in situ hybridization data showing mRNA localization of various nicotinic receptor subunits (*Chrna2, Chrna3, Chrna4, Chrna5, Chrna7, Chrnb2, Chrnb3*) and muscarinic receptors (*Chrm1–4*) for the (**a**) prefrontal cortex and (**b**) primary visual cortex. Scale bar: 500 μm; inset scale bar: 100 μm. Drawing adapted from Allen Mouse Brain Atlas. *PFC* prefrontal cortex, *V1* primary visual area, *CC* corpus callosum, *CA* hippocampus. Image Credit: Allen Institute

adult mouse cortex reveals laminar differences in localization of AChE-expressing neurons as illustrated in Fig. 5. In the prefrontal cortex, there is high intensity of such neurons in the deeper cortical layers, particularly layer 6. In sensory cortex, this pattern of expression is also evident in primary visual cortex but not in primary auditory cortex (Anderson et al. 2009). Since the level of AChE shapes the kinetics of the cholinergic responses, the greater expression of AChE in layer 6 of the prefrontal and visual cortex suggests that their cholinergic responses are under the strongest temporal control.

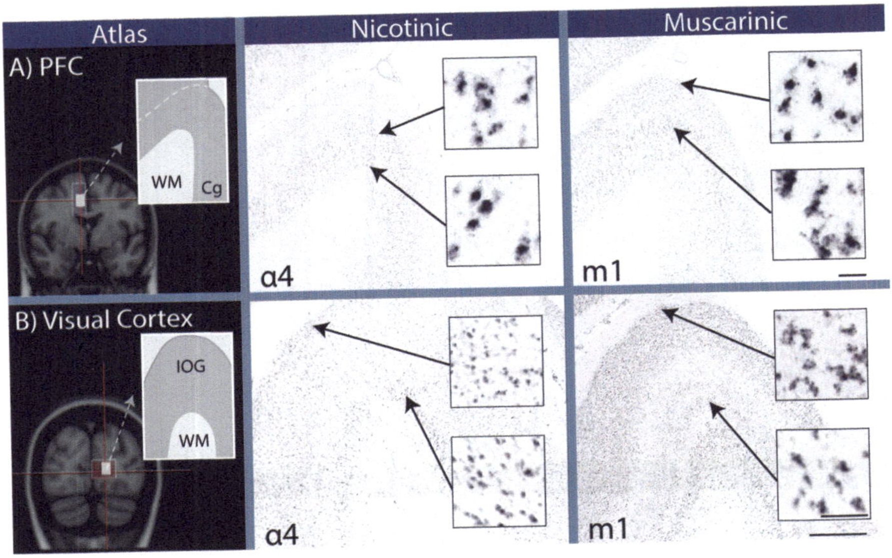

Fig. 4 Human postmortem in situ hybridization data showing mRNA localization of α4 (CHRNA4) and M1 (CHRM1) subunits for the (**a**) prefrontal cortex and (**b**) primary visual cortex. Scale bars: 1000 μm; inset scale bar: 50 μm. Drawing adapted from Allen Human Brain Atlas. *PFC* prefrontal cortex, *Cg* cingulate gyrus, *IOG* inferior occipital gyrus, *WM* white matter. Image Credit: Allen Institute

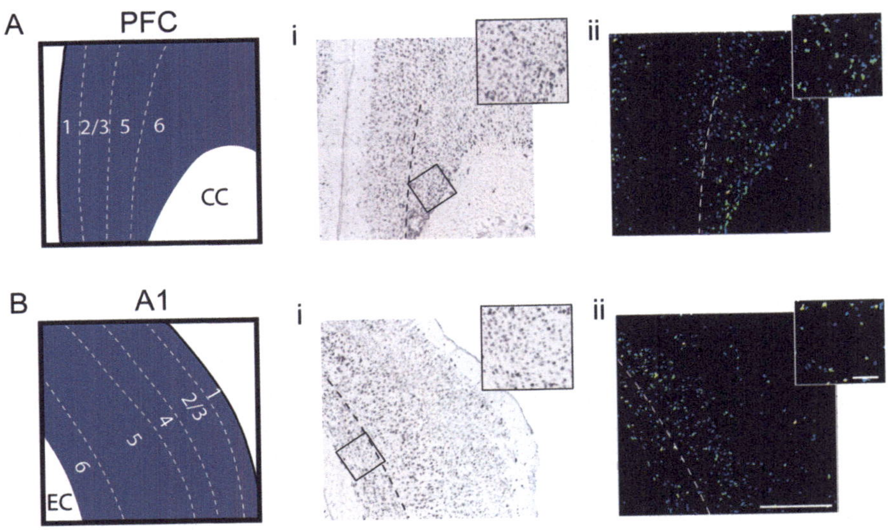

Fig. 5 Mouse brain acetylcholinesterase (AChE) expression in (**a**) prefrontal cortex and (**b**) auditory cortex. (i) In situ hybridization, (ii) Normalized expression. Scale bar: 500 μm; inset scale bar: 100 μm. *CC* corpus callosum, *LV* lateral ventricle, *CPu* caudoputamen, *EC* external capsule. Image credit: Allen Institute

3 Regulation of Cortical Neurons by Endogenous Acetylcholine

The physiological effects of acetylcholine depend on the cell-type specificity of cholinergic receptor expression. While there is substantial pharmacological and electrophysiological work using exogenous agonists, the critical question is what receptors on which cells are recruited by endogenous acetylcholine. To address this question, we turned to research using optogenetics to stimulate cholinergic axon terminals within sensory versus prefrontal cortex. While the in situ hybridization experiments discussed in the previous section give an estimate of the possible subunit expression in different regions, they do not assess the synaptic expression of those receptors in neurons and their activation during endogenous cholinergic release. The availability of optogenetic tools to selectively manipulate cholinergic neurons allows the evaluation of synaptic responses to rapid endogenous acetylcholine release (Zhao et al. 2011; Hedrick et al. 2016; but see Nasirova et al. 2019). This section examines the functional activation of cholinergic receptors in neuronal types across cortical layers in the prefrontal and sensory cortices, with a focus on studies using optogenetic tools to release endogenous acetylcholine and measure postsynaptic responses. A summary of the laminar and cell-type specific effects of endogenous stimulation is shown in Fig. 6. Note that the documentation of excitatory cholinergic effects is typically biased toward the nicotinic receptors, which exert rapid and strong depolarization even from hyperpolarized resting potentials. In contrast, the effects of excitatory muscarinic receptors to depolarize and to accelerate and prolong action potential firing are detectable only near threshold (Hedrick and Waters 2015; Sparks et al. 2017). This state dependence is due to the ability of M1 and M3 muscarinic receptors to inhibit subthreshold voltage-dependent potassium channels (Jentsch 2000).

The superficial layers of cortex appear to be excited by endogenous acetylcholine through nicotinic and excitatory muscarinic receptors, although the latter has not been explored as extensively (Hedrick and Waters 2015; Kimura et al. 2014). Layer 1 cortical interneurons are excited by endogenous acetylcholine, through both α7 and α4β2 nicotinic receptors in the prefrontal cortex but through α4β2 receptor in sensory cortices (Arroyo et al. 2012; Bennett et al. 2012; Hay et al. 2016; Kimura et al. 2014). In layer 2/3, however, there are differences in the effect of acetylcholine between the sensory and prefrontal cortices: direct excitatory effects of acetylcholine are not commonly seen in L2/3 pyramidal neurons in the prefrontal cortex, whereas these neurons show direct nicotinic receptor-mediated EPSCs in the primary visual and auditory cortices (Poorthuis et al. 2013; Hedrick and Waters 2015; Nelson and Mooney 2016; Verhoog et al. 2016). Primarily nicotinic excitation of parvalbumin (PV) and somatostatin (SST) interneurons mediated by α7 and α4β2 nicotinic receptors is observed in layer 2/3 of the rodent prefrontal cortex (Poorthuis et al. 2013; Verhoog et al. 2016; Obermayer et al. 2018, 2019), whereas in the somatosensory cortex, L2/3 SST interneurons show M1/M3 muscarinic excitation (Munoz et al. 2017), and PV interneurons show both nicotinic and muscarinic effects (Dasgupta et al. 2018).

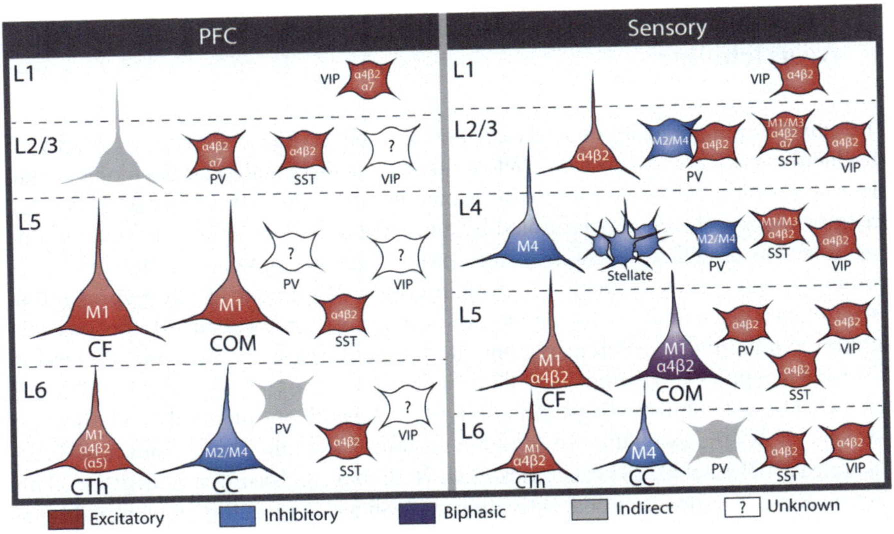

Fig. 6 Schematic comparing layer-specific effects of optogenetically released acetylcholine on the prefrontal cortex (left) and sensory cortex (right). In layer 1, interneurons in neocortex exhibit excitatory responses mediated by nicotinic acetylcholine receptors (Arroyo et al. 2012; Bennett et al. 2012; Hay et al. 2016; Obermayer et al. 2018). Majority of prefrontal layer 2/3 pyramidal neurons lack nicotinic responses (Hedrick and Waters 2015), which are prominent in sensory cortices by contrast (Kimura et al. 2014; Hedrick and Waters 2015; Nelson and Mooney 2016). Layer 2/3 interneurons also appear to be modulated in a region-specific manner with prefrontal interneurons receiving nicotinic excitation (Obermayer et al. 2018; Obermayer et al. 2019; Verhoog et al. 2016; Poorthuis et al. 2013) and sensory cortex interneurons receiving both nicotinic and muscarinic excitation (Dasgupta et al. 2018; Munoz et al. 2017; Letzkus et al. 2011). In sensory cortex, layer 4 pyramidal neurons are primarily inhibited through M4 receptors (Dasgupta et al. 2018), while the interneurons exhibit both inhibitory muscarinic and excitatory nicotinic responses (Dasgupta et al. 2018; Obermayer et al. 2019). In prefrontal cortex, layer 5 pyramidal neurons are excited by M1 muscarinic receptors (Baker et al. 2018), whereas in sensory cortices nicotinic and muscarinic receptors produce excitatory and inhibitory responses, respectively (Dasari et al. 2017; Baker et al. 2018; Joshi et al. 2016). Layer 5 interneurons in prefrontal and sensory cortices show excitatory responses through activation of nicotinic receptors (Nelson and Mooney 2016). Layer 6 pyramidal neurons show similar responses across the cortex, but the responses appear much stronger in the prefrontal cortex (Hedrick and Waters 2015; Hay et al. 2016; Sparks et al. 2017; Venkatesan and Lambe 2020; Tian et al. 2014; Obermayer et al. 2019). Layer 6 pyramidal neurons are excited by both nicotinic and muscarinic receptors (Sparks et al. 2017), and the nicotinic excitation of layer 6 interneurons is also similar across brain regions (Poorthuis et al. 2013; Hay et al. 2016; Obermayer et al. 2018). *PV* parvalbumin, *SST* somatostatin, *VIP* vasoactive intestinal peptide, *CF* corticofugal, *COM* commissural, *CTh* corticothalamic, *CC* corticocortical

The granular layer of sensory cortex appears to be largely inhibited by cholinergic stimulation of inhibitory muscarinic receptors (Dasgupta et al. 2018). In layer 4 of the somatosensory cortex, excitatory neurons including pyramidal and spiny stellate cells show muscarinic M2/M4-mediated inhibition to both exogenously and endogenously released acetylcholine (Eggermann and Feldmeyer 2009; Dasgupta et al.

2018). By contrast, the prefrontal cortex in rodents is agranular, meaning it lacks a defined layer 4 that receives thalamic inputs in contrast to the sensory cortices (Van Aerde and Feldmeyer 2015).

Cholinergic effects on the deeper layers of cortex, containing output projection neurons, are typically excitatory but are more complex. They involve both muscarinic and nicotinic components, with the muscarinic components dominating in layer 5 and both nicotinic and muscarinic components in layer 6 (Hedrick and Waters et al. 2015; Sparks et al. 2017). In layer 5 pyramidal neurons of both the sensory and prefrontal cortices, acetylcholine can cause a transient inhibition due to the opening of SK channels by IP3-mediated release of calcium from internal stores, followed by excitation mediated by M1/M3 muscarinic receptors (Gulledge and Stuart 2005; Gulledge et al. 2009; Dasari et al. 2017; Proulx et al. 2014b). With endogenous acetylcholine release in the prefrontal cortex, both types of response are observed, and both responses are most obvious in neurons that are depolarized to near threshold (Hedrick and Waters 2015). Of note, the muscarinic excitation of prefrontal layer 5 pyramidal neurons shows a projection-specific gradient, with corticofugal neurons showing stronger muscarinic excitation compared to commissural neurons (Baker et al. 2018). In contrast, layer 5 pyramidal neurons in the auditory cortex show both nicotinic and muscarinic effects, with commissural neurons showing α4β2 nicotinic depolarization and muscarinic hyperpolarization and corticofugal neurons showing a prolonged muscarinic depolarization (Joshi et al. 2016). Layer 5 interneurons show α4β2 nicotinic receptor-mediated excitation in both the prefrontal and sensory cortices, with prefrontal PV interneurons showing some evidence of α7-mediated responses to exogenous cholinergic stimulation (Couey et al. 2007; Poorthuis et al. 2013; Nelson and Mooney 2016; Askew et al. 2019).

Layer 6 is the primary output layer of the cortex with a high proportion of corticothalamic neurons (Thomson 2010; Gabbott et al. 2005). Layer 6 pyramidal neurons in the prefrontal and sensory cortices show α4β2 nicotinic receptor-mediated excitation (Hedrick and Waters 2015; Hay et al. 2016; Sparks et al. 2017). However, the nicotinic excitation is potentially stronger in the prefrontal cortex due to the greater expression of the accessory α5 nicotinic receptor subunit in prefrontal layer 6 (Bailey et al. 2010; Proulx et al. 2014a; Tian et al. 2014). Layer 6 pyramidal neurons show projection-dependent cholinergic activation, where corticothalamic neurons are excited via α4β2 nicotinic receptors while corticocortical neurons are inhibited via M2/M4 muscarinic receptors (Yang et al. 2019). Endogenous acetylcholine release activates layer 6 neurons on a rapid timescale via α4β2 nicotinic receptors (Hay et al. 2016; Verhoog et al. 2016; Sparks et al. 2017) with excitatory muscarinic receptors accelerating action potentials and extending the duration of excitation (Sparks et al. 2017). The stronger expression of the α5 nicotinic receptor subunit in prefrontal layer 6 (Fig. 3) appears to contribute to the rapid timescale of cholinergic activation of these neurons, considered critical for attention. Recent work demonstrates that α5 knockout neurons show slower cholinergic activation to optogenetic acetylcholine release (Venkatesan and Lambe 2020). This pattern highlights the importance of temporal control in the prefrontal cortex and is consistent with the higher expression of AChE in prefrontal layer

6 (Fig. 5). While some PV interneurons in the prefrontal cortex receive lesser cholinergic input from the basal forebrain (Sun et al. 2019) and lack direct nicotinic receptor-mediated responses to acetylcholine. SST interneurons in both the prefrontal and somatosensory cortex are directly excited by endogenous acetylcholine through α4β2 nicotinic receptors and mediate lateral inhibition between pyramidal neurons (Poorthuis et al. 2013; Obermayer et al. 2018).

In addition to the direct postsynaptic effects of acetylcholine described above, there are also indirect effects mediated through cholinergic heteroreceptors on excitatory and inhibitory cortical afferents. In both the auditory and prefrontal cortices, thalamocortical axons show α4β2 nicotinic receptor-mediated excitation, which causes glutamate release onto L5 pyramidal neurons (Lambe et al. 2003; Kawai et al. 2007; Lambe et al. 2005). PV interneurons in layer 6 lack direct cholinergic effects but are instead excited indirectly by glutamate release, potentially from layer 6 pyramidal neurons (Kassam et al. 2008; Tian et al. 2016). Acetylcholine can also affect the connection strengths between different neuronal types, notably influencing the strength of inhibition: in the superficial layers of the somatosensory cortex, endogenous acetylcholine enhances pyramidal-to-SST interneuron connection but not pyramidal-to-PV interneuron connections (Urban-Ciecko et al. 2018).

The general pattern of nicotinic receptor neuronal modulation in different cortical layers described above for rodents seems to be preserved in the human cortex as well, with layer 1 and layer 2/3 interneurons showing nicotinic receptor mediated excitation, layer 2/3 pyramidal neurons showing very little activation, and layer 6 pyramidal neurons being strongly excited by acetylcholine in the human frontal and temporal cortices (Verhoog et al. 2016; Obermayer et al. 2018; Poorthuis et al. 2013).

The above studies illustrate that there are significant differences in the pattern of cholinergic excitation between the sensory and prefrontal cortices both in the direction of the cholinergic effects and the receptors that mediate these effects.

4 Behavioral Consequences of Endogenous Acetylcholine Release in Cerebral Cortex

Acetylcholine is critical for the modulation of sensory perception, as well as for cognitive functions ranging from associative learning to attention. This section will describe the cognitive consequences of acetylcholine release in both the sensory and the prefrontal cortex. Across the cerebral cortex, acetylcholine is linked with desynchronization of local field potentials and the increase in higher frequency power associated with cognitive activity. These changes are easier to correlate with precise behavioral outputs in sensory and associative learning tasks than in attention tasks; however, it is thought that the decorrelation of cortical neuronal firing is a fundamental mechanism for the improved behavioral performance which comes with attention on sensory tasks (Cohen and Maunsell 2009).

Acetylcholine efflux in the sensory cortices broadly decorrelates the firing of cortical neurons and desynchronizes the cortical local field potential, which is thought to enhance detection of sensory signals (Goard and Dan 2009; Pinto et al. 2013; Eggermann et al. 2014; Meir et al. 2018). Optogenetic activation of basal forebrain cholinergic neurons in mice desynchronized the local field potential signals in the visual cortex, suppressing the power at lower frequencies (1–5 Hz) and increasing power at higher frequencies (60–100 Hz). This was accompanied by enhanced visual discrimination, while optogenetic inhibition of cholinergic neurons suppressed visual discrimination (Pinto et al. 2013). Subsequently, the same data was used to show that optogenetic activation of cholinergic neurons reduced noise correlations in visual cortical neurons and enhanced the signal amplitude, supporting a role for acetylcholine in enhancing signal-to-noise ratios and improving sensory detection (Minces et al. 2017). There is some evidence that nicotinic excitation of layer 2/3 SST interneurons in the visual cortex is sufficient to reduce neuronal correlations and desynchronize the cortical LFP (Chen et al. 2015). In the somatosensory barrel cortex, it has been shown using calcium imaging of axonal activation that cholinergic axons increase their activity during whisking leading to a suppression of spontaneous activity in layer 2/3 pyramidal neurons (Eggermann et al. 2014). Optogenetic stimulation of cholinergic axons to the somatosensory cortex suppresses ongoing synaptic activity, decorrelates neuronal firing, and results in increased signal-to-noise ratio of sensory responses (Meir et al. 2018).

Cholinergic modulation allows for learning the association of a sensory cue with subsequent reward/punishment. In the auditory cortex, layer 1 interneurons exhibit nicotinic excitation during tone-paired aversive foot shocks, and they inhibit L2/3 PV interneurons causing disinhibition of pyramidal neurons. This disinhibitory circuit is essential for learning fear conditioning (Letzkus et al. 2011). Likewise, cholinergic modulation of the auditory or visual cortex appears necessary for bridging two stimuli that are separated in time (Chubykin et al. 2013; Liu et al. 2015; Guo et al. 2019). Blocking cholinergic receptors inhibits the plasticity of the auditory cortex tonotopic maps following the fear learning (Guo et al. 2019). Specifically, calcium imaging shows cholinergic neurons projecting to the auditory cortex undergo plasticity during associative learning to increase their responses to a conditioned stimulus and exhibit sustained activity bridging the period between the cue and the predicted reinforcement (Guo et al. 2019). A similar role for acetylcholine in associative learning is seen in the visual cortex. Pairing a visual cue with a reward elicits responses in primary visual cortex neurons, which encode the timing of the reward. The activation of basal forebrain cholinergic input to the visual cortex is both necessary and sufficient to cause this associative learning of reward timing representation in V1 neurons (Chubykin et al. 2013; Liu et al. 2015).

Cholinergic axons in the auditory cortex are active during movements, and their activity precedes pupil dilations corresponding to behavioral changes (Reimer et al. 2016). Acetylcholine release depolarizes the cortical neurons, and activating the cholinergic axons during a tone broadens the bandwidth/tuning curve of neurons to different frequencies (Nelson and Mooney 2016). Cholinergic axon activity in the auditory cortex predicts context switching during a working memory task (Kuchibhotla et al. 2017). Adding to the complexity of function within the cholinergic system, it has been suggested that basal forebrain cholinergic neurons contain two populations in vivo, burst firing and regular firing neurons, which may have different behavioral roles (Unal et al. 2012; Laszlovszky et al. 2019). The synchronization of burst firing cholinergic neurons with auditory cortex strongly correlates with the timing of reward/punishment, whereas synchronization of regular firing cholinergic neurons with auditory cortex predicts correct performance on a go-no go task (Laszlovszky et al. 2019).

Attention is dependent on cholinergic modulation of the prefrontal cortex. Removing cholinergic inputs to the prefrontal cortex greatly impairs attention (McGaughy et al. 2002). The prefrontal release of acetylcholine as measured by choline-sensitive microelectrodes reveals that sharp cholinergic transients predict cue detection (Parikh et al. 2007). Optogenetic acetylcholine release in the prefrontal cortex promotes cue evoked gamma oscillations (Howe et al. 2017) known to be critical for attention (Cardin et al. 2009). Bidirectional optogenetic manipulation of prefrontal acetylcholine release affects cue detection with excitation promoting detection and inhibition disrupting detection, and the effects were most evident for cues which were hard to detect (Gritton et al. 2016). Prefrontal cholinergic effects on attention are thought to be strongly dependent on nicotinic receptors, as different nicotinic receptor knockout mice show deficits in a prefrontal-dependent attention task based on visual cues. The nicotinic β2 subunit knockout mice show increased number of omission errors, where they failed to detect the cue. Their poor performance was rescued by the re-expression of β2 subunits in the adult cortex (Guillem et al. 2011). Mice lacking the nicotinic α7 subunit also show increased number of omission errors on the same attention task (Hoyle et al. 2006; Young et al. 2007). Mice lacking the nicotinic α5 subunit also show attention deficits; however, they show poor performance only to the briefest and most challenging durations of cue presentation (Bailey et al. 2010). The latter result highlights the need for cholinergic responses in the prefrontal cortex to have a rapid timescale. This conceptual model is supported by a recent finding that the nicotinic α5 subunit accelerates the timescale of cholinergic responding (Venkatesan and Lambe 2020) as well as the strong expression of AChE by neurons in layer 6 of the prefrontal cortex (Fig. 5), where the α5 subunit is highly expressed.

Interactions between the sensory and prefrontal cortices are essential for the detection of cues and subsequent responding to those cues. The basal forebrain cholinergic system has a key role to play in both these functions. Studies have shown the role of the cholinergic system in the interaction between the prefrontal and auditory cortices. Presentation of auditory stimuli in an oddball task, where mice react to presentation of an unusual tone, results in the generation of oddball-evoked electrical activity in the prefrontal cortex that originates in the basal forebrain and involves the rapid activation of layer 6 neurons (Nguyen and Lin 2014). This finding could support the involvement of the nicotinic α5 subunit expressed in prefrontal layer 6 in rapid cholinergic responding leading to cue detection. Generation of auditory-related electrical activity in the prefrontal cortex is dependent on muscarinic activation of the auditory cortex, and infusion of scopolamine, a muscarinic antagonist, into the auditory cortex reduces the amplitude of tone-evoked potentials in the prefrontal cortex in awake head-fixed mice listening passively (James et al. 2019). This suggests that cholinergic control of attention and sensory processing involves not only the direct action of acetylcholine on neurons in the prefrontal and sensory cortices but also the projections between these two regions.

5 Conclusions and Directions for Future Research

Cholinergic signaling in the sensory and prefrontal cortex is essential for cognitive processes. Despite some differences in cholinergic innervation and receptor targets, there are broad similarities in the modulation of sensory and prefrontal cortex by endogenous acetylcholine. Acetylcholine improves sensory perception and attention, is vital for learned associations, and facilitates cue detection by recruiting the sensory and prefrontal cortex. Our understanding of layer-specific and region-specific cholinergic responses and their behavioral consequences is due to the development of techniques such as in vivo calcium imaging and optogenetic manipulation of endogenous acetylcholine. These techniques allow for a more detailed consideration of cholinergic circuitry, revealing complexity that poses questions for future research. Cortical acetylcholine signaling is under tight control from presynaptic and postsynaptic mechanisms. These include autoinhibition of cholinergic release and breakdown of acetylcholine by AChE, as well as a state dependence of some postsynaptic cholinergic receptors. Such mechanisms for cholinergic tuning are under active investigation. Broadly, there are similarities between rodent and primate cortical cholinergic circuits that underscore the relevance of research in preclinical models which allow us to gain mechanistic insight. Detailed understanding of the key players within the cells and circuits will permit testing of novel treatment targets to enhance cognitive performance in psychiatric and neurological disorders.

Acknowledgments This work was supported by the Canada Research Chair in Developmental Cortical Physiology (EKL) and grants from the Canadian Institutes of Health Research MOP-89825 (EKL) and PJT-153101 (EKL). Fellowship support is from Mary Beatty Fellowship from the University of Toronto (SV) and an Ontario Graduate Scholarship (SP).

References

Ahmed NY, Knowles R, Dehorter N (2019) New insights into cholinergic neuron diversity. Front Mol Neurosci 12:204. https://doi.org/10.3389/fnmol.2019.00204

Allen TG (1999) The role of N-, Q- and R-type Ca2+ channels in feedback inhibition of ACh release from rat basal forebrain neurones. J Physiol 515(Pt 1):93–107. https://doi.org/10.1111/j.1469-7793.1999.093ad.x

Anderson DJ, McKinney M (1988) Muscarinic M2 receptor-mediated cyclic AMP reduction in mechanically dissociated rat cortex. Brain Res 475(1):28–34. https://doi.org/10.1016/0006-8993(88)90195-3

Anderson LA, Christianson GB, Linden JF (2009) Mouse auditory cortex differs from visual and somatosensory cortices in the laminar distribution of cytochrome oxidase and acetylcholinesterase. Brain Res 1252:130–142. https://doi.org/10.1016/j.brainres.2008.11.037

Arroyo S, Bennett C, Aziz D, Brown SP, Hestrin S (2012) Prolonged disynaptic inhibition in the cortex mediated by slow, non-alpha7 nicotinic excitation of a specific subset of cortical interneurons. J Neurosci 32(11):3859–3864. https://doi.org/10.1523/JNEUROSCI.0115-12.2012

Askew CE, Lopez AJ, Wood MA, Metherate R (2019) Nicotine excites VIP interneurons to disinhibit pyramidal neurons in auditory cortex. Synapse 73(9):e22116. https://doi.org/10.1002/syn.22116

Bailey CD, De Biasi M, Fletcher PJ, Lambe EK (2010) The nicotinic acetylcholine receptor alpha5 subunit plays a key role in attention circuitry and accuracy. J Neurosci 30(27):9241–9252. https://doi.org/10.1523/JNEUROSCI.2258-10.2010

Baker AL, O'Toole RJ, Gulledge AT (2018) Preferential cholinergic excitation of corticopontine neurons. J Physiol 596(9):1659–1679. https://doi.org/10.1113/JP275194

Ballinger EC, Ananth M, Talmage DA, Role LW (2016) Basal forebrain cholinergic circuits and signaling in cognition and cognitive decline. Neuron 91(6):1199–1218. https://doi.org/10.1016/j.neuron.2016.09.006

Bennett C, Arroyo S, Berns D, Hestrin S (2012) Mechanisms generating dual-component nicotinic EPSCs in cortical interneurons. J Neurosci 32(48):17287–17296. https://doi.org/10.1523/JNEUROSCI.3565-12.2012

Bloem B, Poorthuis RB, Mansvelder HD (2014a) Cholinergic modulation of the medial prefrontal cortex: the role of nicotinic receptors in attention and regulation of neuronal activity. Front Neural Circuits 8:17. https://doi.org/10.3389/fncir.2014.00017

Bloem B, Schoppink L, Rotaru DC, Faiz A, Hendriks P, Mansvelder HD, van de Berg WD, Wouterlood FG (2014b) Topographic mapping between basal forebrain cholinergic neurons and the medial prefrontal cortex in mice. J Neurosci 34(49):16234–16246. https://doi.org/10.1523/JNEUROSCI.3011-14.2014

Buckley NJ, Bonner TI, Brann MR (1988) Localization of a family of muscarinic receptor mRNAs in rat brain. J Neurosci 8(12):4646–4652

Cardin JA, Carlen M, Meletis K, Knoblich U, Zhang F, Deisseroth K, Tsai LH, Moore CI (2009) Driving fast-spiking cells induces gamma rhythm and controls sensory responses. Nature 459 (7247):663–667. https://doi.org/10.1038/nature08002

Chandler D, Waterhouse BD (2012) Evidence for broad versus segregated projections from cholinergic and noradrenergic nuclei to functionally and anatomically discrete subregions of prefrontal cortex. Front Behav Neurosci 6:20. https://doi.org/10.3389/fnbeh.2012.00020

Chandler DJ, Lamperski CS, Waterhouse BD (2013) Identification and distribution of projections from monoaminergic and cholinergic nuclei to functionally differentiated subregions of prefrontal cortex. Brain Res 1522:38–58. https://doi.org/10.1016/j.brainres.2013.04.057

Chen N, Sugihara H, Sur M (2015) An acetylcholine-activated microcircuit drives temporal dynamics of cortical activity. Nat Neurosci 18(6):892–902. https://doi.org/10.1038/nn.4002

Chubykin AA, Roach EB, Bear MF, Shuler MG (2013) A cholinergic mechanism for reward timing within primary visual cortex. Neuron 77(4):723–735. https://doi.org/10.1016/j.neuron.2012.12.039

Cohen MR, Maunsell JH (2009) Attention improves performance primarily by reducing interneuronal correlations. Nat Neurosci 12(12):1594–1600. https://doi.org/10.1038/nn.2439

Coppola JJ, Disney AA (2018) Is there a canonical cortical circuit for the cholinergic system? Anatomical differences across common model systems. Front Neural Circuits 12:8. https://doi.org/10.3389/fncir.2018.00008

Couey JJ, Meredith RM, Spijker S, Poorthuis RB, Smit AB, Brussaard AB, Mansvelder HD (2007) Distributed network actions by nicotine increase the threshold for spike-timing-dependent plasticity in prefrontal cortex. Neuron 54(1):73–87. https://doi.org/10.1016/j.neuron.2007.03.006

Dani JA, Bertrand D (2007) Nicotinic acetylcholine receptors and nicotinic cholinergic mechanisms of the central nervous system. Annu Rev Pharmacol Toxicol 47:699–729. https://doi.org/10.1146/annurev.pharmtox.47.120505.105214

Dasari S, Hill C, Gulledge AT (2017) A unifying hypothesis for M1 muscarinic receptor signalling in pyramidal neurons. J Physiol 595(5):1711–1723. https://doi.org/10.1113/JP273627

Dasgupta R, Seibt F, Beierlein M (2018) Synaptic release of acetylcholine rapidly suppresses cortical activity by recruiting muscarinic receptors in layer 4. J Neurosci 38(23):5338–5350. https://doi.org/10.1523/JNEUROSCI.0566-18.2018

Do JP, Xu M, Lee SH, Chang WC, Zhang S, Chung S, Yung TJ, Fan JL, Miyamichi K, Luo L, Dan Y (2016) Cell type-specific long-range connections of basal forebrain circuit. Elife 5. https://doi.org/10.7554/eLife.13214

Eckenstein FP, Baughman RW, Quinn J (1988) An anatomical study of cholinergic innervation in rat cerebral cortex. Neuroscience 25(2):457–474. https://doi.org/10.1016/0306-4522(88)90251-5

Eggermann E, Feldmeyer D (2009) Cholinergic filtering in the recurrent excitatory microcircuit of cortical layer 4. Proc Natl Acad Sci U S A 106(28):11753–11758. https://doi.org/10.1073/pnas.0810062106

Eggermann E, Kremer Y, Crochet S, Petersen CCH (2014) Cholinergic signals in mouse barrel cortex during active whisker sensing. Cell Rep 9(5):1654–1660. https://doi.org/10.1016/j.celrep.2014.11.005

Fernandez-Fernandez JM, Wanaverbecq N, Halley P, Caulfield MP, Brown DA (1999) Selective activation of heterologously expressed G protein-gated K+ channels by M2 muscarinic receptors in rat sympathetic neurones. J Physiol 515(Pt 3):631–637. https://doi.org/10.1111/j.1469-7793.1999.631ab.x

Gabbott PL, Warner TA, Jays PR, Salway P, Busby SJ (2005) Prefrontal cortex in the rat: projections to subcortical autonomic, motor, and limbic centers. J Comp Neurol 492(2):145–177. https://doi.org/10.1002/cne.20738

Galvin VC, Arnsten AFT, Wang M (2018) Evolution in neuromodulation – the differential roles of acetylcholine in higher order association vs. primary visual cortices. Front Neural Circuits 12:67. https://doi.org/10.3389/fncir.2018.00067

Gerber U, Stevens DR, McCarley RW, Greene RW (1991) Muscarinic agonists activate an inwardly rectifying potassium conductance in medial pontine reticular formation neurons of the rat in vitro. J Neurosci 11(12):3861–3867

Ghashghaei HT, Barbas H (2001) Neural interaction between the basal forebrain and functionally distinct prefrontal cortices in the rhesus monkey. Neuroscience 103(3):593–614. https://doi.org/10.1016/s0306-4522(00)00585-6

Goard M, Dan Y (2009) Basal forebrain activation enhances cortical coding of natural scenes. Nat Neurosci 12(11):1444–1449. https://doi.org/10.1038/nn.2402

Gotti C, Zoli M, Clementi F (2006) Brain nicotinic acetylcholine receptors: native subtypes and their relevance. Trends Pharmacol Sci 27(9):482–491. https://doi.org/10.1016/j.tips.2006.07.004

Granger AJ, Wang W, Robertson K, El-Rifai M, Zanello A, Bistrong K, Saunders A, Chow B, Nuñez V, Gu C, Sabatini BL (2018) Target-specific co-transmission of acetylcholine and GABA from a subset of cortical VIP+ interneurons. https://doi.org/10.1101/469064

Gritton HJ, Howe WM, Mallory CS, Hetrick VL, Berke JD, Sarter M (2016) Cortical cholinergic signaling controls the detection of cues. Proc Natl Acad Sci U S A 113(8):E1089–E1097. https://doi.org/10.1073/pnas.1516134113

Guillem K, Bloem B, Poorthuis RB, Loos M, Smit AB, Maskos U, Spijker S, Mansvelder HD (2011) Nicotinic acetylcholine receptor beta2 subunits in the medial prefrontal cortex control attention. Science 333(6044):888–891. https://doi.org/10.1126/science.1207079

Gulledge AT, Stuart GJ (2005) Cholinergic inhibition of neocortical pyramidal neurons. J Neurosci 25(44):10308–10320. https://doi.org/10.1523/JNEUROSCI.2697-05.2005

Gulledge AT, Bucci DJ, Zhang SS, Matsui M, Yeh HH (2009) M1 receptors mediate cholinergic modulation of excitability in neocortical pyramidal neurons. J Neurosci 29(31):9888–9902. https://doi.org/10.1523/JNEUROSCI.1366-09.2009

Guo W, Robert B, Polley DB (2019) The cholinergic basal forebrain links auditory stimuli with delayed reinforcement to support learning. Neuron 103(6):1164–1177 e1166. https://doi.org/10.1016/j.neuron.2019.06.024

Hay YA, Lambolez B, Tricoire L (2016) Nicotinic transmission onto layer 6 cortical neurons relies on synaptic activation of non-alpha7 receptors. Cereb Cortex 26(6):2549–2562. https://doi.org/10.1093/cercor/bhv085

Hedrick T, Waters J (2015) Acetylcholine excites neocortical pyramidal neurons via nicotinic receptors. J Neurophysiol 113(7):2195–2209. https://doi.org/10.1152/jn.00716.2014

Hedrick T, Danskin B, Larsen RS, Ollerenshaw D, Groblewski P, Valley M, Olsen S, Waters J (2016) Characterization of channelrhodopsin and archaerhodopsin in cholinergic neurons of Cre-Lox transgenic mice. PLoS One 11(5):e0156596. https://doi.org/10.1371/journal.pone.0156596

Hodge RD, Bakken TE, Miller JA, Smith KA, Barkan ER, Graybuck LT, Close JL, Long B, Johansen N, Penn O, Yao Z, Eggermont J, Hollt T, Levi BP, Shehata SI, Aevermann B, Beller A, Bertagnolli D, Brouner K, Casper T, Cobbs C, Dalley R, Dee N, Ding SL, Ellenbogen RG, Fong O, Garren E, Goldy J, Gwinn RP, Hirschstein D, Keene CD, Keshk M, Ko AL, Lathia K, Mahfouz A, Maltzer Z, McGraw M, Nguyen TN, Nyhus J, Ojemann JG, Oldre A, Parry S, Reynolds S, Rimorin C, Shapovalova NV, Somasundaram S, Szafer A, Thomsen ER, Tieu M, Quon G, Scheuermann RH, Yuste R, Sunkin SM, Lelieveldt B, Feng D, Ng L, Bernard A, Hawrylycz M, Phillips JW, Tasic B, Zeng H, Jones AR, Koch C, Lein ES (2019) Conserved cell types with divergent features in human versus mouse cortex. Nature 573 (7772):61–68. https://doi.org/10.1038/s41586-019-1506-7

Howe WM, Gritton HJ, Lusk NA, Roberts EA, Hetrick VL, Berke JD, Sarter M (2017) Acetylcholine release in prefrontal cortex promotes gamma oscillations and theta-gamma coupling during cue detection. J Neurosci 37(12):3215–3230. https://doi.org/10.1523/JNEUROSCI.2737-16.2017

Hoyle E, Genn RF, Fernandes C, Stolerman IP (2006) Impaired performance of alpha7 nicotinic receptor knockout mice in the five-choice serial reaction time task. Psychopharmacology (Berl) 189(2):211–223. https://doi.org/10.1007/s00213-006-0549-2

Huppe-Gourgues F, Jegouic K, Vaucher E (2018) Topographic organization of cholinergic innervation from the basal forebrain to the visual cortex in the rat. Front Neural Circuits 12:19. https://doi.org/10.3389/fncir.2018.00019

Jain A, Kuryatov A, Wang J, Kamenecka TM, Lindstrom J (2016) Unorthodox acetylcholine binding sites formed by alpha5 and beta3 accessory subunits in alpha4beta2* nicotinic acetylcholine receptors. J Biol Chem 291(45):23452–23463. https://doi.org/10.1074/jbc.M116.749150

James NM, Gritton HJ, Kopell N, Sen K, Han X (2019) Muscarinic receptors regulate auditory and prefrontal cortical communication during auditory processing. Neuropharmacology 144:155–171. https://doi.org/10.1016/j.neuropharm.2018.10.027

Janeczek M, Gefen T, Samimi M, Kim G, Weintraub S, Bigio E, Rogalski E, Mesulam MM, Geula C (2018) Variations in acetylcholinesterase activity within human cortical pyramidal neurons across age and cognitive trajectories. Cereb Cortex 28(4):1329–1337. https://doi.org/10.1093/cercor/bhx047

Jentsch TJ (2000) Neuronal KCNQ potassium channels: physiology and role in disease. Nat Rev Neurosci 1(1):21–30. https://doi.org/10.1038/35036198

Jing M, Zhang P, Wang G, Feng J, Mesik L, Zeng J, Jiang H, Wang S, Looby JC, Guagliardo NA, Langma LW, Lu J, Zuo Y, Talmage DA, Role LW, Barrett PQ, Zhang LI, Luo M, Song Y, Zhu JJ, Li Y (2018) A genetically encoded fluorescent acetylcholine indicator for in vitro and in vivo studies. Nat Biotechnol 36(8):726–737. https://doi.org/10.1038/nbt.4184

Jones BE (2004) Activity, modulation and role of basal forebrain cholinergic neurons innervating the cerebral cortex. Prog Brain Res 145:157–169. https://doi.org/10.1016/S0079-6123(03)45011-5

Joshi A, Kalappa BI, Anderson CT, Tzounopoulos T (2016) Cell-specific cholinergic modulation of excitability of layer 5B principal neurons in mouse auditory cortex. J Neurosci 36(32):8487–8499. https://doi.org/10.1523/JNEUROSCI.0780-16.2016

Kassam SM, Herman PM, Goodfellow NM, Alves NC, Lambe EK (2008) Developmental excitation of corticothalamic neurons by nicotinic acetylcholine receptors. J Neurosci 28(35): 8756–8764. https://www.jneurosci.org/content/28/35/8756.long

Kawai H, Lazar R, Metherate R (2007) Nicotinic control of axon excitability regulates thalamocortical transmission. Nat Neurosci 10(9):1168–1175. https://doi.org/10.1038/nn1956

Keil JM, Qalieh A, Kwan KY (2018) Brain transcriptome databases: a user's guide. J Neurosci 38(10):2399–2412. https://doi.org/10.1523/JNEUROSCI.1930-17.2018

Kim JH, Jung AH, Jeong D, Choi I, Kim K, Shin S, Kim SJ, Lee SH (2016) Selectivity of neuromodulatory projections from the basal forebrain and locus ceruleus to primary sensory cortices. J Neurosci 36(19):5314–5327. https://doi.org/10.1523/JNEUROSCI.4333-15.2016

Kimura R, Safari MS, Mirnajafi-Zadeh J, Kimura R, Ebina T, Yanagawa Y, Sohya K, Tsumoto T (2014) Curtailing effect of awakening on visual responses of cortical neurons by cholinergic activation of inhibitory circuits. J Neurosci 34(30):10122–10133. https://doi.org/10.1523/JNEUROSCI.0863-14.2014

Kohlmeier KA, Ishibashi M, Wess J, Bickford ME, Leonard CS (2012) Knockouts reveal overlapping functions of M(2) and M(4) muscarinic receptors and evidence for a local glutamatergic circuit within the laterodorsal tegmental nucleus. J Neurophysiol 108(10):2751–2766. https://doi.org/10.1152/jn.01120.2011

Kuchibhotla KV, Gill JV, Lindsay GW, Papadoyannis ES, Field RE, Sten TA, Miller KD, Froemke RC (2017) Parallel processing by cortical inhibition enables context-dependent behavior. Nat Neurosci 20(1):62–71. https://doi.org/10.1038/nn.4436

Kuryatov A, Onksen J, Lindstrom J (2008) Roles of accessory subunits in alpha4beta2(∗) nicotinic receptors. Mol Pharmacol 74(1):132–143. https://doi.org/10.1124/mol.108.046789

Lambe EK, Picciotto MR, Aghajanian GK (2003) Nicotine induces glutamate release from thalamocortical terminals in prefrontal cortex. Neuropsychopharmacology 28(2):216–225. https://doi.org/10.1038/sj.npp.1300032

Lambe EK, Olausson P, Horst NK, Taylor JR, Aghajanian GK (2005) Hypocretin and nicotine excite the same thalamocortical synapses in prefrontal cortex: correlation with improved attention in rat. J Neurosci 25(21):5225–5229. https://doi.org/10.1523/JNEUROSCI.0719-05.2005

Laszlovszky T, Dániel S, Tamás FF, Attila G, Adam K, Balázs H (2019) Distinct synchronization, cortical coupling and behavioural function of basal forebrain cholinergic neuron types. https://doi.org/10.1101/703090

Letzkus JJ, Wolff SB, Meyer EM, Tovote P, Courtin J, Herry C, Luthi A (2011) A disinhibitory microcircuit for associative fear learning in the auditory cortex. Nature 480(7377):331–335. https://doi.org/10.1038/nature10674

Levey AI, Kitt CA, Simonds WF, Price DL, Brann MR (1991) Identification and localization of muscarinic acetylcholine receptor proteins in brain with subtype-specific antibodies. J Neurosci 11(10):3218–3226

Lewis DA (1991) Distribution of choline acetyltransferase-immunoreactive axons in monkey frontal cortex. Neuroscience 40(2):363–374. https://doi.org/10.1016/0306-4522(91)90126-9

Li X, Yu B, Sun Q, Zhang Y, Ren M, Zhang X, Li A, Yuan J, Madisen L, Luo Q, Zeng H, Gong H, Qiu Z (2018) Generation of a whole-brain atlas for the cholinergic system and mesoscopic projectome analysis of basal forebrain cholinergic neurons. Proc Natl Acad Sci U S A 115(2):415–420. https://doi.org/10.1073/pnas.1703601115

Lidow MS, Gallager DW, Rakic P, Goldman-Rakic PS (1989) Regional differences in the distribution of muscarinic cholinergic receptors in the macaque cerebral cortex. J Comp Neurol 289(2):247–259. https://doi.org/10.1002/cne.902890206

Liu CH, Coleman JE, Davoudi H, Zhang K, Hussain Shuler MG (2015) Selective activation of a putative reinforcement signal conditions cued interval timing in primary visual cortex. Curr Biol 25(12):1551–1561. https://doi.org/10.1016/j.cub.2015.04.028

Lysakowski A, Wainer BH, Bruce G, Hersh LB (1989) An atlas of the regional and laminar distribution of choline acetyltransferase immunoreactivity in rat cerebral cortex. Neuroscience 28(2):291–336. https://doi.org/10.1016/0306-4522(89)90180-2

McGaughy J, Dalley JW, Morrison CH, Everitt BJ, Robbins TW (2002) Selective behavioral and neurochemical effects of cholinergic lesions produced by intrabasalis infusions of 192 IgG-saporin on attentional performance in a five-choice serial reaction time task. J Neurosci 22(5):1905–1913

Mechawar N, Cozzari C, Descarries L (2000) Cholinergic innervation in adult rat cerebral cortex: a quantitative immunocytochemical description. J Comp Neurol 428(2):305–318. https://doi.org/10.1002/1096-9861(20001211)428:2<305::aid-cne9>3.0.co;2-y

Meir I, Katz Y, Lampl I (2018) Membrane potential correlates of network decorrelation and improved SNR by cholinergic activation in the somatosensory cortex. J Neurosci 38 (50):10692–10708. https://doi.org/10.1523/JNEUROSCI.1159-18.2018

Mesulam MM, Geula C (1991) Acetylcholinesterase-rich neurons of the human cerebral cortex: cytoarchitectonic and ontogenetic patterns of distribution. J Comp Neurol 306(2):193–220. https://doi.org/10.1002/cne.903060202

Mesulam MM, Hersh LB, Mash DC, Geula C (1992) Differential cholinergic innervation within functional subdivisions of the human cerebral cortex: a choline acetyltransferase study. J Comp Neurol 318(3):316–328. https://doi.org/10.1002/cne.903180308

Metherate R, Intskirveli I, Kawai HD (2012) Nicotinic filtering of sensory processing in auditory cortex. Front Behav Neurosci 6:44. https://doi.org/10.3389/fnbeh.2012.00044

Minces V, Pinto L, Dan Y, Chiba AA (2017) Cholinergic shaping of neural correlations. Proc Natl Acad Sci U S A 114(22):5725–5730. https://doi.org/10.1073/pnas.1621493114

Miwa JM, Lester HA, Walz A (2012) Optimizing cholinergic tone through lynx modulators of nicotinic receptors: implications for plasticity and nicotine addiction. Physiology (Bethesda) 27 (4):187–199. https://doi.org/10.1152/physiol.00002.2012

Morishita H, Miwa JM, Heintz N, Hensch TK (2010) Lynx1, a cholinergic brake, limits plasticity in adult visual cortex. Science 330(6008):1238–1240. https://doi.org/10.1126/science.1195320

Mrzljak L, Goldman-Rakic PS (1993) Low-affinity nerve growth factor receptor (p75NGFR)- and choline acetyltransferase (ChAT)-immunoreactive axons in the cerebral cortex and hippocampus of adult macaque monkeys and humans. Cereb Cortex 3(2):133–147. https://doi.org/10.1093/cercor/3.2.133

Mrzljak L, Pappy M, Leranth C, Goldman-Rakic PS (1995) Cholinergic synaptic circuitry in the macaque prefrontal cortex. J Comp Neurol 357(4):603–617. https://doi.org/10.1002/cne.903570409

Munoz W, Tremblay R, Levenstein D, Rudy B (2017) Layer-specific modulation of neocortical dendritic inhibition during active wakefulness. Science 355(6328):954–959. https://doi.org/10.1126/science.aag2599

Nasirova N, Quina LA, Agosto-Marlin IM, Ramirez JM, Lambe EK, Turner EE (2019) Dual recombinase fate mapping reveals a transient cholinergic phenotype in multiple populations of developing glutamatergic neurons. J Comp Neurol. https://doi.org/10.1002/cne.24753

Nelson A, Mooney R (2016) The basal forebrain and motor cortex provide convergent yet distinct movement-related inputs to the auditory cortex. Neuron 90(3):635–648. https://doi.org/10.1016/j.neuron.2016.03.031

Nguyen DP, Lin SC (2014) A frontal cortex event-related potential driven by the basal forebrain. Elife 3:e02148. https://doi.org/10.7554/eLife.02148

Obermayer J, Verhoog MB, Luchicchi A, Mansvelder HD (2017) Cholinergic modulation of cortical microcircuits is layer-specific: evidence from rodent, monkey and human brain. Front Neural Circuits 11:100. https://doi.org/10.3389/fncir.2017.00100

Obermayer J, Heistek TS, Kerkhofs A, Goriounova NA, Kroon T, Baayen JC, Idema S, Testa-Silva G, Couey JJ, Mansvelder HD (2018) Lateral inhibition by Martinotti interneurons is facilitated by cholinergic inputs in human and mouse neocortex. Nat Commun 9(1):4101. https://doi.org/10.1038/s41467-018-06628-w

Obermayer J, Luchicchi A, Heistek TS, de Kloet SF, Terra H, Bruinsma B, Mnie-Filali O, Kortleven C, Galakhova AA, Khalil AJ, Kroon T, Jonker AJ, de Haan R, van den Berg WDJ, Goriounova NA, de Kock CPJ, Pattij T, Mansvelder HD (2019) Prefrontal cortical ChAT-VIP interneurons provide local excitation by cholinergic synaptic transmission and control attention. Nat Commun 10(1):5280. https://doi.org/10.1038/s41467-019-13244-9

Palomero-Gallagher N, Zilles K (2019) Cortical layers: cyto-, myelo-, receptor- and synaptic architecture in human cortical areas. Neuroimage 197:716–741. https://doi.org/10.1016/j.neuroimage.2017.08.035

Parikh V, Kozak R, Martinez V, Sarter M (2007) Prefrontal acetylcholine release controls cue detection on multiple timescales. Neuron 56(1):141–154. https://doi.org/10.1016/j.neuron.2007.08.025

Pinto L, Goard MJ, Estandian D, Xu M, Kwan AC, Lee SH, Harrison TC, Feng G, Dan Y (2013) Fast modulation of visual perception by basal forebrain cholinergic neurons. Nat Neurosci 16 (12):1857–1863. https://doi.org/10.1038/nn.3552

Poorthuis RB, Bloem B, Verhoog MB, Mansvelder HD (2013) Layer-specific interference with cholinergic signaling in the prefrontal cortex by smoking concentrations of nicotine. J Neurosci 33(11):4843–4853. https://doi.org/10.1523/JNEUROSCI.5012-12.2013

Proulx E, Piva M, Tian MK, Bailey CD, Lambe EK (2014a) Nicotinic acetylcholine receptors in attention circuitry: the role of layer VI neurons of prefrontal cortex. Cell Mol Life Sci 71 (7):1225–1244. https://doi.org/10.1007/s00018-013-1481-3

Proulx E, Suri D, Heximer SP, Vaidya VA, Lambe EK (2014b) Early stress prevents the potentiation of muscarinic excitation by calcium release in adult prefrontal cortex. Biol Psychiatry 76 (4):315–323. https://doi.org/10.1016/j.biopsych.2013.10.017

Reimer J, McGinley MJ, Liu Y, Rodenkirch C, Wang Q, McCormick DA, Tolias AS (2016) Pupil fluctuations track rapid changes in adrenergic and cholinergic activity in cortex. Nat Commun 7:13289. https://doi.org/10.1038/ncomms13289

Rossner S, Kues W, Witzemann V, Schliebs R (1993) Laminar expression of m1-, m3- and m4-muscarinic cholinergic receptor genes in the developing rat visual cortex using in situ hybridization histochemistry. Effect of monocular visual deprivation. Int J Dev Neurosci 11 (3):369–378. https://doi.org/10.1016/0736-5748(93)90008-2

Salas R, Orr-Urtreger A, Broide RS, Beaudet A, Paylor R, De Biasi M (2003) The nicotinic acetylcholine receptor subunit alpha 5 mediates short-term effects of nicotine in vivo. Mol Pharmacol 63(5):1059–1066. https://doi.org/10.1124/mol.63.5.1059

Sarter M (2015) Behavioral-cognitive targets for cholinergic enhancement. Curr Opin Behav Sci 4:22–26. https://doi.org/10.1016/j.cobeha.2015.01.004

Sarter M, Parikh V, Howe WM (2009) Phasic acetylcholine release and the volume transmission hypothesis: time to move on. Nat Rev Neurosci 10(5):383–390. https://doi.org/10.1038/nrn2635

Seeger T, Alzheimer C (2001) Muscarinic activation of inwardly rectifying K(+) conductance reduces EPSPs in rat hippocampal CA1 pyramidal cells. J Physiol 535(Pt 2):383–396. https://doi.org/10.1111/j.1469-7793.2001.00383.x

Shapiro MS, Loose MD, Hamilton SE, Nathanson NM, Gomeza J, Wess J, Hille B (1999) Assignment of muscarinic receptor subtypes mediating G-protein modulation of Ca(2+) channels by using knockout mice. Proc Natl Acad Sci U S A 96(19):10899–10904. https://doi.org/10.1073/pnas.96.19.10899

Smiley JF, Morrell F, Mesulam MM (1997) Cholinergic synapses in human cerebral cortex: an ultrastructural study in serial sections. Exp Neurol 144(2):361–368. https://doi.org/10.1006/exnr.1997.6413

Sparks DW, Tian MK, Sargin D, Venkatesan S, Intson K, Lambe EK (2017) Opposing cholinergic and serotonergic modulation of layer 6 in prefrontal cortex. Front Neural Circuits 11:107. https://doi.org/10.3389/fncir.2017.00107

Sun Q, Li X, Ren M, Zhao M, Zhong Q, Ren Y, Luo P, Ni H, Zhang X, Zhang C, Yuan J, Li A, Luo M, Gong H, Luo Q (2019) A whole-brain map of long-range inputs to GABAergic interneurons in the mouse medial prefrontal cortex. Nat Neurosci 22(8):1357–1370. https://doi.org/10.1038/s41593-019-0429-9

Tapia L, Kuryatov A, Lindstrom J (2007) Ca2+ permeability of the (alpha4)3(beta2)2 stoichiometry greatly exceeds that of (alpha4)2(beta2)3 human acetylcholine receptors. Mol Pharmacol 71 (3):769–776. https://doi.org/10.1124/mol.106.030445

Thomson AM (2010) Neocortical layer 6, a review. Front Neuroanat 4:13. https://doi.org/10.3389/fnana.2010.00013

Tian MK, Bailey CD, Lambe EK (2014) Cholinergic excitation in mouse primary vs. associative cortex: region-specific magnitude and receptor balance. Eur J Neurosci 40(4):2608–2618. https://doi.org/10.1111/ejn.12622

Tian MK, Schmidt EF, Lambe EK (2016) Serotonergic suppression of mouse prefrontal circuits implicated in task attention. eNeuro 3(5). https://doi.org/10.1523/ENEURO.0269-16.2016

Turrini P, Casu MA, Wong TP, De Koninck Y, Ribeiro-da-Silva A, Cuello AC (2001) Cholinergic nerve terminals establish classical synapses in the rat cerebral cortex: synaptic pattern and age-related atrophy. Neuroscience 105(2):277–285. https://doi.org/10.1016/s0306-4522(01)00172-5

Unal CT, Golowasch JP, Zaborszky L (2012) Adult mouse basal forebrain harbors two distinct cholinergic populations defined by their electrophysiology. Front Behav Neurosci 6:21. https://doi.org/10.3389/fnbeh.2012.00021

Urban-Ciecko J, Jouhanneau JS, Myal SE, Poulet JFA, Barth AL (2018) Precisely timed nicotinic activation drives SST inhibition in neocortical circuits. Neuron 97(3):611–625 e615. https://doi.org/10.1016/j.neuron.2018.01.037

van Aerde KI, Feldmeyer D (2015) Morphological and physiological characterization of pyramidal neuron subtypes in rat medial prefrontal cortex. Cereb Cortex 25(3):788–805. https://doi.org/10.1093/cercor/bht278

Venkatesan S, Lambe EK (2020) *Chrna5* is essential for a rapid and protected response to optogenetic release of endogenous acetylcholine in prefrontal cortex. https://doi.org/10.1101/2020.05.10.087569

Verhoog MB, Obermayer J, Kortleven CA, Wilbers R, Wester J, Baayen JC, De Kock CPJ, Meredith RM, Mansvelder HD (2016) Layer-specific cholinergic control of human and mouse cortical synaptic plasticity. Nat Commun 7:12826. https://doi.org/10.1038/ncomms12826

von Engelhardt J, Eliava M, Meyer AH, Rozov A, Monyer H (2007) Functional characterization of intrinsic cholinergic interneurons in the cortex. J Neurosci 27(21):5633–5642. https://doi.org/10.1523/JNEUROSCI.4647-06.2007

Wada E, McKinnon D, Heinemann S, Patrick J, Swanson LW (1990) The distribution of mRNA encoded by a new member of the neuronal nicotinic acetylcholine receptor gene family (alpha 5) in the rat central nervous system. Brain Res 526(1):45–53. https://doi.org/10.1016/0006-8993(90)90248-a

Wang J, Lindstrom J (2018) Orthosteric and allosteric potentiation of heteromeric neuronal nicotinic acetylcholine receptors. Br J Pharmacol 175(11):1805–1821. https://doi.org/10.1111/bph.13745

Wang J, Kuryatov A, Sriram A, Jin Z, Kamenecka TM, Kenny PJ, Lindstrom J (2015) An accessory agonist binding site promotes activation of alpha4beta2∗ nicotinic acetylcholine receptors. J Biol Chem 290(22):13907–13918. https://doi.org/10.1074/jbc.M115.646786

Wilking JA, Stitzel JA (2015) Natural genetic variability of the neuronal nicotinic acetylcholine receptor subunit genes in mice: Consequences and confounds. Neuropharmacology 96 (Pt B):205–212. https://doi.org/10.1016/j.neuropharm.2014.11.022

Yang D, Robert G, Guanxiao Q, Gabriele R, Dirk F (2019) Cell type-specific modulation of layer 6A excitatory microcircuits by acetylcholine in rat barrel cortex. https://doi.org/10.1101/701318

Young JW, Crawford N, Kelly JS, Kerr LE, Marston HM, Spratt C, Finlayson K, Sharkey J (2007) Impaired attention is central to the cognitive deficits observed in alpha 7 deficient mice. Eur Neuropsychopharmacol 17(2):145–155. https://doi.org/10.1016/j.euroneuro.2006.03.008
Zaborszky L, Gombkoto P, Varsanyi P, Gielow MR, Poe G, Role LW, Ananth M, Rajebhosale P, Talmage DA, Hasselmo ME, Dannenberg H, Minces VH, Chiba AA (2018) Specific basal forebrain-cortical cholinergic circuits coordinate cognitive operations. J Neurosci 38 (44):9446–9458. https://doi.org/10.1523/JNEUROSCI.1676-18.2018
Zhang W, Basile AS, Gomeza J, Volpicelli LA, Levey AI, Wess J (2002) Characterization of central inhibitory muscarinic autoreceptors by the use of muscarinic acetylcholine receptor knock-out mice. J Neurosci 22(5):1709–1717
Zhao S, Ting JT, Atallah HE, Qiu L, Tan J, Gloss B, Augustine GJ, Deisseroth K, Luo M, Graybiel AM, Feng G (2011) Cell type-specific channelrhodopsin-2 transgenic mice for optogenetic dissection of neural circuitry function. Nat Methods 8(9):745–752. https://doi.org/10.1038/nmeth.1668
Zilles K, Palomero-Gallagher N (2017) Multiple transmitter receptors in regions and layers of the human cerebral cortex. Front Neuroanat 11:78. https://doi.org/10.3389/fnana.2017.00078

Cholinergic Signaling Dynamics and Cognitive Control of Attention

Vinay Parikh and Debra A. Bangasser

Contents

Abstract The central cholinergic system is one of the most important modulator neurotransmitter system implicated in diverse behavioral processes. Activation of the basal forebrain cortical cholinergic input system represents a critical step in cortical information processing. This chapter explores recent developments illustrating cortical cholinergic transmission mediate defined cognitive operations, which is contrary to the traditional view that acetylcholine acts as a slowly acting neuromodulator that influences arousal cortex-wide. Specifically, we review the evidence that phasic cholinergic signaling in the prefrontal cortex is a causal mediator of signal detection. In addition, studies that support the neuromodulatory role of cholinergic inputs in top-down attentional control are summarized. Finally, we review new findings that reveal sex differences and hormonal regulation of the cholinergic-attention system.

V. Parikh (✉) and D. A. Bangasser
Department of Psychology and Neuroscience Program, Temple University, Philadelphia, PA, USA
e-mail: vinay.parikh@temple.edu

Curr Topics Behav Neurosci (2020) 45: 71–88
https://doi.org/10.1007/7854_2020_133
Published Online: 24 May 2020

Keywords Acetylcholine · Attention · Muscarinic receptors · Nicotinic receptors · Sex differences

1 Introduction

Cholinergic inputs to the entire cortical mantle originate in the nucleus basalis of Meynert (nBM), substantia innominata (SI), the horizontal nucleus of the diagonal band (HDB), and the preoptic nucleus (collectively termed basal forebrain, BF). The BF corticopetal cholinergic system constitutes the most rostral component of neuromodulatory input systems, and its anatomical organization reflects its ability to orchestrate cortical information processing. Of the many different behavioral and cognitive processes that relate to the central cholinergic system, fundamental aspects of attention are closely linked to the activity of cortical cholinergic inputs (Ballinger et al. 2016; Sarter et al. 2005, 2016). Therefore, there is considerable interest in the dynamics of cortical cholinergic signaling and cholinergic regulation of attentional processes and capacities and developing procholinergic therapies to treat cognitive deficits in psychiatric and neurological disorders.

Historically, the organization of the BF cortical projection system was described as a diffuse and undifferentiated projection system with widespread cortical innervation, which corresponds to the notion that acetylcholine (ACh) influences the excitability of neurons cortex-wide to modulate global states of arousal and wakefulness. Earlier studies that focused on slow and regionally nonspecific changes in ACh efflux (volume transmission) supported this view (Bartolini and Pepeu 1967; Descarries 1998; Phillis 1968). However, advancement in electrochemical approaches to monitor neurochemical events at high temporal and spatial resolution in the past decade led to the characterization of phasic ACh release that was linked to specific cognitive events (Howe et al. 2013; Parikh et al. 2007). These developments, along with the refinement of neuroanatomical tools that revealed a highly organized topographic arrangement of cortical target-specific groups of BF cholinergic neurons, challenged previous conceptualization and support modality-/region-specific function of ACh (Lean et al. 2019; Zaborszky et al. 2015, 2018).

Cholinergic signaling is elicited by presynaptic release of ACh that activates two classes of ACh receptors, nicotinic (nAChR) and muscarinic (mAChR), in a spatially and temporally selected fashion due to the constraints imposed by the potent ACh-metabolizing enzyme acetylcholinesterase (AChE). nAChRs are a family of ligand-gated ionotropic receptors that mediate fast synaptic transmission by altering cation channel currents. Neuronal nAChRs are pentameric structures that are formed from a combination of five membrane-spanning units consisting of nine isoforms of α subunits (α2–α10) and three isoforms of β subunits (β2–β4) and arranged either as a heteromeric or homomeric assemblies (Gotti et al. 2009). Within the mammalian cortex, homomeric α7 and heteromeric $\alpha4\beta2$ are the most predominant and widely distributed nAChRs. mAChRs are metabotropic receptors that, following activation by ACh, transduce their signaling via heterotrimeric G proteins. The M1 family of mAChRs include M1, M3, and M5 and signals via Gq proteins, while M2 and M4 mAChRs belong to the M2 family that are coupled to Gi proteins (Thiele 2013).

Efforts to develop cholinomimetic drugs as cognition enhancers have largely focused on boosting cholinergic transmission. Although psychopharmacological research to augment cholinergic signaling have generally focused on AChE inhibitors, the procognitive therapeutic efficacy of these drugs in human subjects have remained limited (Pepeu and Giovannini 2009). It is suggested that higher baseline ACh levels, as a result of AChE blockade, would result in generalized activation of cholinergic auto- and hetero-receptors that may uncouple presynaptic and postsynaptic mechanisms and, consequently, produce complex changes in the local cortical networks (Hasselmo and Sarter 2011; Pepeu and Giovannini 2009). Likewise, pharmacological studies that focused on nonselectively modulating mAChRs and nAChRs largely reported complex effects on cognition (Hasselmo and Sarter 2011; Sarter et al. 2009a). This chapter explores recent developments in our understanding of the cholinergic mechanisms of attention. Specifically, the evidence that phasic cholinergic signaling in the prefrontal cortex (PFC) is a causal mediator of signal detection will be discussed. Moreover, studies that support the tonic neuromodulatory role of cholinergic inputs in top-down attentional control, and those that provide insights into the potential cellular substrates that integrate the phasic and neuromodulatory cholinergic signaling modes, will be reviewed. Because sex differences exist in the central cholinergic system, we will also highlight new findings that reveal sex differences in cholinergic-attention system. In conclusion, the framework to develop procholinergic therapies by targeting specific components of cortical cholinergic signaling will be briefly presented.

2 Cortical ACh and Attentional Performance

Substantial evidence from lesion and microdialysis studies supported the hypothesis that cortical cholinergic projections are necessary for performance in tasks that assess a range of attentional functions. A plethora of studies conducted in rodents demonstrated that selective lesions of BF cholinergic neurons and their cortical inputs produced by the immunotoxin 192-IgG saporin impair performance in various tasks of attention. For instance, cortical cholinergic deafferentation of rats trained in an operant sustained attention task (SAT) disrupted animals' ability to detect the signal (correctly respond on signal trials), while sparing response accuracies on non-signal trials (McGaughy et al. 1996, 2000). In a cross-modal divided attention task that requires the processing of a visual and auditory conditioned stimulus, selective cholinergic lesions resulted in a speed-accuracy tradeoff under conditions of modality uncertainty, with longer correct response latencies in bimodal than in unimodal blocks of trials (Botly and De Rosa 2009; Newman and McGaughy 2008; Turchi and Sarter 1997). Additionally, removal of cholinergic inputs from the medial

PFC reduced choice accuracies under conditions of increased attentional load and increased perseverative responding in animals performing the five-choice serial reaction time task (Dalley et al. 2004; Maddux et al. 2007). It is noteworthy that lesions of BF cholinergic neurons did not robustly impact performance of subjects trained in tasks that did not explicitly tax attentional processes, illustrating the specificity of cognitive impairments produced by cortical cholinergic deafferentation (Baxter et al. 1995; Frick et al. 2004; Vuckovich et al. 2004).

Studies employing in vivo microdialysis procedures in rats reproducibly demonstrated performance-associated increases in ACh release in the frontal and parietal cortex in operant tasks of attention (Arnold et al. 2002; Dalley et al. 2001; Himmelheber et al. 1997; Kozak et al. 2006; Passetti et al. 2000). More importantly, the levels of ACh release in attentional task-performing animals varied as a function of the demands on attention but did not correlate with levels of attentional performance. Such increases in cortical ACh efflux were not observed in animals performing various operant control procedures that do not explicitly tax attentional processes.

Although this research substantiated the claim made by lesions studies concerning the necessary role of BF cholinergic system in attentional performance, it remained limited in explaining the precise cognitive operations that are supported by cortical cholinergic activity. Microdialysis studies typically required 5–10 min of sample collection to detect ACh concentration in the dialysate using HPLC, which took over tens of trials to generate a single data point. Thus, the low temporal resolution of ACh release data limited the demonstration of specific attention task events or the behavioral/cognitive operations that are responsible for the increases in cortical cholinergic signaling. Consequently, conclusions based on microdialysis data were consistent with the conventional characteristics of ACh as a slowly acting cortex-wide neuromodulator optimizing input processing by regulating arousal states. As noted above, the presence of ubiquitous and highly potent ACh-metabolizing enzyme AChE, and the presence of nAChRs that mediate faster cholinergic signaling, suggests that the functions of forebrain cholinergic system are not sufficiently described by such notions. With the advent of electrochemical approaches and refinement in the design of enzyme-based microelectrodes, the measurement of cholinergic transmission on a faster time-scale became possible. The evidence generated from research based on these technical breakthroughs, which show rapid (phasic) changes in cholinergic transmission transients in specific behavioral contexts, led to the revision of previous conceptualizations of BF cholinergic system as discussed next.

3 Prefrontal Cholinergic Mechanisms of Signal Detection and Attentional Control

3.1 Phasic ACh Release Mediates Cue Detection

Considerable progress has been made in the development of enzyme-based biosensors to measure electrochemically rapid (on the time-scale of milliseconds to seconds) changes in extracellular choline levels, as a marker for cortical ACh release (Giuliano et al. 2008; Parikh et al. 2004). These approaches allow real-time monitoring of cholinergic signaling on a trial-by-trial basis in tasks of attention and have advanced our understanding of the specific role of phasic cholinergic signaling in signal (cue) detection. *Detection* here implies a cognitive-attentional process that relates to the entry of information concerning the presence of a stimulus (signal or cue) into a processing stream that allows the subject to report the existence of a signal by an arbitrary response established by the experimenter (Posner et al. 1980). This process is distinct from *orientating* that mostly reflects a process of aligning sensory response to the salient cue.

One of the initial studies that employed choline-sensitive biosensors to record cholinergic activity from the medial PFC was conducted in *awake* rats, performing a cued-appetitive response task (Parikh et al. 2007). This study demonstrated phasic cholinergic signals (cholinergic transients) evoked by "detected" cues (visual stimulus) that generated a distinct shift from ongoing behavior (e.g., grooming) toward the monitoring of the reward ports, followed by port approach and reward retrieval in response to reward delivery. The onset of the cholinergic transient was highly correlated with the onset of the behavioral shift. Moreover, prefrontal cholinergic transients were specifically associated with detected cues and did not occur with other task events such as reward delivery and reward retrieval. In trials involving missed cues, where the animal oriented to the cue but failed to initiate any response, cholinergic signals were not observed. Removal of cholinergic inputs to the recording region by locally infusing cholino-immunotoxin 192-IgG saporin, completely abolished cue-evoked phasic cholinergic signals in detected trials confirming that signals originated from cholinergic terminals. Collectively, these findings suggested that transient or phasic increases in prefrontal cholinergic activity mediate cue-evoked cognitive operations in attention-demanding contexts.

Additional experiments indicated that variation in the time interval between cue and reward delivery caused variation of the timing of the peak amplitude of cue-evoked cholinergic signals (Parikh et al. 2007; Parikh and Sarter 2008). This was an important observation as it indicated that cholinergic transients do not merely reflect sensory encoding of the cue. If that was the case, variations of cue-reward intervals should not affect the timing of the cholinergic transients. The variation of the timing of cue-evoked cholinergic transients indicates that phasic ACh release in the PFC is associated with a cognitive operation (cue detection), the timing of which is a function of cue-reward intervals.

A subsequent study that recorded PFC cholinergic activity in rats performing an operant SAT reported cholinergic transients during "hits," i.e., correct responses on signal trials (Howe et al. 2013). Surprisingly, phasic cholinergic signals were observed only in 40% of hits. The cholinergic transients that were generated during hits were preceded either by correct rejections (correct responses on non-signal trials) or misses (incorrect responses on signal trials). Hits that were not associated with cholinergic transients were those preceded by hits. These findings indicated that phasic cholinergic signals mediate signal detection specifically in situations that involves a shift from monitoring to cue-directed behavior (shift hits). Additional evidence from fMRI studies conducted in humans performing the SAT task illustrated increase BOLD activation in the right rostrolateral/orbital PFC and right BF during shift hits and that this activation was associated with faster reaction times (Howe et al. 2013; Sarter et al. 2016).

Another study that combined optogenetics with electrochemistry tested the hypothesis that cholinergic transients have the capacity to cause signal detection even in the absence of signals (Gritton et al. 2016). Photostimulation of channel rhodopsin-expressed BF cholinergic neurons and prefrontal cholinergic terminals generated optogenetically evoked cholinergic transients and increased hit rates in SAT-performing mice. Moreover, suppression of phasic cholinergic activity by photostimulating halorhodopsin-expressed BF cholinergic neurons resulted in reduced hits without affecting correct rejections. Collectively, these findings indicate that phasic cholinergic signaling, specifically in the PFC, is not only associated exclusively with cue detection but are actually the causal mediators of shift hits (i.e., shifts from monitoring to signal detection). This view aligns with the lesion studies (discussed earlier) that show the detrimental effects of cortical cholinergic deafferentation were linked to detection performance (i.e., hit rates on signal trials and not correct rejections).

3.2 *Top-Down Control of Attention and Cholinergic Neuromodulation*

The ability to maintain stable task performance in the face of challenges or distractors requires attentional effort (Sarter et al. 2006). Cholinergic neuromodulation of the prefrontal efferent projections is conceptualized to enhance stimulus processing and to suppress the processing of irrelevant stimuli, distractors, or noise in a top-down fashion (Sarter et al. 2005). This hypothesis was supported by previous microdialysis studies that reported sustained increases in cholinergic activity during attentional challenges. For instance, steady increases in ACh efflux in medial PFC of SAT-performing rats were observed when animals moved from non-performing (baseline) stage to the performing (task) stage; however, ACh levels increased further with the presentation of visual distractors despite a reduction in hits (Kozak et al. 2006; St Peters et al. 2011). Human fMRI studies conducted in subjects performing SAT reported comparable increases in right PFC activity from baseline to SAT and then to the distracting condition (Berry et al. 2017; Demeter et al. 2008, 2011). Furthermore, SAT-associated ACh release in the medial PFC was attenuated in sign-tracking rats that show poor attentional control (Paolone et al. 2013).

Extracellular ACh efflux measured using microdialysis reflects a slower (tonic) component of cholinergic signaling that ranges from hundreds of seconds to tens of minutes. Tonic cholinergic activity is proposed to reflect a top-down neuromodulatory role of BF-cholinergic neurons to regulate cortical detection circuitry in an attempt to maintain task performance under conditions of distraction (Sarter and Lustig 2019). Although the dissociation between phasic and neuromodulatory (tonic) components of cholinergic signaling appears to be distinct in terms of cognitive operations; the two modes may interact to support overall attentional performance. This notion is supported by a previous in vivo amperometry study that reported a positive correlation between the magnitude of slower (timescale of minutes) session-related increases in tonic cholinergic activity and the amplitudes of phasic cholinergic signals in animals performing the cued-appetitive response task (Parikh et al. 2007). Given the constraints imposed by AChE on cholinergic signaling, the view that neuromodulatory/tonic cholinergic activity is driven by "volume transmission" is debated (Sarter et al. 2009b). It remains to be seen whether cholinergic neuromodulation is a consequence of sustained activity of BF cholinergic neurons, local presynaptic regulation in the cortical microcircuits, or another population of BF cholinergic neurons that produce tonic discharges (Sarter and Kim 2015; Sarter et al. 2014; Unal et al. 2012).

4 Cellular Regulation of Cholinergic Signaling Modes

4.1 High-Affinity Choline Transporters (CHTs)

Cholinergic terminals recover choline from the synaptic cleft following ACh degradation by AChE, through a hemicholinium-3 (HC-3)-sensitive high-affinity choline transporter (CHT). Because cholinergic synapses rely heavily on choline for ACh production, the capacity to import choline into presynaptic cholinergic compartments via CHTs dictates the rate of ACh synthesis and release (Ferguson and Blakely 2004; Sarter and Parikh 2005). CHT-mediated choline uptake was enhanced in the synaptosomes isolated from the medial PFC of SAT-performing rats; such increases in choline uptake were not observed in animals that completed a behavioral control session (Apparsundaram et al. 2005). The same study also reported attention performance-associated increases in the densities of CHTs on the surface membrane of prefrontal synaptosomes relative to the intracellular pools (outward CHT trafficking). Another study found a decline in the capacity to generate prefrontal cholinergic transients following sustained BF stimulation in CHT heterozygous mice (Parikh et al. 2013). Moreover, these mutants displayed high vulnerability to the effects of visual distractors in SAT and disrupted trafficking of subcellular CHTs. Likewise, a recent fMRI study that involved human subjects expressing a I89V variant of CHT (low CHT capacity) did not find increases in right prefrontal activity in these subjects during increases in attentional demands that is typically seen in normal subjects (Berry et al. 2015). Taken together, these interesting findings point toward an important role of CHT function in regulating presynaptic cholinergic neuromodulation and in sustaining phasic cholinergic signaling under situations that impose increased demands on BF cholinergic neurons, such as top-down attentional control.

4.2 nAChRs

Substantial evidence indicates that the administration of nicotine and nAChR agonists, specifically those that activate α4β2 nAChRs, exert beneficial effects on attention and related cognitive abilities (Allison and Shoaib 2013; Howe et al. 2010; Newhouse et al. 2004; Sarter et al. 2009a; Stolerman et al. 2000; Wilens and Decker 2007). α4β2 nAChRs situated on thalamic glutamatergic projections in the medial PFC are an important component of attention circuitry and that stimulation of these receptors increase glutamatergic activity (Lambe et al. 2003; Lucas-Meunier et al. 2009). Moreover, neuropharmacological studies employing in vivo amperometry demonstrated that the stimulation of α4β2 nAChRs produces transient increases in glutamate and ACh release in the medial PFC and that thalamocortical glutamatergic terminals are necessary for the generation of cholinergic transients (Parikh et al. 2008, 2010). Moreover, systemic administration of a full α4β2 nAChR agonist S38232 improved attentional performance following the presentation of distractor in rats (Howe et al. 2010). As noted above, attention control requires cholinergic neuromodulation, and it is possible that α4β2 nAChR activation facilitates phasic cholinergic signaling by tonically modulating glutamatergic-cholinergic interactions (Hasselmo and Sarter 2011). Although α7 nAChR agonists have also been reported to augment prefrontal glutamatergic transmission, they did not produce faster cholinergic transients as observed with the stimulation of α4β2 nAChRs (Bortz et al. 2013; Parikh et al. 2010). It is possible that α7 nAChRs recruit other ascending modulators such as monoamines which impact the dynamics of BF cholinergic signaling in a different way resulting in more complex effects on attention.

4.3 mAChRs

Systemic administration of mAChR antagonist scopolamine has consistently been shown to produce attentional impairments indicating that mAChRs may be important for cholinergic mediation of attention (Callahan et al. 1993; Chudasama et al. 2004; Young et al. 2013). However, the beneficial effects of mAChR agonists on cognitive processes have remained complex and could not be reliably demonstrated in clinics presumably due to lack of the availability of specific ligands targeting specific mAChR subtypes. It has been suggested that postsynaptic M1 receptors localized on cortical pyramidal neurons enhance voltage-dependent Ca^{2+} influx and action potential output in response to phasic release of ACh (Dasari et al. 2017). Moreover, a recent study reported that cue-evoked cholinergic transients in the medial PFC of animals performing the Pavlovian cued-approach task triggered theta-gamma coupling, and this synchronization and cue detection was disrupted following M1 receptor blockade (Howe et al. 2017). Thus, M1 receptor activation may regulate phasic ACh-induced prefrontal network synchrony required for cue detection.

5 Sex Differences and the Cholinergic Mediation of Attention

5.1 Neurochemical Sex Differences

The synthesis, release, and postsynaptic effects of many neurotransmitters systems are influenced by biological sex, and the BF corticopetal cholinergic system is no exception. As noted, a critical ACh-producing region within this circuit is the nBM. In rats, although a sex difference is not always observed (Gibbs 1996), there are reports that the nBM of females has more neurons containing the cholinergic synthetic enzyme, choline acetyltransferase (ChAT), than the nBM of males (Takase et al. 2007, 2009). The sex difference in ChAT neurons may be specific to the nBM, as it does not occur in the HDB (Takase et al. 2009), suggesting that the regions within the BF corticopetal cholinergic system are differentially influenced by sex. An increase in ChAT in the female nBM could facilitate their production of ACh relative to males. Consistent with this idea, the females have higher tonic ACh release than males in the mPFC (Takase et al. 2007, 2009). This sex differences in cortical ACh is observed across the circadian release profile for ACh (Takase et al. 2009). The enhanced tonic cortical ACh release in females may facilitate their top-down attentional control relative to males. However, sex differences in phasic ACh release have not been assessed. Given the greater capacity of females to synthesis of ACh, it is possible that a similar sex difference in phasic cholinergic signals would be detected, but further studies are needed.

Sex differences in ACh production and release likely result from circulating ovarian hormones. Cholinergic neurons in the BF contain estradiol receptors (ER), including ERα and the G-protein coupled ER (Gibbs 1996; Miettinen et al. 2002). Although ER levels in the BF cholinergic neurons are comparable in male and female rats (Gibbs 1996), the higher circulating levels of estradiol in females could preferentially influence their BF. One mechanism by which estradiol could influence cholinergic neurons is via the regulation of ChAT. Estradiol administration to ovariectomized female rats increases ChAT levels in nBM, but not the HDB (Gibbs 1997; Gibbs et al. 1994). In contrast to the role of estradiol, manipulation of testosterone in males does not affect ChAT in the nBM (Nakamura et al. 2002). Collectively, these studies suggest an increased nBM production and release of ACh in females that is driven by estradiol regulation of ChAT.

As noted, ACh exerts its effects through nAChRs and mAChRs. In humans, women have more β2-containing nAChRs and mAChRs in the frontal cortex than men (Cosgrove et al. 2012; Yoshida et al. 2000). These receptors also appear to be regulated by estrogens. Estradiol potentiates the human alpha4beta2 subtype of the nicotinic receptor (Curtis et al. 2002). In rats, mAChR binding is highest in females in the proestrous stage of the estrous cycle, which is the stage when estradiol levels are highest (van Huizen et al. 1994). Additional evidence for estrogenic regulation of cholinergic receptors comes from studies associating the loss of estrogens in menopause with a reduction in nAChRs and mAChRs (Norbury et al. 2007; Tinkler and Voytko 2005). Interestingly, postmenopausal women receiving estrogen replacement therapy have higher mAChR density in the lateral frontal cortex than untreated postmenopausal women (Norbury et al. 2007), indicating that restoring estrogen levels can mitigate against the negative effect of hormone loss on cholinergic receptor levels. When these findings are considered with the aforementioned studies on ACh production and release, it appears that, compared to males, the basal forebrain corticopetal system of females has a greater capacity for producing and responding to ACh, which could improve attention in females.

5.2 Behavioral Sex Differences

Psychiatric disorders with attention dysregulation often occur at different rates in men and women. For example, men are more likely to be diagnosed with ADHD and schizophrenia (Mendrek and Mancini-Marïe 2016; Ramtekkar et al. 2010). These disorders can also present differently in men and women, such that men with schizophrenia, for example, have greater deficits in cognitive processes, including attention (Goldstein et al. 1998; Mendrek and Mancini-Marïe 2016; Zhang et al. 2012). In aging populations, there is evidence that women have higher rates of Alzheimer's disease than men (Gao et al. 1998; Mazure and Swendsen 2016). This sex difference has been attributed to a loss of estradiol in women, and there is some evidence that hormone replacement therapy reduces Alzheimer's disease risk, especially when hormone replacement therapy is initiated within a short period of oophorectomy or natural menopause (Mielke et al. 2014; Rocca et al. 2011; Whitmer et al. 2011).

In healthy populations, there is also evidence for sex differences in certain aspects of attention. For example, women outperform men on a divided attention paradigm and their enhanced capacity to rapidly switch attention is thought to explain their better ability to multitask than men (Seçer and Yılmazoğulları 2016; Stoet et al. 2013). In rodents, females also do better at certain tasks of attention than males. For example, auditory distractors are less disruptive in female than male mice in an interval timing task (Buhusi et al. 2017). However, sex differences in attention may be specific to certain attentional processes because they are not observed in every attention task. For example, male and female rats perform similarly under baseline parameters in task of spatial divided attention (Bayless et al. 2012; Jentsch and Taylor 2003). When the task is made more difficult (e.g., by increasing the intertrial interval, decreasing the visual stimulus), females make more vigilance errors, while males make more errors of inhibitory control (Bayless et al. 2012, Jentsch and Taylor 2003). Similarly, performance on the SAT is comparable between male and female rats, even on the signal trials that require the release of ACh in the mPFC (Bangasser et al. 2017; Cole et al. 2016). These studies indicate that sex differences in attention differ based on the attentional process examined and often do not emerge until tested under challenging conditions.

There are some reports of estradiol regulating attentional processes. On a task of divided attention, a loss of estrogens impaired performance when conditions were challenging and this decrement was rescued by the administration of estradiol (Barnes et al. 2006). In contrast, performance in the sustained attention did not change across the estrous cycle (Cole et al. 2016), and ovariectomy did not impair performance on the task and, surprisingly, prevented a decrease in performance across the session (McGaughy and Sarter 1999). However, if BF cholinergic neurons were damaged with a selective neurotoxin, high levels of estradiol improved aspects of performance on the sustained attention task (McGaughy and Sarter 1999). These data suggest that when task parameters are easy the effects of estradiol on attention are difficult to detect; however, when the system is challenged, estradiol improves attention. In support of this, we challenged male and female rats with the stress neuropeptide, corticotropin-releasing factor (CRF), and assessed their performance on the sustained attention task (Cole et al. 2016). We found a significant dose-dependent impairment in all aspects of attention that was similar between the sexes. However, when the estrous cycle stage was assessed in females, we found that CRF impaired attention during estrous cycle phases with low levels of ovarian hormones but had little effect during phases with high levels of ovarian hormones (Cole et al. 2016). Functional connectivity analysis on brain networks activated (as measured with cFOS) by CRF revealed that female in the proestrous phase of their cycle that is characterized by high ovarian hormone levels had higher connectivity between the nBM and mPFC than females in the phase of their cycle with low ovarian hormones and males (Wiersielis et al. 2016). This finding indicates that estradiol may promote stress resilience by increasing the coupling of brain regions within the of the BF corticopetal cholinergic system. The mechanism by which this occurs, however, remains to be determined.

In sum, females appear to produce and release more ACh in the BF corticopetal system, and this effect is linked to estradiol. When it comes to behavior, which is more complex and often involves many regions, neurotransmitter, and hormones systems, there tends to be a bias toward females being better than males in certain aspects of attention, but this does not occur for all endpoints tested. When the system is challenged, however, estradiol can help promote resilience to attention deficits.

This finding suggests that treatment with estrogens may be a method to improve attention in people diagnosed with psychiatric disorders. In support of this idea, the selective estrogen receptor modulator, raloxifene, improved attention/processing speed for both men and women with schizophrenia (Weickert et al. 2015). More work is needed, but understanding sex differences and hormonal regulation of the BF corticopetal cholinergic system will likely lead to novel therapies to improve cognition in psychiatric patients.

6 Conclusions

The presented evidence in support of the view that cortical cholinergic signaling mediates discrete components of attentional processing challenges the traditional conceptualizations that view ACh as a slow neuromodulator of cortical arousal. The findings that phasic ACh release mediates the detection of signals in attention-demanding contexts have major implications in understanding the role of cholinergic dysfunction in the manifestation of cognitive symptoms of neuropsychiatric disorders and age-related dementias. Dysregulated phasic cholinergic transients could disrupt attentional abilities of patients suffering from schizophrenia and attention-deficit hyperactivity disorder (Sarter and Paolone 2011; Sarter et al. 2012). Abnormalities in the orchestration of phasic cholinergic signaling may precede global and structural decline in cholinergic function and consequently the loss of cholinergic neurons in Alzheimer's disease (Mesulam 2004).

The development of procholinergic drugs to improve cognitive symptoms of psychiatric and neurological conditions may benefit tremendously by moving away from previous views concerning volume transmission of ACh and not focusing on drugs that produce generalized increase in cholinergic transmission (such as AChE inhibitors). As discussed above, the new evidence from neuropharmacology and behavioral studies indicate that drugs that specifically amplify cholinergic transients via tonic neuromodulation of cholinergic synapses (e.g., $\alpha4\beta2$ nAChR agonists) may improve attentional control. Likewise, M1-selective mAChR agonists may exert beneficial effects on cue detection by enhancing the efficiency of phasic ACh for synchronizing the activity of prefrontal networks. At the presynaptic level, drugs that influence molecular mechanisms to enhance the capacity of cholinergic synapses to sustain phasic cholinergic signaling (e.g., choline transporter-mediated choline uptake mechanisms) may enhance attentional performance. Finally, research on the hormonal regulation of cholinergic transmission is just beginning to answer specific questions concerning sex differences in the cholinergic-attention system. This research will greatly benefit the development of procholinergic drugs for sex-specific treatment of the cognitive symptoms of psychiatric disorders.

Acknowledgments The authors' research is supported by grants from the National Institute on Aging (AG046580) to VP and National Science Foundation (NSF CAREER IOS-15524) to DAB.

References

Allison C, Shoaib M (2013) Nicotine improves performance in an attentional set shifting task in rats. Neuropharmacology 64:314–320

Apparsundaram S, Martinez V, Parikh V, Kozak R, Sarter M (2005) Increased capacity and density of choline transporters situated in synaptic membranes of the right medial prefrontal cortex of attentional task-performing rats. J Neurosci 25:3851–3856

Arnold HM, Burk JA, Hodgson EM, Sarter M, Bruno JP (2002) Differential cortical acetylcholine release in rats performing a sustained attention task versus behavioral control tasks that do not explicitly tax attention. Neuroscience 114:451–460

Ballinger EC, Ananth M, Talmage DA, Role LW (2016) Basal forebrain cholinergic circuits and signaling in cognition and cognitive decline. Neuron 91:1199–1218

Bangasser DA, Wicks B, Waxler DE, Eck SR (2017) Touchscreen sustained attention task (SAT) for rats. J Vis Exp 127:e56219

Barnes P, Staal V, Muir J, Good MA (2006) 17-β estradiol administration attenuates deficits in sustained and divided attention in young ovariectomized rats and aged acyclic female rats. Behav Neurosci 120:1225–1234

Bartolini A, Pepeu G (1967) Investigations into the acetylcholine output from the cerebral cortex of the cat in the presence of hyoscine. Br J Pharmacol Chemother 31:66–73

Baxter MG, Bucci DJ, Gorman LK, Wiley RG, Gallagher M (1995) Selective immunotoxic lesions of basal forebrain cholinergic cells: effects on learning and memory in rats. Behav Neurosci 109:714–722

Bayless DW, Darling JS, Stout WJ, Daniel JM (2012) Sex differences in attentional processes in adult rats as measured by performance on the 5-choice serial reaction time task. Behav Brain Res 235:48–54

Berry AS, Blakely RD, Sarter M, Lustig C (2015) Cholinergic capacity mediates prefrontal engagement during challenges to attention: evidence from imaging genetics. NeuroImage 108:386–395

Berry AS, Sarter M, Lustig C (2017) Distinct frontoparietal networks underlying attentional effort and cognitive control. J Cogn Neurosci 29:1212–1225

Bortz DM, Mikkelsen JD, Bruno JP (2013) Localized infusions of the partial alpha 7 nicotinic receptor agonist SSR180711 evoke rapid and transient increases in prefrontal glutamate release. Neuroscience 255:55–67

Botly LC, De Rosa E (2009) Cholinergic deafferentation of the neocortex using 192 IgG-saporin impairs feature binding in rats. J Neurosci 29:4120–4130

Buhusi M, Bartlett MJ, Buhusi CV (2017) Sex differences in interval timing and attention to time in C57Bl/6J mice. Behav Brain Res 324:96–99

Callahan MJ, Kinsora JJ, Harbaugh RE, Reeder TM, Davis RE (1993) Continuous ICV infusion of scopolamine impairs sustained attention of rhesus monkeys. Neurobiol Aging 14:147–151

Chudasama Y, Dalley JW, Nathwani F, Bouger P, Robbins TW (2004) Cholinergic modulation of visual attention and working memory: dissociable effects of basal forebrain 192-IgG-saporin lesions and intraprefrontal infusions of scopolamine. Learn Mem 11:78–86

Cole RD, Kawasumi Y, Parikh V, Bangasser DA (2016) Corticotropin releasing factor impairs sustained attention in male and female rats. Behav Brain Res 296:30–34

Cosgrove KP, Esterlis I, McKee SA et al (2012) Sex differences in availability of β2∗-nicotinic acetylcholine receptors in recently abstinent tobacco smokers. Arch Gen Psychiatry 69:418–427

Curtis L, Buisson B, Bertrand S, Bertrand D (2002) Potentiation of human α4β2 neuronal nicotinic acetylcholine receptor by estradiol. Mol Pharmacol 61:127

Dalley JW, McGaughy J, O'Connell MT, Cardinal RN, Levita L, Robbins TW (2001) Distinct changes in cortical acetylcholine and noradrenaline efflux during contingent and noncontingent performance of a visual attentional task. J Neurosci 21:4908–4914

Dalley JW, Theobald DE, Bouger P, Chudasama Y, Cardinal RN, Robbins TW (2004) Cortical cholinergic function and deficits in visual attentional performance in rats following 192 IgG-saporin-induced lesions of the medial prefrontal cortex. Cereb Cortex 14:922–932

Dasari S, Hill C, Gulledge AT (2017) A unifying hypothesis for M1 muscarinic receptor signalling in pyramidal neurons. J Physiol 595:1711–1723

Demeter E, Sarter M, Lustig C (2008) Rats and humans paying attention: cross-species task development for translational research. Neuropsychology 22:787–799

Demeter E, Hernandez-Garcia L, Sarter M, Lustig C (2011) Challenges to attention: a continuous arterial spin labeling (ASL) study of the effects of distraction on sustained attention. NeuroImage 54:1518–1529

Descarries L (1998) The hypothesis of an ambient level of acetylcholine in the central nervous system. J Physiol Paris 92:215–220

Ferguson SM, Blakely RD (2004) The choline transporter resurfaces: new roles for synaptic vesicles? Mol Interv 4:22–37

Frick KM, Kim JJ, Baxter MG (2004) Effects of complete immunotoxin lesions of the cholinergic basal forebrain on fear conditioning and spatial learning. Hippocampus 14:244–254

Gao S, Hendrie HC, Hall KS, Hui S (1998) The relationships between age, sex, and the incidence of dementia and Alzheimer disease: a meta-analysis. Arch Gen Psychiatry 55:809–815

Gibbs RB (1996) Expression of estrogen receptor-like immunoreactivity by different subgroups of basal forebrain cholinergic neurons in gonadectomized male and female rats. Brain Res 720:61–68

Gibbs RB (1997) Effects of estrogen on basal forebrain cholinergic neurons vary as a function of dose and duration of treatment. Brain Res 757:10–16

Gibbs RB, Wu D, Hersh LB, Pfaff DW (1994) Effects of estrogen replacement on the relative levels of choline Acetyltransferase, trkA, and nerve growth factor messenger RNAs in the basal forebrain and hippocampal formation of adult rats. Exp Neurol 129:70–80

Giuliano C, Parikh V, Ward JR, Chiamulera C, Sarter M (2008) Increases in cholinergic neurotransmission measured by using choline-sensitive microelectrodes: enhanced detection by hydrolysis of acetylcholine on recording sites? Neurochem Int 52:1343–1350

Goldstein JM, Seidman LJ, Goodman JM, Koren D, Lee H et al (1998) Are there sex differences in neuropsychological functions among patients with schizophrenia? Am J Psychiatr 155:1358–1364

Gotti C, Clementi F, Fornari A, Gaimarri A, Guiducci S et al (2009) Structural and functional diversity of native brain neuronal nicotinic receptors. Biochem Pharmacol 78:703–711

Gritton HJ, Howe WM, Mallory CS, Hetrick VL, Berke JD, Sarter M (2016) Cortical cholinergic signaling controls the detection of cues. Proc Natl Acad Sci U S A 113:E1089–E1097

Hasselmo ME, Sarter M (2011) Modes and models of forebrain cholinergic neuromodulation of cognition. Neuropsychopharmacology 36:52–73

Himmelheber AM, Sarter M, Bruno JP (1997) Operant performance and cortical acetylcholine release: role of response rate, reward density, and non-contingent stimuli. Brain Res Cogn Brain Res 6:23–36

Howe WM, Ji J, Parikh V, Williams S, Mocaer E et al (2010) Enhancement of attentional performance by selective stimulation of alpha4beta2(∗) nAChRs: underlying cholinergic mechanisms. Neuropsychopharmacology 35:1391–1401

Howe WM, Berry AS, Francois J, Gilmour G, Carp JM et al (2013) Prefrontal cholinergic mechanisms instigating shifts from monitoring for cues to cue-guided performance: converging electrochemical and fMRI evidence from rats and humans. J Neurosci 33:8742–8752

Howe WM, Gritton HJ, Lusk NA, Roberts EA, Hetrick VL et al (2017) Acetylcholine release in prefrontal cortex promotes gamma oscillations and theta-gamma coupling during cue detection. J Neurosci 37:3215–3230

Jentsch JD, Taylor JR (2003) Sex-related differences in spatial divided attention and motor impulsivity in rats. Behav Neurosci 117:76–83

Kozak R, Bruno JP, Sarter M (2006) Augmented prefrontal acetylcholine release during challenged attentional performance. Cereb Cortex 16:9–17

Lambe EK, Picciotto MR, Aghajanian GK (2003) Nicotine induces glutamate release from thalamocortical terminals in prefrontal cortex. Neuropsychopharmacology 28:216–225

Lean GA, Liu YJ, Lyon DC (2019) Cell type specific tracing of the subcortical input to primary visual cortex from the basal forebrain. J Comp Neurol 527:589–599

Lucas-Meunier E, Monier C, Amar M, Baux G, Fregnac Y, Fossier P (2009) Involvement of nicotinic and muscarinic receptors in the endogenous cholinergic modulation of the balance between excitation and inhibition in the young rat visual cortex. Cereb Cortex 19:2411–2427

Maddux JM, Kerfoot EC, Chatterjee S, Holland PC (2007) Dissociation of attention in learning and action: effects of lesions of the amygdala central nucleus, medial prefrontal cortex, and posterior parietal cortex. Behav Neurosci 121:63–79

Mazure CM, Swendsen J (2016) Sex differences in Alzheimer's disease and other dementias. Lancet Neurol 15:451–452

McGaughy J, Sarter M (1999) Effects of ovariectomy, 192 IgG-saporin-induced cortical cholinergic deafferentation, and administration of estradiol on sustained attention performance in rats. Behav Neurosci 113:1216–1232

McGaughy J, Kaiser T, Sarter M (1996) Behavioral vigilance following infusions of 192 IgG-saporin into the basal forebrain: selectivity of the behavioral impairment and relation to cortical AChE-positive fiber density. Behav Neurosci 110:247–265

McGaughy J, Everitt BJ, Robbins TW, Sarter M (2000) The role of cortical cholinergic afferent projections in cognition: impact of new selective immunotoxins. Behav Brain Res 115:251–263

Mendrek A, Mancini-Marïe A (2016) Sex/gender differences in the brain and cognition in schizophrenia. Neurosci Biobehav Rev 67:57–78

Mesulam M (2004) The cholinergic lesion of Alzheimer's disease: pivotal factor or side show? Learn Mem 11:43–49

Mielke MM, Vemuri P, Rocca WA (2014) Clinical epidemiology of Alzheimer's disease: assessing sex and gender differences. Clin Epidemiol 6:37–48

Miettinen RA, Kalesnykas G, Koivisto EH (2002) Estimation of the total number of cholinergic neurons containing estrogen receptor-alpha in the rat basal forebrain. J Histochem Cytochem 50:891–902

Nakamura N, Fujita H, Kawata M (2002) Effects of gonadectomy on immunoreactivity for choline acetyltransferase in the cortex, hippocampus, and basal forebrain of adult male rats. Neuroscience 109:473–485

Newhouse PA, Potter A, Singh A (2004) Effects of nicotinic stimulation on cognitive performance. Curr Opin Pharmacol 4:36–46

Newman LA, McGaughy J (2008) Cholinergic deafferentation of prefrontal cortex increases sensitivity to cross-modal distractors during a sustained attention task. J Neurosci 28:2642–2650

Norbury R, Travis MJ, Erlandsson K, Waddington W, Ell PJ, Murphy DGM (2007) Estrogen therapy and brain muscarinic receptor density in healthy females: a SPET study. Horm Behav 51:249–257

Paolone G, Angelakos CC, Meyer PJ, Robinson TE, Sarter M (2013) Cholinergic control over attention in rats prone to attribute incentive salience to reward cues. J Neurosci 33:8321–8335

Parikh V, Sarter M (2008) Cholinergic mediation of attention: contributions of phasic and tonic increases in prefrontal cholinergic activity. Ann N Y Acad Sci 1129:225–235

Parikh V, Pomerleau F, Huettl P, Gerhardt GA, Sarter M, Bruno JP (2004) Rapid assessment of in vivo cholinergic transmission by amperometric detection of changes in extracellular choline levels. Eur J Neurosci 20:1545–1554

Parikh V, Kozak R, Martinez V, Sarter M (2007) Prefrontal acetylcholine release controls cue detection on multiple timescales. Neuron 56:141–154

Parikh V, Man K, Decker MW, Sarter M (2008) Glutamatergic contributions to nicotinic acetylcholine receptor agonist-evoked cholinergic transients in the prefrontal cortex. J Neurosci 28:3769–3780

Parikh V, Ji J, Decker MW, Sarter M (2010) Prefrontal beta2 subunit-containing and alpha7 nicotinic acetylcholine receptors differentially control glutamatergic and cholinergic signaling. J Neurosci 30:3518–3530
Parikh V, St Peters M, Blakely RD, Sarter M (2013) The presynaptic choline transporter imposes limits on sustained cortical acetylcholine release and attention. J Neurosci 33:2326–2337
Passetti F, Dalley JW, O'Connell MT, Everitt BJ, Robbins TW (2000) Increased acetylcholine release in the rat medial prefrontal cortex during performance of a visual attentional task. Eur J Neurosci 12:3051–3058
Pepeu G, Giovannini MG (2009) Cholinesterase inhibitors and beyond. Curr Alzheimer Res 6:86–96
Phillis JW (1968) Acetylcholine release from the cerebral cortex: its role in cortical arousal. Brain Res 7:378–389
Posner MI, Snyder CR, Davidson BJ (1980) Attention and the detection of signals. J Exp Psychol 109:160–174
Ramtekkar UP, Reiersen AM, Todorov AA, Todd RD (2010) Sex and age differences in attention-deficit/hyperactivity disorder symptoms and diagnoses: implications for DSM-V and ICD-11. J Am Acad Child Adolesc Psychiatry 49:217–28e1–3
Rocca WA, Grossardt BR, Shuster LT (2011) Oophorectomy, menopause, estrogen treatment, and cognitive aging: clinical evidence for a window of opportunity. Brain Res 1379:188–198
Sarter M, Kim Y (2015) Interpreting chemical neurotransmission in vivo: techniques, time scales, and theories. ACS Chem Neurosci 6:8–10
Sarter M, Lustig C (2019) Cholinergic double duty: cue detection and attentional control. Curr Opin Psychol 29:102–107
Sarter M, Parikh V (2005) Choline transporters, cholinergic transmission and cognition. Nat Rev Neurosci 6:48–56
Sarter M, Hasselmo ME, Bruno JP, Givens B (2005) Unraveling the attentional functions of cortical cholinergic inputs: interactions between signal-driven and cognitive modulation of signal detection. Brain Res Brain Res Rev 48:98–111
Sarter M, Gehring WJ, Kozak R (2006) More attention must be paid: the neurobiology of attentional effort. Brain Res Rev 51:145–160
Sarter M, Paolone G (2011) Deficits in attentional control: cholinergic mechanisms and circuitry-based approaches. Behav Neurosci 125:825–835
Sarter M, Parikh V, Howe WM (2009a) nAChR agonist-induced cognition enhancement: integration of cognitive and neuronal mechanisms. Biochem Pharmacol 78:658–667
Sarter M, Parikh V, Howe WM (2009b) Phasic acetylcholine release and the volume transmission hypothesis: time to move on. Nat Rev Neurosci 10:383–390
Sarter M, Lustig C, Taylor SF (2012) Cholinergic contributions to the cognitive symptoms of schizophrenia and the viability of cholinergic treatments. Neuropharmacology 62:1544–1553
Sarter M, Lustig C, Howe WM, Gritton H, Berry AS (2014) Deterministic functions of cortical acetylcholine. Eur J Neurosci 39:1912–1920
Sarter M, Lustig C, Berry AS, Gritton H, Howe WM, Parikh V (2016) What do phasic cholinergic signals do? Neurobiol Learn Mem 130:135–141
Seçer I, Yılmazoğulları Y (2016) Are attentional resources a mediator for sex differences in memory? Int J Psychol 51:117–122
St Peters M, Demeter E, Lustig C, Bruno JP, Sarter M (2011) Enhanced control of attention by stimulating mesolimbic-corticopetal cholinergic circuitry. J Neurosci 31:9760–9771
Stoet G, O'Connor DB, Conner M, Laws KR (2013) Are women better than men at multi-tasking? BMC Psychol 1:18
Stolerman IP, Mirza NR, Hahn B, Shoaib M (2000) Nicotine in an animal model of attention. Eur J Pharmacol 393:147–154
Takase K, Mitsushima D, Funabashi T, Kimura F (2007) Sex difference in the 24-h acetylcholine release profile in the premotor/supplementary motor area of behaving rats. Brain Res 1154:105–115

Takase K, Kimura F, Yagami T, Mitsushima D (2009) Sex-specific 24-h acetylcholine release profile in the medial prefrontal cortex: simultaneous measurement of spontaneous locomotor activity in behaving rats. Neuroscience 159:7–15

Thiele A (2013) Muscarinic signaling in the brain. Annu Rev Neurosci 36:271–294

Tinkler GP, Voytko ML (2005) Estrogen modulates cognitive and cholinergic processes in surgically menopausal monkeys. Prog Neuro-Psychopharmacol Biol Psychiatry 29:423–431

Turchi J, Sarter M (1997) Cortical acetylcholine and processing capacity: effects of cortical cholinergic deafferentation on crossmodal divided attention in rats. Brain Res Cogn Brain Res 6:147–158

Unal CT, Golowasch JP, Zaborszky L (2012) Adult mouse basal forebrain harbors two distinct cholinergic populations defined by their electrophysiology. Front Behav Neurosci 6:21

van Huizen F, March D, Cynader MS, Shaw C (1994) Muscarinic receptor characteristics and regulation in rat cerebral cortex: changes during development, aging and the oestrous cycle. Eur J Neurosci 6:237–243

Vuckovich JA, Semel ME, Baxter MG (2004) Extensive lesions of cholinergic basal forebrain neurons do not impair spatial working memory. Learn Mem 11:87–94

Weickert TW, Weinberg D, Lenroot R, Catts SV, Wells R et al (2015) Adjunctive raloxifene treatment improves attention and memory in men and women with schizophrenia. Mol Psychiatry 20:685

Whitmer RA, Quesenberry CP, Zhou J, Yaffe K (2011) Timing of hormone therapy and dementia: the critical window theory revisited. Ann Neurol 69:163–169

Wiersielis KR, Wicks B, Simko H, Cohen SR, Khantsis S et al (2016) Sex differences in corticotropin releasing factor-evoked behavior and activated networks. Psychoneuroendocrinology 73:204–216

Wilens TE, Decker MW (2007) Neuronal nicotinic receptor agonists for the treatment of attention-deficit/hyperactivity disorder: focus on cognition. Biochem Pharmacol 74:1212–1223

Yoshida T, Kuwabara Y, Sasaki M, Fukumura T, Ichimiya A et al (2000) Sex-related differences in the muscarinic acetylcholinergic receptor in the healthy human brain--a positron emission tomography study. Ann Nucl Med 14:97–101

Young JW, Geyer MA, Rissling AJ, Sharp RF, Eyler LT et al (2013) Reverse translation of the rodent 5C-CPT reveals that the impaired attention of people with schizophrenia is similar to scopolamine-induced deficits in mice. Transl Psychiatry 3:e324

Zaborszky L, Csordas A, Mosca K, Kim J, Gielow MR et al (2015) Neurons in the basal forebrain project to the cortex in a complex topographic organization that reflects corticocortical connectivity patterns: an experimental study based on retrograde tracing and 3D reconstruction. Cereb Cortex 25:118–137

Zaborszky L, Gombkoto P, Varsanyi P, Gielow MR, Poe G et al (2018) Specific basal forebrain-cortical cholinergic circuits coordinate cognitive operations. J Neurosci 38:9446–9458

Zhang XY, Chen DC, Xiu MH, Yang FD, Haile CN et al (2012) Gender differences in never-medicated first-episode schizophrenia and medicated chronic schizophrenia patients. J Clin Psychiatry 73:1025–1033

Involvement of Nicotinic Receptors in Working Memory Function

Veronica C. Galvin, Amy F. T. Arnsten, and Min Wang

Contents

Abstract The prefrontal cortex underlies our high order cognitive abilities and is the target of projections from many neuromodulatory nuclei. The dorsolateral prefrontal cortex is particularly critical for rule representation and working memory, or the ability to hold information "in mind" in the absence of sensory input. Emerging evidence supports a prominent and permissive role for acetylcholine in these excitatory circuits, through actions at cholinergic nicotinic receptors. Here we review the involvement of acetylcholine in working memory via actions at nicotinic receptors.

Keywords Acetylchoine · Nicotinic receptor · The prefrontal cortex · Working memory

V. C. Galvin, A. F. T. Arnsten, and M. Wang (✉)
Department of Neuroscience, Yale School of Medicine, New Haven, CT, USA
e-mail: veronica.galvin@yale.edu; amy.arnsten@yale.edu; min.wang@yale.edu

Curr Topics Behav Neurosci (2020) 45: 89–100
https://doi.org/10.1007/7854_2020_142
Published Online: 26 May 2020

1 Introduction

The prefrontal cortex (PFC) is the most newly evolved cortical region, showing the greatest expansion across primate evolution, and subserves our higher cognitive functions. The PFC is innervated by many neuromodulators, which act to gate or fine-tune the activity of PFC networks. One of these neuromodulators, acetylcholine (ACh), plays a particularly important role in dorsal PFC areas involved in attention, top down control of actions, and working memory. This chapter will focus on the cholinergic mechanisms influencing the attention, working memory, and rule representation functions of the dorsolateral PFC (dlPFC). All of these functions require the ability to maintain information in mind – e.g., a goal or rule for future actions – and to protect this information from the distraction of intervening events. Recent research shows that cholinergic mechanisms play a key role in these dlPFC functions and provide a physiological link between arousal state and strength of cognitive operations.

2 Cholinergic Nuclei and Projections

ACh is synthesized and released in eight primary nuclei in the primate brain. Four of these nuclei reside in the brainstem and midbrain and project to the thalamus, the dopamine-producing midbrain nuclei, interpeduncular brain stem nuclei, and the superior colliculi. Two of these brainstem and midbrain cholinergic nuclei, the pedunculopontine nucleus (PPT) and the laterodorsal tegmental nucleus (LDT), play a critical role in arousal and sleep circuitry, with dense projections to the brainstem reticular formation and to thalamic relay nuclei (Steriade et al. 1990; Steriade et al. 1988; Yeomans 2012). The remaining four cholinergic nuclei comprise the basal forebrain (BF) and project to olfactory bulb, hippocampus, amygdala, and cortex. The four distinct nuclei comprising the BF are distinguished based on projection patterns. The nucleus basalis of Meynert (CH4) contains >90% cholinergic neurons and innervates the entire cortical mantle and amygdala. This nucleus can itself be divided into four distinct subregions based on innervation targets. The horizontal limb of the diagonal band (Ch3) contains ~1% cholinergic neurons and heavily innervates the olfactory bulb. The vertical limb of the diagonal band (CH2) and the medial septum (CH1) express ~70% and 10% cholinergic neurons respectively, and both primarily target the hippocampal formation and the hypothalamus (Mesulam et al. 1983).

ACh acts via two classes of receptors, nicotinic and muscarinic. The muscarinic class of receptors are metabotropic, of which there are five types (M1–M5) that can be split into two groups based on their G-protein coupling to either $G\alpha_q$ (M1, M3, M5) or $G\alpha_i$ (M2, M4). The nicotinic class of receptors are ionotropic receptors comprised of either homomeric alpha (1–9) subunits or a heterogeneous combination of alpha and beta (1–4) subunits. These nicotinic receptors differ in their affinity for acetylcholine, their sensitization following ACh binding, and their permeability to different ion species (for review see Albuquerque et al. 1997). The two most studied of these nicotinic receptors, the homomeric α7 receptor (Nicα7R) and the heteromeric α4β2 receptor (Nicα4β2R), both influence working memory microcircuits in the dlPFC and will be the focus of this chapter. We will also discuss the emerging field of muscarinic actions in dlPFC, with focus on M1.

3 Cholinergic Regulation of Arousal State

Accumulating evidence points to a key role for ACh projections in conscious wakefulness, where studies in cats have shown high firing rates in the ACh brain stem nuclei during conscious wakefulness and paradoxical or REM sleep but very low firing rates during deep/slow wave sleep (Kayama et al. 1992; Steriade et al. 1990). Studies in rats have also implicated an important role for ACh in circadian rhythm and sleep onset, where injection of the nicotinic antagonist α-bungarotoxin into the superchiasmatic nucleus blocked the effects of light exposure in the pineal gland, which rapidly reverses high levels of serotonin N-acetyltransferase (Zatz and Brownstein 1981). These studies are supported by research in patients with temporal lobe epilepsy where imaging and intracranial electroencephalography has indicated a key role for the cholinergic midbrain pedunculopontine tegmental nucleus in loss of consciousness during seizure (Andrews et al. 2019; Englot et al. 2010). While the percentage of cholinergic cells within the BF appears to be much lower in rodents (Gritti et al. 1997) (for review of species difference see Coppola and Disney 2018) these circuits have also been shown to play a critical role in sleep-wake cycles and PFC arousal in mice (Xu et al. 2015).

4 Acetylcholine in PFC and Attention

Consistent with its role in cortical arousal, ACh has been shown to play an important role in attentional modulation (Voytko et al. 1994), e.g., in the primary visual cortex (Herrero et al. 2008), and in the PFC. Studies in rodents have found an increase in ACh release in both PFC and parietal cortex during visually detected cues, and PFC guided attention can cause release of ACh in parietal cortex, but parietal guided cannot initiate ACh release in PFC, highlighting the importance of cholinergic tone in PFC for top-down cognitive processes (Nelson et al. 2005; Parikh et al. 2007).

Attention can be driven either by bottom-up processes (such as salient visual stimuli in the environment) or from top-down processes in the PFC directing attentional resources (Buschman and Miller 2007), the latter of which mediates cholinergic release in other cortical areas. Evidence in humans suggests that this PFC-initiated ACh release in visual areas is only critical for top-down attentional modulation, but not bottom-up salient cue detection (Rokem et al. 2010). The actions of ACh following the transient release in PFC during cued detection tasks in rodents are through both Nicα7R and Nicα4β2R, where Nicα4β2R are critical for regulating the amplitude of ACh release, and Nicα7R contribute to the decay of such transients in PFC (Parikh et al. 2010). This distinct contribution of different nicotinic receptors for cholinergic release and modulation of neuronal activity is also seen in working memory circuitry in the dlPFC.

5 Acetylcholine in Working Memory

The dlPFC is critical for working memory. It has been appreciated for almost a century that lesions to the frontal lobe produce marked, permanent deficits on working memory tasks (Jacobsen 1936), with lesions restricted to the dlPFC region surrounding the principle sulcus specifically impairing visual spatial working memory (Goldman and Rosvold 1970). The importance of ACh for working memory in primates was clearly shown in 2011 by Croxson and colleagues, where depletion of ACh selectively in PFC caused a dramatic reduction in performance of a visuospatial working memory task while showing no deficit in other PFC-dependent tasks. This deficit in working memory performance was equitable to the behavioral effect of complete dlPFC tissue ablation (Croxson et al. 2011; Goldman and Rosvold 1970; Goldman et al. 1971). Recent work has focused on uncovering the physiological basis for these important actions.

The neural microcircuits that underlie spatial working memory are thought to reside in deep layer III of dlPFC and are comprised of glutamatergic pyramidal neurons and GABAergic interneurons (Goldman-Rakic 1995; Kritzer and Goldman-Rakic 1995). This lamina also shows the greatest expansion across primate evolution, showing particularly significant expansion of pyramidal cell dendrites and increases in dendritic spine density (Elston et al. 2005; Elston et al. 2006; Elston and Fujita 2014). The excitatory neurons in these circuits show extensive horizontal projections, allowing for connections between distal columns of layer III neurons (Gonzalez-Burgos et al. 2000). Neurons within dlPFC increase their firing across the delay epoch in a working memory task (Funahashi et al. 1989). These neurons, termed “Delay cells,” show this persistent activity for particular spatial locations, with distinct spatially tuned preferred locations in visual space. The extensive horizontal projections of these neurons in deep layer III are thought to create networks of interconnected neurons with similar spatial tuning, forming a circuit able to support recurrent excitation to maintain representation of visuospace in the absence of continued sensory input, sculpted by GABAergic inhibition to limit activity to preferred directions (Goldman-Rakic 1995).

The excitatory connections between these similarly tuned pyramidal neurons critically rely on NMDA receptors (NMDARs), and particularly rely on NMDARs containing NR2B subunits. Unlike classic excitatory synapses, these glutamatergic connections express the slower closing, NR2B subunits in the adult post-synaptic density (PSD), and computational modeling has shown that the slower kinetics of

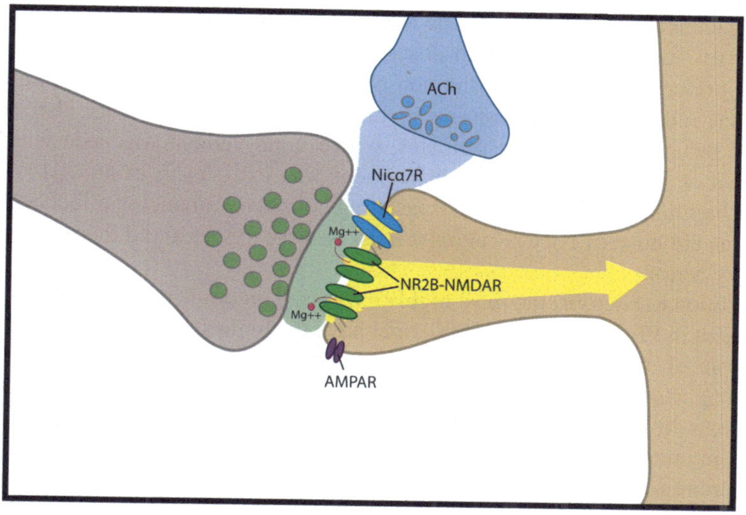

Fig. 1 Nicotinic α7 receptors are permissive for NMDA receptor activation in dlPFC circuits underlying working memory. The recurrent excitatory connections between glutamatergic pyramidal neurons underlying spatial working memory require NR2B-NMDA receptor activation. NMDA receptors require both agonist binding and membrane depolarization for receptor activation. Glutamatergic AMPA receptors play a minimal role in these circuits, and instead ACh plays this permissive depolarization role in dlPFC for NMDA receptor activation via actions at Nicα7 receptors present within the PSD of dendritic spines

NR2B are critical for maintaining persistent activity through recurrent excitation (Wang et al. 2013). Blockade of NMDAR in general, or selective targeting of either NMDAR-NR2B or NMDAR-NR2A, markedly reduces Delay cell firing. Importantly, blockade of AMPAR had only subtle effects on Delay cell firing, indicating that the classic, permissive role of AMPAR must be carried out by other receptors in or near the PSD. As detailed below, ACh appears to play this critical role in activating these NMDAR-dependent circuits for delay-related activity in dlPFC.

Nicα7Rs are highly expressed in dlPFC layer III, and immunoEM localized these receptors within the PSD of presumed glutamatergic synapses on dendritic spines (Yang et al. 2013). Activation of Nicα7R using selective agonists caused a significant increase in delay-related firing for the preferred direction and not for other non-preferred directions, enhancing spatial tuning. Application of Nicα7R selective antagonists had the reverse effect, dramatically and significantly eroding delay-related firing and reducing spatial tuning. Consistent with a permissive effect of Nicα7R for NMDAR activation (Fig. 1), these experiments showed that delay activity was increased with direct application of NMDA, but not under conditions when Nicα7R were blocked (Yang et al. 2013). Thus, Nicα7R stimulation was needed for excitatory NMDA actions.

The enhancing effect of activating Nicα7R showed an inverted-U dose-response, both at the cellular level on delay-related activity and at the behavioral level when given systemically to primates. While low doses of Nicα7R agonists enhance Delay cell activity, iontophoresis of high doses to Delay cells causes increased firing for both the preferred direction and the non-preferred directions, an increase in overall firing that reduces the signal-to-noise ratio. This inverted-U was also seen in systemic behavioral studies in primates, where improvement was seen with lower doses, but at higher doses this enhancement was lost (Yang et al. 2013). These findings highlight the difficulties in translating cognitive-enhancing effects to human studies, as the dose of Nicα7R agonist must be very low to avoid these nonspecific excitatory actions.

In addition to Nicα7R, the most highly expressed nicotinic receptor, Nicα4β2s are also present in PFC (Quik et al. 2000), and have the highest affinity for nicotine, and expression of these receptors in the dopamine reward pathway underlie the reinforcing and addicting properties of nicotine (Picciotto et al. 1998). In PFC, experiments investigating the role of these receptors in working memory found a key role in attention-related activity and resistance to distractors (Sun et al. 2017). Iontophoretic activation of these receptors in dlPFC with a selective agonist did not significantly increase delay-related firing, but could prevent the significant reduction in delay activity seen during presentation of distracting stimuli during the delay period. Blockade of these receptors also does not significantly impact Delay cell activity, but does significantly reduce activity of neurons showing elevated persistent activity across the entire duration of each trial, termed "Fixation" cells (Sun et al. 2017). These neurons are thought to represent visual attention during spatial working memory tasks, required for the duration of each trial. A critical role for ACh actions at Nicα4β2Rs for attention is also supported by studies in rodents, where attentional deficits in α4β2 KO mice were restored by lentiviral expression of α4β2 exclusively in PFC (Guillem et al. 2011). Thus, the interplay of attention and working memory processes is intimately intertwined in these circuits, e.g., protecting the contents of working memory from distraction. The circuit basis for Nicα4β2R actions is unknown, but may provide clues regarding the mechanism by which dlPFC can suppress irrelevant information from the contents of working memory.

6 Cholinergic Modulation of Cognitive Control

There is also recent evidence that ACh acts on muscarinic receptors in dlPFC during both spatial working memory and cognitive control tasks in primates (Major et al. 2015, 2018; Vijayraghavan et al. 2018; Zhou et al. 2011). Both behavioral performance on a working memory task and Delay cell activity are impaired in primates administered the general muscarinic antagonist scopolamine (Zhou et al. 2011). To determine if these effects of systemic muscarinic antagonism were due to local actions on receptors in dlPFC, further studies have investigated the role of local modulation of muscarinic receptors. Activity during the delay period in a cognitive control task was also found to be reduced when scopolamine was administered locally to dlPFC via iontophoresis (Major et al. 2015). In these experiments, subjects must make an eye saccade either toward or away from a target based on a rule indicated by a particular cue. Neurons in dlPFC with preferential activity for one rule or saccade direction showed reduced firing following scopolamine application (Major et al. 2015). These neurons also showed reduced rule selectivity following high doses of the general muscarinic agonist, carbachol (Major et al. 2018), either via exciting or suppressing activity, suggesting an inverted-U effect of cholinergic actions in PFC for muscarinic activation similar to nicotinic receptors (Yang et al. 2013). As carbachol also stimulates nicotinic a7 receptors (Li et al. 2010), these effects of destruction of rule representation seen with carbachol application may also be influenced by local nicotinic actions.

Muscarinic M1 receptors are the most highly expressed in cortex (Levey 1993) and have been localized postsynaptically on neurons in primate dlPFC (Mrzljak et al. 1993). Thus, these receptors may be the prominent substrate in PFC for the actions of general muscarinic drugs. To investigate this, Vijayraghavan and colleagues assessed alterations in neuronal firing in dlPFC during the above-mentioned cognitive control task using M1-selective compounds, with mixed effects (Vijayraghavan et al. 2018). Both the M1 selective positive allosteric modulator used, an M1 selective agonist, and M1 selective antagonist all inhibited a proportion of neurons and excited a proportion of neurons. These results seemed to be similar for both presumed pyramidal broad spiking neurons and presumed GABAergic narrow spiking cells. The proportion of inhibited neurons over excited neurons increased dose dependently with M1 receptor stimulation, but showed no significant change across doses with M1 receptor antagonist application. These data suggest there may be high endogenous ACh tone in primate PFC during engaged cognitive behavior in young adult monkeys, and thus additional M1 agonist application overstimulates receptors and reduces rule selectivity, as seen with general muscarinic antagonism with carbachol (Major et al. 2018; Vijayraghavan et al. 2018). Given the therapeutic potential of M1R agonists, this remains an area of continued research.

7 Relevance to Disease

In addition to deepening our knowledge of the biological basis for healthy cognitive functioning, understanding the role of ACh in cognitive processes such as working memory is also critical for our understanding of psychiatric disorders with cognitive deficits. One disease with prominent working memory deficits and PFC atrophy is schizophrenia. Onset of this disorder appears in late adolescence to early adulthood, and among other symptoms is characterized by significant cognitive dysfunction, including working memory deficits. Our understanding of ACh actions is particularly relevant to schizophrenia, as the gene locus for the nicotinic α7 subunit has been found in genetic association studies to represent increased risk (Bakanidze et al. 2013), and there is evidence of reduced receptor protein for the α7 subunit in PFC of patients (Guan et al. 1999). Additionally, schizophrenic patients show a significantly higher rate of tobacco smoking than the general population (Hughes et al. 1986) which may indicate a mechanism for self-medicating to boost dlPFC circuits critical for working memory and attention in the brain. Nicα7R agonists and PAMs have been a focus for therapeutic drug development, but the narrow inverted-U dose-response has made this challenging.

The cholinergic system has also been implicated in the pathophysiology of Alzheimer's disease (AD), where over-progression of the disease pyramidal cells in association cortex is most susceptible to degeneration, leading to significant cognitive impairments through advancing stages (Bussiere et al. 2003; Morrison and Hof 2007). AD is associated with a loss of BF volume in advancing stages of cognitive deterioration (Grothe et al. 2012) and BF neurons exhibit both amyloid β deposits and neurofibrillary tangle pathology in various stages of AD (Sassin et al. 2000). The known association of cholinergic degradation in AD led to the development of currently available treatments for the disease (such as galantamine, rivastigmine, and donepezil), all of which primarily inhibit the activity of the enzyme that degrades ACh, acetylcholinesterase (Galimberti and Scarpini 2016). These drugs all boost cognition in patients initially, but do not halt or alter the progression of the disease, and as such are only able to boost remaining neurons based on remaining ACh release, but as BF cells and cortical neurons themselves die, these drugs lose efficacy (Bullock and Dengiz 2005).

8 Conclusion

ACh plays a critical role in conscious arousal and PFC-dependent processes. As detailed here, one of these PFC-dependent cognitive processes in dlPFC, working memory, requires ACh release for circuit function. In these dlPFC circuits, ACh acts on both muscarinic M1 and nicotinic α7 and $\alpha4\beta2$ receptors to regulate the neurons underlying working memory, with permissive actions for Delay cell activation at Nicα7Rs and resistance to distractors via actions at Nic$\alpha4\beta2$Rs. ACh also acts on Nic$\alpha4\beta2$Rs to contribute to attention, significantly modulating Fixation cell activity during cognitive tasks. Understanding the intricacies of cholinergic modulation of PFC not only advances our understanding of cognitive processes in the brain, but also strengthens our understanding of disorders of the brain, and closer to finding effective treatments and potential causes. Highlighted here is the relevance of our understanding of cholinergic actions in working memory for schizophrenia, though this information is also relevant for many other disorders with cognitive symptoms.

References

Albuquerque EX, Alkondon M, Pereira EF, Castro NG, Schrattenholz A, Barbosa CT et al (1997) Properties of neuronal nicotinic acetylcholine receptors: pharmacological characterization and modulation of synaptic function. J Pharmacol Exp Ther 280(3):1117–1136

Andrews JP, Yue Z, Ryu JH, Neske G, McCormick DA, Blumenfeld H (2019) Mechanisms of decreased cholinergic arousal in focal seizures: in vivo whole-cell recordings from the pedunculopontine tegmental nucleus. Exp Neurol 314:74–81. https://doi.org/10.1016/j.expneurol.2018.11.008

Bakanidze G, Roinishvili M, Chkonia E, Kitzrow W, Richter S, Neumann K et al (2013) Association of the nicotinic receptor alpha7 subunit gene (CHRNA7) with schizophrenia and visual backward masking. Front Psych 4:133. https://doi.org/10.3389/fpsyt.2013.00133

Bullock R, Dengiz A (2005) Cognitive performance in patients with Alzheimer's disease receiving cholinesterase inhibitors for up to 5 years. Int J Clin Pract 59(7):817–822. https://doi.org/10.1111/j.1368-5031.2005.00562.x

Buschman TJ, Miller EK (2007) Top-down versus bottom-up control of attention in the prefrontal and posterior parietal cortices. Science 315(5820):1860–1862. https://doi.org/10.1126/science.1138071

Bussiere T, Giannakopoulos P, Bouras C, Perl DP, Morrison JH, Hof PR (2003) Progressive degeneration of nonphosphorylated neurofilament protein-enriched pyramidal neurons predicts cognitive impairment in Alzheimer's disease: stereologic analysis of prefrontal cortex area 9. J Comp Neurol 463(3):281–302. https://doi.org/10.1002/cne.10760

Coppola JJ, Disney AA (2018) Is there a canonical cortical circuit for the cholinergic system? Anatomical differences across common model systems. Front Neural Circuits 12:8. https://doi.org/10.3389/fncir.2018.00008

Croxson PL, Kyriazis DA, Baxter MG (2011) Cholinergic modulation of a specific memory function of prefrontal cortex. Nat Neurosci 14(12):1510–1512. https://doi.org/10.1038/nn.2971

Elston GN, Fujita I (2014) Pyramidal cell development: postnatal spinogenesis, dendritic growth, axon growth, and electrophysiology. Front Neuroanat 8:78. https://doi.org/10.3389/fnana.2014.00078

Elston GN, Benavides-Piccione R, Defelipe J (2005) A study of pyramidal cell structure in the cingulate cortex of the macaque monkey with comparative notes on inferotemporal and primary visual cortex. Cereb Cortex 15(1):64–73. https://doi.org/10.1093/cercor/bhh109

Elston GN, Benavides-Piccione R, Elston A, Zietsch B, Defelipe J, Manger P et al (2006) Specializations of the granular prefrontal cortex of primates: implications for cognitive processing. Anat Rec A Discov Mol Cell Evol Biol 288(1):26–35. https://doi.org/10.1002/ar.a.20278

Englot DJ, Yang L, Hamid H, Danielson N, Bai X, Marfeo A et al (2010) Impaired consciousness in temporal lobe seizures: role of cortical slow activity. Brain 133(Pt 12):3764–3777. https://doi.org/10.1093/brain/awq316

Funahashi S, Bruce CJ, Goldman-Rakic PS (1989) Mnemonic coding of visual space in the monkey's dorsolateral prefrontal cortex. J Neurophysiol 61(2):331–349

Galimberti D, Scarpini E (2016) Old and new acetylcholinesterase inhibitors for Alzheimer's disease. Expert Opin Investig Drugs 25(10):1181–1187. https://doi.org/10.1080/13543784.2016.1216972

Goldman PS, Rosvold HE (1970) Localization of function within the dorsolateral prefrontal cortex of the rhesus monkey. Exp Neurol 27(2):291–304

Goldman PS, Rosvold HE, Vest B, Galkin TW (1971) Analysis of the delayed-alternation deficit produced by dorsolateral prefrontal lesions in the rhesus monkey. J Comp Physiol Psychol 77(2):212–220

Goldman-Rakic PS (1995) Cellular basis of working memory. Neuron 14(3):477–485

Gonzalez-Burgos G, Barrionuevo G, Lewis DA (2000) Horizontal synaptic connections in monkey prefrontal cortex: an in vitro electrophysiological study. Cereb Cortex 10(1):82–92

Gritti I, Mainville L, Mancia M, Jones BE (1997) GABAergic and other noncholinergic basal forebrain neurons, together with cholinergic neurons, project to the mesocortex and isocortex in the rat. J Comp Neurol 383(2):163–177

Grothe M, Heinsen H, Teipel SJ (2012) Atrophy of the cholinergic basal forebrain over the adult age range and in early stages of Alzheimer's disease. Biol Psychiatry 71(9):805–813. https://doi.org/10.1016/j.biopsych.2011.06.019
Guan ZZ, Zhang X, Blennow K, Nordberg A (1999) Decreased protein level of nicotinic receptor alpha7 subunit in the frontal cortex from schizophrenic brain. Neuroreport 10(8):1779–1782
Guillem K, Bloem B, Poorthuis RB, Loos M, Smit AB, Maskos U et al (2011) Nicotinic acetylcholine receptor beta2 subunits in the medial prefrontal cortex control attention. Science 333(6044):888–891. https://doi.org/10.1126/science.1207079
Herrero JL, Roberts MJ, Delicato LS, Gieselmann MA, Dayan P, Thiele A (2008) Acetylcholine contributes through muscarinic receptors to attentional modulation in V1. Nature 454 (7208):1110–1114. https://doi.org/10.1038/nature07141
Hughes JR, Hatsukami DK, Mitchell JE, Dahlgren LA (1986) Prevalence of smoking among psychiatric outpatients. Am J Psychiatry 143(8):993–997. https://doi.org/10.1176/ajp.143.8.993
Jacobsen CF (1936) Studies of cerebral function in primates. Comp Psychol Monogr 13:1–68
Kayama Y, Ohta M, Jodo E (1992) Firing of 'possibly' cholinergic neurons in the rat laterodorsal tegmental nucleus during sleep and wakefulness. Brain Res 569(2):210–220
Kritzer MF, Goldman-Rakic PS (1995) Intrinsic circuit organization of the major layers and sublayers of the dorsolateral prefrontal cortex in the rhesus monkey. J Comp Neurol 359 (1):131–143. https://doi.org/10.1002/cne.903590109
Levey AI (1993) Immunological localization of m1-m5 muscarinic acetylcholine receptors in peripheral tissues and brain. Life Sci 52(5-6):441–448
Li YZ, Liu XH, Rong F, Hu S, Sheng ZY (2010) Carbachol inhibits TNF-alpha-induced endothelial barrier dysfunction through alpha 7 nicotinic receptors. Acta Pharmacol Sin 10:1389–1394. https://doi.org/10.1038/aps.2010.165
Major AJ, Vijayraghavan S, Everling S (2015) Muscarinic attenuation of mnemonic rule representation in macaque dorsolateral prefrontal cortex during a pro- and anti-saccade task. J Neurosci 35(49):16064–16076. https://doi.org/10.1523/JNEUROSCI.2454-15.2015
Major AJ, Vijayraghavan S, Everling S (2018) Cholinergic overstimulation attenuates rule selectivity in macaque prefrontal cortex. J Neurosci 38(5):1137–1150. https://doi.org/10.1523/JNEUROSCI.3198-17.2017
Mesulam MM, Mufson EJ, Levey AI, Wainer BH (1983) Cholinergic innervation of cortex by the basal forebrain: cytochemistry and cortical connections of the septal area, diagonal band nuclei, nucleus basalis (substantia innominata), and hypothalamus in the rhesus monkey. J Comp Neurol 214(2):170–197. https://doi.org/10.1002/cne.902140206
Morrison JH, Hof PR (2007) Life and death of neurons in the aging cerebral cortex. Int Rev Neurobiol 81:41–57. https://doi.org/10.1016/S0074-7742(06)81004-4
Mrzljak L, Levey AI, Goldman-Rakic PS (1993) Association of m1 and m2 muscarinic receptor proteins with asymmetric synapses in the primate cerebral cortex: morphological evidence for cholinergic modulation of excitatory neurotransmission. Proc Natl Acad Sci U S A 90 (11):5194–5198
Nelson CL, Sarter M, Bruno JP (2005) Prefrontal cortical modulation of acetylcholine release in posterior parietal cortex. Neuroscience 132(2):347–359. https://doi.org/10.1016/j.neuroscience.2004.12.007
Parikh V, Kozak R, Martinez V, Sarter M (2007) Prefrontal acetylcholine release controls cue detection on multiple timescales. Neuron 56(1):141–154. https://doi.org/10.1016/j.neuron.2007.08.025
Parikh V, Ji J, Decker MW, Sarter M (2010) Prefrontal beta2 subunit-containing and alpha7 nicotinic acetylcholine receptors differentially control glutamatergic and cholinergic signaling. J Neurosci 30(9):3518–3530. https://doi.org/10.1523/JNEUROSCI.5712-09.2010
Picciotto MR, Zoli M, Rimondini R, Lena C, Marubio LM, Pich EM et al (1998) Acetylcholine receptors containing the beta2 subunit are involved in the reinforcing properties of nicotine. Nature 391(6663):173–177. https://doi.org/10.1038/34413

Quik M, Polonskaya Y, Gillespie A, Jakowec M, Lloyd GK, Langston JW (2000) Localization of nicotinic receptor subunit mRNAs in monkey brain by in situ hybridization. J Comp Neurol 425 (1):58–69
Rokem A, Landau AN, Garg D, Prinzmetal W, Silver MA (2010) Cholinergic enhancement increases the effects of voluntary attention but does not affect involuntary attention. Neuropsychopharmacology 35(13):2538–2544. https://doi.org/10.1038/npp.2010.118
Sassin I, Schultz C, Thal DR, Rub U, Arai K, Braak E, Braak H (2000) Evolution of Alzheimer's disease-related cytoskeletal changes in the basal nucleus of Meynert. Acta Neuropathol 100 (3):259–269
Steriade M, Pare D, Parent A, Smith Y (1988) Projections of cholinergic and non-cholinergic neurons of the brainstem core to relay and associational thalamic nuclei in the cat and macaque monkey. Neuroscience 25(1):47–67
Steriade M, Datta S, Pare D, Oakson G, Curro Dossi RC (1990) Neuronal activities in brain-stem cholinergic nuclei related to tonic activation processes in thalamocortical systems. J Neurosci 10 (8):2541–2559
Sun Y, Yang Y, Galvin VC, Yang S, Arnsten AF, Wang M (2017) Nicotinic alpha4beta2 cholinergic receptor influences on dorsolateral prefrontal cortical neuronal firing during a working memory task. J Neurosci 37(21):5366–5377. https://doi.org/10.1523/JNEUROSCI.0364-17.2017
Vijayraghavan S, Major AJ, Everling S (2018) Muscarinic M1 receptor overstimulation disrupts working memory activity for rules in primate prefrontal cortex. Neuron 98(6):1256–1268. e1254. https://doi.org/10.1016/j.neuron.2018.05.027
Voytko ML, Olton DS, Richardson RT, Gorman LK, Tobin JR, Price DL (1994) Basal forebrain lesions in monkeys disrupt attention but not learning and memory. J Neurosci 14(1):167–186
Wang M, Yang Y, Wang CJ, Gamo NJ, Jin LE, Mazer JA et al (2013) NMDA receptors subserve persistent neuronal firing during working memory in dorsolateral prefrontal cortex. Neuron 77 (4):736–749. https://doi.org/10.1016/j.neuron.2012.12.032
Xu M, Chung S, Zhang S, Zhong P, Ma C, Chang WC et al (2015) Basal forebrain circuit for sleep-wake control. Nat Neurosci 18(11):1641–1647. https://doi.org/10.1038/nn.4143
Yang Y, Paspalas CD, Jin LE, Picciotto MR, Arnsten AF, Wang M (2013) Nicotinic alpha7 receptors enhance NMDA cognitive circuits in dorsolateral prefrontal cortex. Proc Natl Acad Sci U S A 110(29):12078–12083. https://doi.org/10.1073/pnas.1307849110
Yeomans JS (2012) Muscarinic receptors in brain stem and mesopontine cholinergic arousal functions. Handb Exp Pharmacol 208:243–259. https://doi.org/10.1007/978-3-642-23274-9_11
Zatz M, Brownstein MJ (1981) Injection of alpha-bungarotoxin near the suprachiasmatic nucleus blocks the effects of light on nocturnal pineal enzyme activity. Brain Res 213(2):438–442
Zhou X, Qi XL, Douglas K, Palaninathan K, Kang HS, Buccafusco JJ et al (2011) Cholinergic modulation of working memory activity in primate prefrontal cortex. J Neurophysiol 106 (5):2180–2188. https://doi.org/10.1152/jn.00148.2011

Nicotinic Receptors Underlying Nicotine Dependence: Evidence from Transgenic Mouse Models

Cassandra D. Gipson and Christie D. Fowler

Contents

Abstract Nicotine underlies the reinforcing properties of tobacco cigarettes and e-cigarettes. After inhalation and absorption, nicotine binds to various nicotinic acetylcholine receptor (nAChR) subtypes localized on the pre- and postsynaptic membranes of cells, which subsequently leads to the modulation of cellular function and neurotransmitter signaling. In this chapter, we begin by briefly reviewing the current understanding of nicotine's actions on nAChRs and highlight considerations regarding nAChR subtype localization and pharmacodynamics. Thereafter, we discuss the seminal discoveries derived from genetically modified mouse models, which have greatly contributed to our understanding of nicotine's effects on the reward-related mesolimbic pathway and the aversion-related habenulo-interpeduncular pathway. Thereafter, emerging areas of research focusing on modulation of nAChR expression and/or function are considered. Taken together, these

C. D. Gipson
Department of Family and Community Medicine, University of Kentucky, Lexington, KY, USA

C. D. Fowler (✉)
Department of Neurobiology and Behavior, University of California, Irvine, CA, USA
e-mail: cdfowler@uci.edu

Curr Topics Behav Neurosci (2020) 45: 101–122
https://doi.org/10.1007/7854_2020_134
Published Online: 29 May 2020

discoveries have provided a foundational understanding of various genetic, neurobiological, and behavioral factors underlying the motivation to use nicotine and related dependence processes, which are thereby advancing drug discovery efforts to promote long-term abstinence.

Keywords Addiction · Brain · e-cigarette · Nicotine · Nicotinic acetylcholine receptors (nAChRs) · Tobacco cigarette

1 Introduction

Nicotine is the primary active constituent in tobacco-containing products, which is responsible for maintaining smoking behavior in humans (Stolerman and Jarvis 1995). Recently, nicotine has also been formulated for vapor inhalation via e-cigarette devices (Electronic Nicotine Delivery Systems, or ENDS (St Helen et al. 2016)). Concomitant with a decrease in combustible tobacco cigarette use, the use of e-cigarettes, especially among adolescents, has drastically risen in recent years (Wang et al. 2018). Indeed, from 2017 to 2018, there was a rapid increase in vaping prevalence among adolescents aged ~13–18 years old, with nicotine vaping rates translating to roughly an additional 1.3 million adolescent users in 2018 compared to 2017 (Miech et al. 2019). Although e-cigarettes may have value as a nicotine replacement strategy for current tobacco smokers (Hajek et al. 2019), the increasing patterns of e-cigarette use among adolescents have become of high concern and warrant further investigation. As well, among current smokers, some studies show that e-cigarettes are not liked as much as tobacco cigarettes (Strasser et al. 2016), and, therefore, additional research is needed to determine the ability of e-cigarettes to accomplish nicotine replacement and harm reduction to act as a quit aid (Rennie et al. 2016; Selya et al. 2018).

Chronic exposure to nicotine or nicotine-containing products is associated with detrimental health effects, including enhanced brain injury and/or stroke risk (Sifat et al. 2018); altered blood-brain barrier permeability (Hawkins et al. 2004); promotion of tumor growth via nicotine and its carcinogenic metabolites cotinine, N′-nitrosonornicotine (NNN), and 4-(methylnitrosamino)-1-(3-pyridyl)-1-butanone (NNK (Ginzkey et al. 2012, 2013; Jacob et al. 2009; Nakada et al. 2012); and early onset of menopause in women (Bellavia et al. 2016), among others. Of importance, recent research has found that e-cigarette smoke may be carcinogenic and lead to increased risk of lung and bladder cancer, as well as heart disease, due to DNA damage (Lee et al. 2018). Given the harmful health effects associated with chronic use of nicotine-containing products, understanding the mechanisms that drive nicotine use is essential.

Nicotine is an agonist at nicotinic acetylcholine receptors (nAChRs), the structure and function of which were discovered in the early 1980s using ligand-binding assays (Clarke et al. 1985; Patrick and Stallcup 1977). The nAChRs are members of a large family of cys-loop homologous receptors, which also includes muscle acetylcholine receptors, GABAA/C, glycine, and serotonin type 3 receptors (Miller and Smart 2010). The nAChRs are pentameric ion channels, whereby either homomeric or heteromeric subunits combine together to form a central pore (Zoli et al. 2015). The subunit composition and stoichiometry of nAChRs determine its unique pharmacological binding profile, as well as its susceptibility to desensitization (Picciotto et al. 2008).

Decades of research have uncovered important neural mechanisms that drive nicotine self-administration behavior. Specifically, nicotine engenders self-administration through activation of high-affinity β2 subunit-containing (denoted β2∗) nAChRs localized on dopamine-containing cell bodies within the ventral tegmental area (VTA (Klink et al. 2001; Picciotto et al. 1998)) and by altering glutamatergic and GABAergic tone in the VTA (Mansvelder et al. 2002). The net result of nicotine-induced activation of nAChRs is increased levels of extracellular dopamine within the nucleus accumbens (NAc (Pontieri et al. 1996)), which is hypothesized to contribute to its reinforcing effects. Although activation of nAChRs can lead to motivated behavior, it is also through desensitization/inactivation that nAChRs can alter acetylcholine signaling and neuronal function (Colombo et al. 2013), which may contribute to modulation of nicotine-motivated behavior. Importantly, both acetylcholine and nicotine lead to nAChR desensitization; however, nicotine leads to prolonged inactivation of these receptors, with a slower rate of recovery than the endogenous ligand (Giniatullin et al. 2005). Interestingly, α4β2 nAChR desensitization occurs following cigarette smoking, which is correlated with reductions in cigarette craving (Brody et al. 2006).

It is important to note that nAChRs are expressed neuronally, on both pre- and postsynaptic terminals as well as on postsynaptic somatodendrites (Albuquerque et al. 2009). Additionally, some nAChRs (e.g., homomeric α7) are localized on astrocytes and microglia in the brain (Graham et al. 2003; Jensen et al. 1997; Noda and Kobayashi 2017; Shen and Yakel 2012; Shytle et al. 2004), which have important functions at glutamatergic neuronal synapses that impact synaptic plasticity (Wang et al. 2013). Subunit composition of nAChRs in the brain can vary depending on region- and cell type-specific localization, the topography of which continues to be studied (Gaimarri et al. 2007; Gotti et al. 2006; Hendrickson et al. 2013), and is important in understanding neurobehavioral processes modulated by nAChRs.

In this chapter, we describe the current state of the knowledge regarding nAChR subtype expression in a brain region- and pathway-specific manner as it relates to nicotine dependence and comorbid pathologies. To accomplish this goal, we address current tools in the field that have allowed for exploration of the role of nAChR function and expression in addiction-related processes, with a focus on findings garnered from transgenic mouse models. As well, we illuminate novel areas of research focusing on modulating nAChR expression and/or function, which may have important implications for nicotine dependence processes. Given the evolving landscape of nicotine-containing product use (Fowler et al. 2017), a better understanding of the neural processes underlying the motivation to use nicotine is needed to enhance drug discovery efforts to promote cessation from nicotine-containing products.

2 nAChR Function and Signaling

As noted above, the nAChR receptor composition plays an important role in response to pharmacological agents. When an agonist (e.g., acetylcholine or nicotine) is bound to nAChRs, the receptors are first activated and then can desensitize, followed by recovery once the agonist is unbound. The EC_{50} value, which represents nAChR activation for the concentration of agonist producing half-maximal response amplitude, varies based on subunit composition. For example, the acetylcholine EC_{50} value is 513 for rat α7 (Papke and Porter Papke 2002) but is 14 for rat α3β4 (Bohler et al. 2001). Conversely, the measure for desensitization, or the concentration of agonist required to reduce the amplitude of the response by 50%, is termed the IC_{50}. Based on subunit composition, desensitization-induced inhibition of receptors can vary when activated by the same agonist; for instance, the rat α4β2 IC_{50} with nicotine is <0.01 (Paradiso and Steinbach 2003), whereas rat α7 is 1.3 (Fenster et al. 1997; Giniatullin et al. 2005). These values represent different rates of activation and inhibition, which have profound effects on nAChR modulation of neuronal function. It should be noted that the same receptor subtype can desensitize at different rates based on the agonist present. For example, the rat α7 IC_{50} value is 1.3 for nicotine but is $>10,000$ for acetylcholine (Fenster et al. 1997; Papke and Porter Papke 2002). This is important because recovery rates are dependent upon the agonist present, with the recovery rate from nicotine taking longer than acetylcholine in some cases (e.g., α4β2 (Paradiso and Steinbach 2003)). Given the important role of nAChR subtypes such as α4β2 in the reinforcing effects of nicotine (Changeux et al. 1998) and in modulating dopamine release (Mansvelder and McGehee 2002), different rates of desensitization and recovery of different nAChRs likely play key roles in the neural circuitry underlying nicotine addiction.

There are generally two binding sites for neuronal heteromeric nAChRs, each of which is formed by a pocket between subunits extracellularly at the ligand-binding N-terminal domain (Karlin 2002; Sine et al. 2002). Neuronally, α7 nAChRs are mainly homomeric and have five potential binding sites between the α subunits (Drisdel and Green 2000). Recently, additional binding sites for heteromeric nAChRs have been identified which are dependent on subunit composition (Jain et al. 2016). When a ligand is bound, the channel opens within microseconds (Albuquerque et al. 2009), indicative of the rapid responsivity of these channels. A sequence of events occurs to alter the conformational state of the channel in order to open. Through computer-generated modeling, it has been determined that when acetylcholine or nicotine is bound, hydrogen bonds among amino acids rearrange near the binding pocket. Subsequently, the C-loop moves toward the central pore, which then allows the Cys-Cys pair to interact with the bound ligand and results in the ligand being trapped deep within the pore between the subunits (Gao et al. 2005; Hansen et al. 2005).

As mentioned above, neuronal nAChRs can be expressed somatodendritically, presynaptically or postsynaptically (Broide and Leslie 1999; McGehee et al. 1995; Wonnacott 1997). Nicotine binds to nAChRs located in the brain, which have identified subunits of α2–7, and β2–4 (Boulter et al. 1987; Couturier et al. 1990; Picciotto et al. 2008). Somatodendritically expressed nAChRs play a modulatory role in the neurotransmission of other systems in response to nicotine (Wonnacott et al. 2006), such as dopamine (Nisell et al. 1994). nAChRs that are expressed somatodendritically and modulate dopamine release appear to contain α6 and β3 subunits and differ in pharmacological response to nicotine and epibatidine compared to those expressed on terminals in the striatum (Reuben et al. 2000). Presynaptic nAChRs are also important in modulating neurotransmitter release. For instance, α7 nAChRs are Ca^{2+}-permeable, rapidly desensitize following activation, and are expressed on many cell types including at glutamatergic terminals in brain regions such as the hippocampus and NAc (Fabian-Fine et al. 2001; Kaiser and Wonnacott 2000). Thus, nAChRs can enhance release of neurotransmitters from synaptic terminals and may provide a feedforward mechanism by which cholinergic signaling gates neurotransmitter release. Postsynaptically, nAChRs modulate many functions within the brain, including the flow of auditory information in the thalamus. Specifically, β2-containing heteromeric nAChRs are located on neurons within the medial geniculate body and receive cholinergic input from the pontomesencephalic tegmentum, and these receptors undergo an age-related decline in their expression and function (Sottile et al. 2017). As well, postsynaptic nAChR subunits such as α6, α7, β2, and/or β4 within the laterodorsal tegmentum undergo changes in subunit composition due to age, which results in differential nicotine-induced neuronal excitability (Christensen and Kohlmeier 2016; Kaneda 2017; Kaneda et al. 2016; Shinohara et al. 2014; Taoka et al. 2016).

For decades, research has shown that chronic exposure to nicotine significantly alters expression and function of neuronal nAChRs (Fenster et al. 1999a, b; Gentry and Lukas 2002; Quick and Lester 2002). One nAChR subunit heavily involved in nicotine-induced striatal dopamine release and nicotine self-administration behavior, β2 (see a more thorough description below), is upregulated and desensitized after chronic nicotine exposure, as measured via binding assays in intact oocytes (Fenster et al. 1999b). Following chronic activation due to nicotine, different subunit-containing nAChRs appear to desensitize at different rates, which is thought to underlie their ability to modulate different neurotransmitter systems. Further, changes in receptor expression with chronic nicotine appear to be cell type- and nAChR subtype-dependent (Benwell et al. 1988; Lallai et al. 2019; Marks et al. 1992; Perry et al. 1999). Both neuronal and non-neuronal cholinergic signaling involve some of the same subtypes of nAChRs and are associated with pathologies such as lung cancer (Mucchietto et al. 2016). One such subunit is α5, which is expressed in non-neuronal tissues (Chini et al. 1992) such as the lung, pancreas, stomach, and gliomas (Jia et al. 2016; Yoshikawa et al. 2005; Zia et al. 1997), and may mediate nicotine-induced lung cancer cell proliferation (Ma et al. 2014; Sun and

Ma 2015). This subunit forms functional complexes with α4β2 or α3β4 subunits, and polymorphism of the human α5 gene, *CHRNA5*, is associated with nicotine dependence and lung cancer (Bierut et al. 2008; Chen et al. 2009; Saccone et al. 2007, 2009). Importantly, the variant of α5, characterized by a change in the 398th amino acid from aspartic acid to asparagine (D398N), has been associated with a reduction in the function of the human α3β4α5 nAChR (George et al. 2012), which has important implications for smoking cessation outcomes as well as other health pathologies.

3 nAChRs Mediating Nicotine Reinforcement

Genetically modified mouse models have allowed for the interrogation of receptors and circuits underlying complex behaviors. In the tobacco and nicotine field, groundbreaking initial studies by Picciotto, Changeux, and colleagues (Picciotto et al. 1995, 1998) have provided an important foundation for the further progression of these animal models. Beginning with knockout mice, subsequent approaches have incorporated various genetic and technical tools to achieve more select manipulation of target protein or neurotransmitter function. These advances include, but are not limited to, humanized knockin genes, modified receptors, Cre driver lines with floxed viral approaches, optogenetic and chemicogenetic expression of receptors in a cell type-specific manner, promoter-driven fluorescent reporter lines, and, most recently, CRISPR-Cas9 directed genetic modifications. Findings derived thus far from such approaches within each circuit are discussed in the following paragraphs.

3.1 Mesolimbic Pathway

The positive rewarding effects of nicotine involve the brain's mesolimbic pathway (Kenny and Markou 2005; Rice and Cragg 2004), consisting of dopaminergic projections from the VTA. The VTA integrating circuits and projection regions contain various nAChR subtypes expressed on dopaminergic, glutamatergic, and GABAergic neurons (Charpantier et al. 1998; Klink et al. 2001; Mameli-Engvall et al. 2006; Mansvelder and McGehee 2002). For instance, inhibitory GABAergic projections from the rostromedial tegmental nucleus (RMTg) express terminal α4β2∗ nAChRs. VTA dopaminergic cells projecting to both the NAc and prefrontal cortex (PFC) express α4α6β2, α4β2, and α6β2 nAChRs, allowing for regulation of dopamine signaling through either somatic or presynaptic expression. These VTA dopaminergic neurons may also co-express glutamate or GABA, and it has been recently shown that heteromeric nAChRs mediate excitatory signaling in the dopaminergic-glutamate co-expressing cells (Yan et al. 2018). Within the NAc, the dopaminergic terminal nAChRs become activated by cholinergic interneurons and modulate dopamine's activation of GABAergic medium spiny neurons expressing dopamine D1 or D2 receptors. Intra-VTA glutamatergic circuits also appear to modulate GABAergic signaling via axoaxonic connections onto RMTg terminals. Further, glutamatergic projections from other brain regions, such as the PFC and subiculum, express presynaptic α7 nAChRs and have been found to terminate on the soma of dopaminergic neurons. Moreover, expression of the α2, α5, and β3 nAChR subunits has also been localized within the VTA. Together, this complicated pattern of nAChR expression makes defining the specific subtype contribution to nicotine reward and reinforcement challenging. However, significant advances have been made in this regard.

Initial studies in knockout mice have supported pharmacological findings implicating nAChRs expressing the β2 nAChR subunit in mediating reward- and reinforcement-related processes. In the striatum, nicotine application induces a robust increase in dopamine release, which can be blocked by administration of the nAChR antagonist mecamylamine (Mifsud et al. 1989). However, this nicotine-mediated increase in dopamine release was absent in the striatum of mice lacking the β2 nAChR subunit (Picciotto et al. 1998). To examine the involvement of this subunit on nicotine reinforcement, mice were assessed in an intravenous nicotine self-administration protocol, a technique with high translational validity to patterns of nicotine consumption in humans. Interestingly, while the wild-type mice exhibited sustained nicotine self-administration behavior, the β2 knockout mice did not self-administer nicotine (Picciotto et al. 1998). A further study revealed similar findings with a lack of sustained self-administration behavior in the absence of the β2 nAChR subunit with nicotine infusions directly into the VTA (Maskos et al. 2005). More recently, viral-mediated re-expression of the β2 nAChR subunit in the VTA of the knockout mice was shown to "rescue" the behavioral phenotype, in which this site-specific re-expression led to the mice acquiring nicotine self-administration (Orejarena et al. 2012). Additional support from studies with β2 knockout mice demonstrates that the β2∗ nAChR is necessary for the formation of a conditioned place preference to a nicotine-paired environment and the discriminative stimulus properties of nicotine (Shoaib et al. 2002; Walters et al. 2006). In a cutting-edge approach, Mourot and colleagues used a viral technique to express light-controllable β2∗ nAChRs in the VTA, and, during light exposure, the VTA β2∗ nAChRs became inhibited, which thereby was sufficient to prevent the formation of a nicotine-induced conditioned place preference (Durand-de Cuttoli et al. 2018).

In addition to the β2 subunit, lack of sustained nicotine self-administration has also been found in mice with knockout of the α4 and α6 nAChR subunits, and, importantly, the behavioral phenotype could be restored with re-expression of these subunits in the VTA of each respective knockout line (Exley et al. 2011; Maskos et al. 2005; Picciotto et al. 1998; Pons et al. 2008). Further, dopaminergic neuron-specific deletion of the α4 subunit was found to prevent the formation of a nicotine-mediated conditioned place preference (McGranahan et al. 2011). In a complementary approach, transgenic α4 and α6 nAChR hypersensitive knockin mice were generated, in which a single point mutation renders the receptor subtype more responsive to nicotine. For the α4 subunit, this genetic modification led to an enhancement of the rewarding effects of nicotine, as assessed with conditioned place preference (Tapper et al. 2004), and, for the α6 subunit, mice exhibited a potentiation of nicotine-mediated locomotor effects and increased glutamatergic transmission with VTA neurons (Berry et al. 2015). As further evidence for these specific receptor subtypes, pharmacological administration of the relatively selective α4β2 nAChR antagonist, DHβE, also decreased nicotine self-administration in rats (Corrigall and Coen 1989; Harvey et al. 1996; Watkins et al. 1999). These findings are paralleled by studies demonstrating that DHβE attenuates the stimulatory effects of nicotine on brain reward systems (Harrison et al. 2002). Together, these findings support the notion that α4β2 and/or α4α6β2 nAChRs on dopaminergic circuits in the VTA mediate the reinforcing properties of nicotine.

The involvement of the α7 nAChR in nicotine dependence has been somewhat controversial. As noted above, glutamatergic axons containing presynaptic α7 nAChRs terminate on the soma of dopaminergic neurons in the VTA, suggesting a regulatory role for downstream dopaminergic signaling. Initial pharmacological studies demonstrated that administration of the α7-selective antagonist, methyllycaconitine, attenuates nicotine self-administration in rats (Markou and Paterson 2001), a finding that was further substantiated with site-specific VTA injections in wild-type mice (Besson et al. 2012). However, studies in α7 nAChR knockout mice failed to find differences with intravenous nicotine self-administration and nicotine-mediated conditioned place preference compared to wild-type littermates (Pons et al. 2008). More recently, Granon and colleagues (Besson et al. 2012) were able to establish a dose-dependent effect with intra-VTA nicotine self-administration, in which the α7 nAChR knockout mice exhibited decreased self-administration at a low, but not high, nicotine dose. Further, when administered a peripheral injection of nicotine, nicotine-induced dopamine outflow in the NAc was sustained over a longer period of time in the α7 knockout mice (120 min), as compared to the wild-type mice (15 min) (Besson et al. 2012). In consideration of α7 nAChRs' presynaptic circuit localization, lower affinity for nicotine, and rapid recovery from desensitization, the receptor's effects on the mechanisms underlying nicotine reinforcement appear to be more nuanced.

3.2 *Habenulo-Interpeduncular Pathway*

As a drug of abuse, nicotine is distinctive in that the aversive properties appear to sharply contrast the rewarding properties of the drug, thereby limiting the range of doses that promote reinforcement and drug consumption. Nicotine's aversive effects are mediated by the medial habenula (MHb), a brain structure that directly projects to the interpeduncular nucleus (IPN). The MHb-IPN circuit has been characterized as containing the densest expression of cholinergic fibers and various nAChR subunits within the brain, including the α5, α3, and β4 nAChR subunits (Marks et al. 1992; Villani et al. 1983). The aversive signaling of this circuit has been demonstrated in several studies with genetically modified rodents. For instance, α5 nAChR subunit knockout mice exhibit a high level of motivation to consume large quantities of nicotine, and viral-mediated re-expression of α5 subunits within this pathway restores nicotine intake to wild-type levels (Fowler et al. 2011). In addition, while wild-type mice exhibit inhibitory motivational effects at high doses of nicotine, the α5 nAChR knockout mice continue to exhibit reward-related effects, as assessed with both conditioned place preference and intracranial self-stimulation (Fowler et al. 2013; Jackson et al. 2010). The conclusions drawn from the knockout mice are supported by complementary studies using viral-mediated knockdown of the α5 nAChR subunit in rats, in which decreased expression of α5 nAChR subunits selectively in the habenula similarly increases nicotine intake and also decreased the inhibitory effects of higher nicotine doses on the activity of the brain reward circuitry (Fowler et al. 2011). Presynaptic α5* nAChRs on MHb terminals appear to facilitate glutamate release from cholinergic and glutamatergic co-expressing axons in the IPN (Fowler et al. 2011; Girod and Role 2001), which is thought to mediate this effect. Further, chronic nicotine appears to mitigate the activation of a subpopulation of α5-expressing neurons in the IPN, which subsequently provide negative feedback onto habenular terminals and mitigate nicotine reward, as assessed with conditioned place preference (Ables et al. 2017). The presence of the α5 nAChR subunit in α4β2, α3β2, and α3β4 nAChR receptors has been shown to alter nicotine binding and/or desensitization kinetics in vitro (Ramirez-Latorre et al. 1996; Wang et al. 1996), and all of these subtypes are expressed in the MHb-IPN pathway. Furthermore, the β4 nAChR subunit has also been shown to mediate aversive processing for nicotine. Under conditions of β4 nAChR subunit overexpression, mice consume less nicotine solution (Frahm et al. 2011), thereby suggesting that an α5β4* nAChR subtype may underlie an inhibitory motivational signal for nicotine in the MHb-IPN pathway. These findings in mouse models are further supported by human genome-wide association studies demonstrating that allelic variation in the *CHRNA3-CHRNA5-CHRNB4* gene cluster, which encodes α3, α5, and β4, respectively, increases vulnerability to developing tobacco dependence (Bierut et al. 2008; Kapoor et al. 2012; Wang et al. 2009). Recently, the non-synonymous SNP in the α5 gene that has been implicated in nicotine dependence in humans was inserted into the genome of rats to generate a transgenic humanized α5SNP model (Forget et al. 2018). The behavior of the α5SNP rat closely parallels the mouse knockout model, in which greater levels of nicotine are self-administered at high doses. In addition, an increase in nicotine-induced reinstatement was found in the α5SNP rats (Forget et al. 2018), suggesting a role for this genetic variant in relapse-related behavior.

4 nAChRs in Other Aspects of Nicotine Dependence

4.1 *Nicotine Enhancement of Cue Association*

Nicotine administration has been shown to enhance the acquisition of certain learned behaviors, such as contextual fear conditioning and trace cued fear conditioning. These findings may underlie nicotine's cue-related conditioning effects with drug use, in that later exposure to the cue during abstinence may promote drug relapse. Nicotine's enhancing effect on contextual fear conditioning is prevented in mice with knockout of the β2, but not α7, nAChR subunit (Davis and Gould 2007; Portugal et al. 2008). These effects likely involve the hippocampus since systemic or site-specific hippocampal administration of the β2 nAChR antagonist DHβE mitigates contextual fear learning in wild-type, but not β2 subunit knockout, mice (Davis and Gould 2007; Portugal et al. 2008). An enhancement of nicotine-mediated cued, but not trace or contextual, fear conditioning was also found in female, but not male, α2 nAChR subunit knockout mice (Lotfipour et al. 2013). Interestingly, mice with a hypersensitive α2∗ nAChR exhibit impaired contextual fear conditioning, an effect which could be rescued with pretreatment of nicotine (Lotfipour et al. 2017).

4.2 *Nicotine Withdrawal*

Following chronic nicotine administration, wild-type mice exhibit a range of behaviors indicative of the withdrawal state, including somatic signs (such as shaking, paw tremors, writhing), increased anxiety-like behavior in the elevated plus maze, increased brain reward thresholds, learning deficits in a contextual fear conditioning paradigm, and development of a conditioned place aversion to a withdrawal-associated environment. Studies with the β2 knockout mouse indicate that β2∗ nAChRs are involved in withdrawal-related anxiety-like behavior and conditioned place aversion, but not in the expression of somatic withdrawal signs (Jackson et al. 2008; Salas et al. 2004). Further, α7 nAChRs have been implicated in the initial expression of withdrawal symptomology, including anhedonia and somatic signs, but the α7 subunit knockout mice do not differ from wild-type mice at later time points (e.g., 24+ h) (Grabus et al. 2005; Salas et al. 2007; Stoker et al. 2012). Moreover, decreased somatic withdrawal signs have been found in α2, α5, and β4 nAChR subunit knockout mice, as compared to their respective wild-type littermates (Lotfipour et al. 2013; Salas et al. 2004, 2009). Interestingly, all of these subunits exhibit selectively dense expression in the MHb-IPN pathway, which has also been specifically implicated in somatic aspects of nicotine withdrawal. Administration of the general nAChR antagonist mecamylamine into the MHb-IPN pathway is sufficient to precipitate withdrawal, whereas injections into the cortex, VTA, or hippocampus are ineffective (Salas et al. 2009), and re-exposure to nicotine during withdrawal results in increased activity of MHb and IPN neurons (Arvin et al. 2019; Gorlich et al. 2013). Further, injections of antagonists for α4β2∗ or α6β2∗, but not α3β4∗, nAChRs in the MHb decrease the expression of anxiety-related behavior under conditions of nicotine withdrawal in mice (Pang et al. 2016). Together, these findings suggest that nAChRs are involved in various aspects of nicotine withdrawal based on their localization and expression patterns within the brain.

5 Modulators of nAChRs Influencing Expression and Function

The expression and function of nAChRs may be modulated at various points from protein translation to membrane insertion to subsequent function. Early receptor binding studies in humans found increased expression of nAChRs in chronic smokers (Benwell et al. 1988; Perry et al. 1999), suggesting a change in cellular activation following prolonged nicotine exposure. Given that chronic agonist receptor activation typically results in receptor downregulation, this finding was unexpected, although it was also evidenced in more controlled rodent studies (Marks et al. 1983, 1992). The likely mechanism underlying receptor upregulation was recently elucidated as it was found that nicotine and nAChR ligands can act as "chaperones" for α4 and β2 nAChR subunits (Henderson et al. 2014; Kuryatov et al. 2005; Srinivasan et al. 2011), thereby allowing for increased expression of the high-affinity nAChR subtype in the membrane. As the nAChR subunit protein is translated in the endoplasmic reticulum, the chaperone mechanism is thought to facilitate transport by promoting the trafficking of the protein to the plasma membrane and subsequent insertion of the assembled nAChR.

Intracellular proteins have also been shown to stabilize nAChR subunits in the endoplasmic reticulum and regulate subunit assembly into specific nAChR subtypes, resulting in either an increase or decrease in nAChR subtype-specific membrane expression (Dau et al. 2013; Wanamaker and Green 2007). For instance, α7 nAChRs are selectively targeted to the dendritic membrane by Ric-3, thus facilitating receptor expression (Alexander et al. 2010). In contrast, members of the Ly-6/neurotoxin gene superfamily, which includes lynx1 and lynx2, have been demonstrated to decrease receptor expression by acting as inhibitory chaperones during protein translation and trafficking, and, moreover, lynx proteins also bind directly to the extracellular face of nAChRs on the cell membrane, resulting in a decrease in ligand-binding efficiency and increase in the desensitization rate for nAChRs containing the α4, α3, α5, α7, β2, and/or β4 subunits (George et al. 2017; Ibanez-Tallon et al. 2002; Lyukmanova et al. 2011; Miwa et al. 1999; Nichols et al. 2014). In the cortex, lynx1 is expressed in both glutamatergic and GABAergic neurons, whereas lynx2 has been mainly localized in glutamatergic neurons (Demars and Morishita 2014). Lynx1 also appears to exhibit preferential binding affinity to the α:α interface, which would allow for increased interaction with the stoichiometry present in the lower sensitivity $\alpha4_3\beta2_2$ nAChRs (Nichols et al. 2014). In addition to intracellular proteins, other endogenous factors may interact with nAChRs to modulate function. For instance, estradiol has been shown to bind to the C-terminal tail of the α4 subunit to potentiate the activation of α4* nAChRs in the presence of acetylcholine, an effect that was selective for α4 as differences were not found with the α3 subunit (Curtis et al. 2002). More recently, phosphorylation sites have been identified on α4β2* nAChRs, suggesting a direct role for the receptor in mediating calcium-/calmodulin-dependent protein kinase II and protein kinase A intracellular signaling (Miller et al. 2018). Together, these nAChR subtype-specific interactions, along with cell type-specific expression patterns, may allow for selective modulation of various aspects of cholinergic signaling, thereby permitting each endogenous modulator to differentially regulate neural processes.

6 Beyond Nicotine Dependence

Although heavily involved in processes of nicotine dependence, nAChRs have also been implicated as mechanisms underlying other disease states, including Alzheimer's disease (AD (Lombardo and Maskos 2015)), schizophrenia (specifically, α7 (Jones 2018)), Parkinson's disease (PD (Jurado-Coronel et al. 2016)), and overeating/weight gain (Shariff et al. 2016), among others. Discovery of these mechanisms has led to multiple phase II clinical trials for nAChR compounds that have pro-cognitive effects (although many of these attempts have failed, see Lewis et al. 2017). Varenicline, a full agonist at α7 and a partial agonist at α4β2 nAChRs, is prescribed as a smoking cessation agent but also has efficacy in decreasing sucrose consumption and producing pro-cognitive effect in rodent models (Potasiewicz et al. 2018; Shariff et al. 2016). Interestingly, varenicline may improve cognitive function in patients with schizophrenia (Shim et al. 2012). In AD, medications have been developed that inhibit breakdown of the enzymes that metabolize acetylcholine (inhibition of acetylcholinesterase and/or butyrylcholinesterase), such as donepezil or rivastigmine. Additionally, drug development efforts have included compounds that act as positive allosteric modulators at α7 nAChRs in addition to AChE inhibition, including galantamine. Galantamine slows progression of plaque formation preclinically (Bhattacharya et al. 2014) and has shown efficacy in improving cognition and global functioning in patients with AD (Deardorff et al. 2015). Although statistically significant, these benefits are modest, and thus additional drugs are needed. Taken together, these studies illustrate a need for refinement of medications that target nAChRs for indications beyond nicotine dependence.

7 Conclusions

Since the mid-1990s, significant advances have been made with transgenic animal models to allow for better interrogation of specific nAChRs and circuits underlying nicotine dependence. Studies have built upon prior findings to reveal integral roles for various subunits in the mechanisms underlying nicotine's actions in the brain, with relevance to addiction. The α4α6β2∗ nAChRs in the mesolimbic pathway appear to be important in mediating the reinforcing properties of nicotine, whereas the α5 and β4 nAChR subunits in the MHb-IPN mitigate the aversive properties of higher nicotine doses that thereby limit drug intake. In addition to these effects on drug consumption, nAChRs have also been implicated in other aspects of the dependence processes, including withdrawal, cue-associated learning, and psychiatric comorbidity. This foundation holds the promise to provide the field with a basis for new discoveries to formulate more efficacious therapeutics. For instance, in consideration of the involvement of the $\alpha4\beta2*$ nAChRs in nicotine reinforcement, it is perhaps not surprising that varenicline has similar or greater effectiveness in promoting smoking cessation compared to nicotine replacement therapy and other approved therapeutics, such as bupropion (Gonzales et al. 2006). Drug development efforts are also focused on modulating the MHb-IPN circuit to enhance nicotine-mediated aversion and thus decrease further drug intake (Fowler and Kenny 2014; Jin et al. 2014). For instance, GLP-1 receptors have been shown to alter nicotine intake via modulation of the MHb-IPN circuit (Tuesta et al. 2017), and a GLP-1 receptor agonist, liraglutide, is currently being tested for smoking cessation in a clinical trial (Ashare 2019). In another approach to minimize nicotine entry into the brain, NicA2-J1 has been developed as a reengineered nicotine-degrading enzyme (Kallupi et al. 2018). Interestingly, while NicA2-J1 does not appear to induce significant differences in nicotine intake, decreased withdrawal and relapse-related behaviors were found in rats (Kallupi et al. 2018). Therefore, the field will certainly continue to advance by better defining the various genetic, behavioral, and biological mechanisms underlying addiction so that long-term abstinence can be readily achieved by those seeking to quit tobacco and e-cigarettes.

References

Ables JL et al (2017) Retrograde inhibition by a specific subset of interpeduncular alpha5 nicotinic neurons regulates nicotine preference. Proc Natl Acad Sci U S A 114:13012–13017. https://doi.org/10.1073/pnas.1717506114

Albuquerque EX, Pereira EF, Alkondon M, Rogers SW (2009) Mammalian nicotinic acetylcholine receptors: from structure to function. Physiol Rev 89:73–120. https://doi.org/10.1152/physrev.00015.2008

Alexander JK, Sagher D, Krivoshein AV, Criado M, Jefford G, Green WN (2010) Ric-3 promotes alpha7 nicotinic receptor assembly and trafficking through the ER subcompartment of dendrites. J Neurosci 30:10112–10126. https://doi.org/10.1523/JNEUROSCI.6344-09.2010

Arvin MC et al (2019) Chronic nicotine exposure alters the neurophysiology of habenulo-interpeduncular circuitry. J Neurosci., in press. https://doi.org/10.1523/JNEUROSCI.2816-18.2019

Ashare R (2019) Daily liraglutide for nicotine dependence, NCT03712098. https://ClinicalTrials.gov/show/NCT03712098

Bellavia A, Wolk A, Orsini N (2016) Differences in age at death according to smoking and age at menopause. Menopause 23:108–110. https://doi.org/10.1097/GME.0000000000000501

Benwell ME, Balfour DJ, Anderson JM (1988) Evidence that tobacco smoking increases the density of (-)-[3H]nicotine binding sites in human brain. J Neurochem 50:1243–1247

Berry JN, Engle SE, McIntosh JM, Drenan RM (2015) alpha6-containing nicotinic acetylcholine receptors in midbrain dopamine neurons are poised to govern dopamine-mediated behaviors and synaptic plasticity. Neuroscience 304:161–175. https://doi.org/10.1016/j.neuroscience.2015.07.052

Besson M et al (2012) Alpha7-nicotinic receptors modulate nicotine-induced reinforcement and extracellular dopamine outflow in the mesolimbic system in mice. Psychopharmacology 220:1–14. https://doi.org/10.1007/s00213-011-2422-1

Bhattacharya S, Haertel C, Maelicke A, Montag D (2014) Galantamine slows down plaque formation and behavioral decline in the 5XFAD mouse model of Alzheimer's disease. PLoS One 9:e89454. https://doi.org/10.1371/journal.pone.0089454

Bierut LJ et al (2008) Variants in nicotinic receptors and risk for nicotine dependence. Am J Psychiatry 165:1163–1171

Bohler S, Gay S, Bertrand S, Corringer PJ, Edelstein SJ, Changeux JP, Bertrand D (2001) Desensitization of neuronal nicotinic acetylcholine receptors conferred by N-terminal segments of the beta 2 subunit. Biochemistry 40:2066–2074

Boulter J, Connolly J, Deneris E, Goldman D, Heinemann S, Patrick J (1987) Functional expression of two neuronal nicotinic acetylcholine receptors from cDNA clones identifies a gene family. Proc Natl Acad Sci U S A 84:7763–7767

Brody AL et al (2006) Cigarette smoking saturates brain alpha 4 beta 2 nicotinic acetylcholine receptors. Arch Gen Psychiatry 63:907–915. https://doi.org/10.1001/archpsyc.63.8.907

Broide RS, Leslie FM (1999) The alpha7 nicotinic acetylcholine receptor in neuronal plasticity. Mol Neurobiol 20:1–16. https://doi.org/10.1007/BF02741361

Changeux JP et al (1998) Brain nicotinic receptors: structure and regulation, role in learning and reinforcement Brain research. Brain Res Rev 26:198–216

Charpantier E, Barneoud P, Moser P, Besnard F, Sgard F (1998) Nicotinic acetylcholine subunit mRNA expression in dopaminergic neurons of the rat substantia nigra and ventral tegmental area. Neuroreport 9:3097–3101

Chen X et al (2009) Variants in nicotinic acetylcholine receptors alpha5 and alpha3 increase risks to nicotine dependence. Am J Med Genet Part B Neuropsychiatr Genet 150B:926–933. https://doi.org/10.1002/ajmg.b.30919

Chini B, Clementi F, Hukovic N, Sher E (1992) Neuronal-type alpha-bungarotoxin receptors and the alpha 5-nicotinic receptor subunit gene are expressed in neuronal and nonneuronal human cell lines. Proc Natl Acad Sci U S A 89:1572–1576. https://doi.org/10.1073/pnas.89.5.1572

Christensen MH, Kohlmeier KA (2016) Age-related changes in functional postsynaptic nicotinic acetylcholine receptor subunits in neurons of the laterodorsal tegmental nucleus, a nucleus important in drug addiction. Addict Biol 21:267–281. https://doi.org/10.1111/adb.12194

Clarke PB, Schwartz RD, Paul SM, Pert CB, Pert A (1985) Nicotinic binding in rat brain: autoradiographic comparison of [3H]acetylcholine, [3H]nicotine, and [125I]-alpha-bungarotoxin. J Neurosci 5:1307–1315

Colombo SF, Mazzo F, Pistillo F, Gotti C (2013) Biogenesis, trafficking and up-regulation of nicotinic ACh receptors. Biochem Pharmacol 86:1063–1073. https://doi.org/10.1016/j.bcp.2013.06.023

Corrigall WA, Coen KM (1989) Nicotine maintains robust self-administration in rats on a limited-access schedule. Psychopharmacology 99:473–478

Couturier S et al (1990) A neuronal nicotinic acetylcholine receptor subunit (alpha 7) is developmentally regulated and forms a homo-oligomeric channel blocked by alpha-BTX. Neuron 5:847–856

Curtis L, Buisson B, Bertrand S, Bertrand D (2002) Potentiation of human alpha4beta2 neuronal nicotinic acetylcholine receptor by estradiol. Mol Pharmacol 61:127–135

Dau A, Komal P, Truong M, Morris G, Evans G, Nashmi R (2013) RIC-3 differentially modulates alpha4beta2 and alpha7 nicotinic receptor assembly, expression, and nicotine-induced receptor upregulation. BMC Neurosci 14:47. https://doi.org/10.1186/1471-2202-14-47

Davis JA, Gould TJ (2007) beta2 subunit-containing nicotinic receptors mediate the enhancing effect of nicotine on trace cued fear conditioning in C57BL/6 mice. Psychopharmacology 190:343–352. https://doi.org/10.1007/s00213-006-0624-8

Deardorff WJ, Feen E, Grossberg GT (2015) The use of cholinesterase inhibitors across all stages of Alzheimer's disease. Drugs Aging 32:537–547. https://doi.org/10.1007/s40266-015-0273-x

Demars MP, Morishita H (2014) Cortical parvalbumin and somatostatin GABA neurons express distinct endogenous modulators of nicotinic acetylcholine receptors. Mol Brain 7:75. https://doi.org/10.1186/s13041-014-0075-9

Drisdel RC, Green WN (2000) Neuronal alpha-bungarotoxin receptors are alpha7 subunit homomers. J Neurosci 20:133–139

Durand-de Cuttoli R et al (2018) Manipulating midbrain dopamine neurons and reward-related behaviors with light-controllable nicotinic acetylcholine receptors. Elife:7. https://doi.org/10.7554/eLife.37487

Exley R et al (2011) Distinct contributions of nicotinic acetylcholine receptor subunit alpha4 and subunit alpha6 to the reinforcing effects of nicotine. Proc Natl Acad Sci U S A 108:7577–7582. https://doi.org/10.1073/pnas.1103000108

Fabian-Fine R, Skehel P, Errington ML, Davies HA, Sher E, Stewart MG, Fine A (2001) Ultrastructural distribution of the alpha7 nicotinic acetylcholine receptor subunit in rat hippocampus. J Neurosci 21:7993–8003

Fenster CP, Rains MF, Noerager B, Quick MW, Lester RA (1997) Influence of subunit composition on desensitization of neuronal acetylcholine receptors at low concentrations of nicotine. J Neurosci 17:5747–5759

Fenster CP, Hicks JH, Beckman ML, Covernton PJ, Quick MW, Lester RA (1999a) Desensitization of nicotinic receptors in the central nervous system. Ann N Y Acad Sci 868:620–623

Fenster CP, Whitworth TL, Sheffield EB, Quick MW, Lester RA (1999b) Upregulation of surface alpha4beta2 nicotinic receptors is initiated by receptor desensitization after chronic exposure to nicotine. J Neurosci 19:4804–4814

Forget B et al (2018) A human polymorphism in CHRNA5 is linked to relapse to nicotine seeking in transgenic rats. Curr Biol 28:3244–3253.e3247. https://doi.org/10.1016/j.cub.2018.08.044

Fowler CD, Kenny PJ (2014) Nicotine aversion: neurobiological mechanisms and relevance to tobacco dependence vulnerability. Neuropharmacology 76(Pt B):533–544. https://doi.org/10.1016/j.neuropharm.2013.09.008

Fowler CD, Lu Q, Johnson PM, Marks MJ, Kenny PJ (2011) Habenular alpha5 nicotinic receptor subunit signalling controls nicotine intake. Nature 471:597–601

Fowler CD, Tuesta L, Kenny PJ (2013) Role of alpha5* nicotinic acetylcholine receptors in the effects of acute and chronic nicotine treatment on brain reward function in mice. Psychopharmacology 229:503–513

Fowler CD et al (2017) Basic science and public policy: informed regulation for nicotine and tobacco products. Nicotin Tob Res. https://doi.org/10.1093/ntr/ntx175

Frahm S et al (2011) Aversion to nicotine is regulated by the balanced activity of beta4 and alpha5 nicotinic receptor subunits in the medial habenula. Neuron 70:522–535. https://doi.org/10.1016/j.neuron.2011.04.013

Gaimarri A, Moretti M, Riganti L, Zanardi A, Clementi F, Gotti C (2007) Regulation of neuronal nicotinic receptor traffic and expression. Brain Res Rev 55:134–143. https://doi.org/10.1016/j.brainresrev.2007.02.005

Gao F et al (2005) Agonist-mediated conformational changes in acetylcholine-binding protein revealed by simulation and intrinsic tryptophan fluorescence. J Biol Chem 280:8443–8451. https://doi.org/10.1074/jbc.M412389200

Gentry CL, Lukas RJ (2002) Regulation of nicotinic acetylcholine receptor numbers and function by chronic nicotine exposure. Curr Drug Targets CNS Neurol Disord 1:359–385

George AA, Lucero LM, Damaj MI, Lukas RJ, Chen X, Whiteaker P (2012) Function of human alpha3beta4alpha5 nicotinic acetylcholine receptors is reduced by the alpha5(D398N) variant. J Biol Chem 287:25151–25162. https://doi.org/10.1074/jbc.M112.379339

George AA, Bloy A, Miwa JM, Lindstrom JM, Lukas RJ, Whiteaker P (2017) Isoform-specific mechanisms of alpha3beta4*-nicotinic acetylcholine receptor modulation by the prototoxin lynx1. FASEB J 31:1398–1420. https://doi.org/10.1096/fj.201600733R

Giniatullin R, Nistri A, Yakel JL (2005) Desensitization of nicotinic ACh receptors: shaping cholinergic signaling. Trends Neurosci 28:371–378. https://doi.org/10.1016/j.tins.2005.04.009

Ginzkey C et al (2012) Analysis of nicotine-induced DNA damage in cells of the human respiratory tract. Toxicol Lett 208:23–29. https://doi.org/10.1016/j.toxlet.2011.09.029

Ginzkey C, Friehs G, Koehler C, Hackenberg S, Hagen R, Kleinsasser NH (2013) Assessment of nicotine-induced DNA damage in a genotoxicological test battery. Mutat Res 751:34–39. https://doi.org/10.1016/j.mrgentox.2012.11.004

Girod R, Role LW (2001) Long-lasting enhancement of glutamatergic synaptic transmission by acetylcholine contrasts with response adaptation after exposure to low-level nicotine. J Neurosci 21:5182–5190

Gonzales D et al (2006) Varenicline, an alpha4beta2 nicotinic acetylcholine receptor partial agonist, vs sustained-release bupropion and placebo for smoking cessation: a randomized controlled trial. JAMA 296:47–55

Gorlich A, Antolin-Fontes B, Ables JL, Frahm S, Slimak MA, Dougherty JD, Ibanez-Tallon I (2013) Reexposure to nicotine during withdrawal increases the pacemaking activity of cholinergic habenular neurons. Proc Natl Acad Sci U S A 110:17077–17082

Gotti C, Zoli M, Clementi F (2006) Brain nicotinic acetylcholine receptors: native subtypes and their relevance. Trends Pharmacol Sci 27:482–491. https://doi.org/10.1016/j.tips.2006.07.004

Grabus SD, Martin BR, Imad Damaj M (2005) Nicotine physical dependence in the mouse: involvement of the alpha7 nicotinic receptor subtype. Eur J Pharmacol 515:90–93

Graham AJ et al (2003) Differential nicotinic acetylcholine receptor subunit expression in the human hippocampus. J Chem Neuroanat 25:97–113

Hajek P et al (2019) A randomized trial of E-cigarettes versus nicotine-replacement therapy. N Engl J Med 380:629–637. https://doi.org/10.1056/NEJMoa1808779

Hansen SB, Sulzenbacher G, Huxford T, Marchot P, Taylor P, Bourne Y (2005) Structures of Aplysia AChBP complexes with nicotinic agonists and antagonists reveal distinctive binding interfaces and conformations. EMBO J 24:3635–3646. https://doi.org/10.1038/sj.emboj.7600828

Harrison AA, Gasparini F, Markou A (2002) Nicotine potentiation of brain stimulation reward reversed by DHbE and SCH 23390, but not by eticlopride, LY 314582 or MPEP in rats. Psychopharmacology 160:56–66

Harvey SC, Maddox FN, Luetje CW (1996) Multiple determinants of dihydro-beta-erythroidine sensitivity on rat neuronal nicotinic receptor alpha subunits. J Neurochem 67:1953–1959

Hawkins BT, Abbruscato TJ, Egleton RD, Brown RC, Huber JD, Campos CR, Davis TP (2004) Nicotine increases in vivo blood-brain barrier permeability and alters cerebral microvascular tight junction protein distribution. Brain Res 1027:48–58. https://doi.org/10.1016/j.brainres.2004.08.043

Henderson BJ et al (2014) Nicotine exploits a COPI-mediated process for chaperone-mediated up-regulation of its receptors. J Gen Physiol 143:51–66. https://doi.org/10.1085/jgp.201311102

Hendrickson LM, Guildford MJ, Tapper AR (2013) Neuronal nicotinic acetylcholine receptors: common molecular substrates of nicotine and alcohol dependence. Front Psychiatr 4:29. https://doi.org/10.3389/fpsyt.2013.00029

Ibanez-Tallon I, Miwa JM, Wang HL, Adams NC, Crabtree GW, Sine SM, Heintz N (2002) Novel modulation of neuronal nicotinic acetylcholine receptors by association with the endogenous prototoxin lynx1. Neuron 33:893–903

Jackson KJ, Martin BR, Changeux JP, Damaj MI (2008) Differential role of nicotinic acetylcholine receptor subunits in physical and affective nicotine withdrawal signs. J Pharmacol Exp Ther 325:302–312

Jackson KJ, Marks MJ, Vann RE, Chen X, Gamage TF, Warner JA, Damaj MI (2010) Role of alpha5 nicotinic acetylcholine receptors in pharmacological and behavioral effects of nicotine in mice. J Pharmacol Exp Ther 334:137–146

Jacob T, Clouden N, Hingorani A, Ascher E (2009) The effect of cotinine on telomerase activity in human vascular smooth muscle cells. J Cardiovasc Surg 50:345–349

Jain A, Kuryatov A, Wang J, Kamenecka TM, Lindstrom J (2016) Unorthodox acetylcholine binding sites formed by alpha5 and beta3 accessory subunits in alpha4beta2∗ nicotinic acetylcholine receptors. J Biol Chem 291:23452–23463. https://doi.org/10.1074/jbc.M116.749150

Jensen JJ, Winzer-Serhan UH, Leslie FM (1997) Glial regulation of alpha 7-type nicotinic acetylcholine receptor expression in cultured rat cortical neurons. J Neurochem 68:112–120

Jia Y et al (2016) Nicotine inhibits Cisplatin-induced apoptosis via regulating alpha5-nAChR/AKT signaling in human gastric Cancer cells. PloS One 11:e0149120. https://doi.org/10.1371/journal.pone.0149120

Jin X, Bermudez I, Steinbach JH (2014) The nicotinic alpha5 subunit can replace either an acetylcholine-binding or nonbinding subunit in the alpha4beta2∗ neuronal nicotinic receptor. Mol Pharmacol 85:11–17. https://doi.org/10.1124/mol.113.089979

Jones C (2018) alpha7 nicotinic acetylcholine receptor: a potential target in treating cognitive decline in schizophrenia. J Clin Psychopharmacol 38:247–249. https://doi.org/10.1097/JCP.0000000000000859

Jurado-Coronel JC, Avila-Rodriguez M, Capani F, Gonzalez J, Moran VE, Barreto GE (2016) Targeting the nicotinic acetylcholine receptors (nAChRs) in astrocytes as a potential therapeutic target in Parkinson's disease. Curr Pharm Des 22:1305–1311

Kaiser S, Wonnacott S (2000) Alpha-bungarotoxin-sensitive nicotinic receptors indirectly modulate [(3)H]dopamine release in rat striatal slices via glutamate release. Mol Pharmacol 58:312–318

Kallupi M, Xue S, Zhou B, Janda KD, George O (2018) An enzymatic approach reverses nicotine dependence, decreases compulsive-like intake, and prevents relapse. Sci Adv 4:eaat4751. https://doi.org/10.1126/sciadv.aat4751

Kaneda K (2017) The contribution of neuroplasticity induced in cholinergic neurons of the laterodorsal tegmental nucleus to cocaine addiction. Nihon Shinkei Seishin Yakurigaku Zasshi 37:1–7

Kaneda K, Kamii H, Taoka N, Minami M (2016) The role of neuroplasticity in cholinergic neurons of the laterodorsal tegmental nucleus for cocaine addiction. Nihon Arukoru Yakubutsu Igakkai Zasshi 51:259–267

Kapoor M et al (2012) Variants located upstream of CHRNB4 on chromosome 15q25.1 are associated with age at onset of daily smoking and habitual smoking. PloS One 7:e33513

Karlin A (2002) Emerging structure of the nicotinic acetylcholine receptors. Nat Rev Neurosci 3:102–114

Kenny PJ, Markou A (2005) Conditioned nicotine withdrawal profoundly decreases the activity of brain reward systems. J Neurosci 25:6208–6212. https://doi.org/10.1523/JNEUROSCI.4785-04.2005

Klink R, de Kerchove d'Exaerde A, Zoli M, Changeux JP (2001) Molecular and physiological diversity of nicotinic acetylcholine receptors in the midbrain dopaminergic nuclei. J Neurosci 21:1452–1463

Kuryatov A, Luo J, Cooper J, Lindstrom J (2005) Nicotine acts as a pharmacological chaperone to up-regulate human alpha4beta2 acetylcholine receptors. Mol Pharmacol 68:1839–1851. https://doi.org/10.1124/mol.105.012419

Lallai V et al (2019) Nicotine acts on cholinergic signaling mechanisms to directly modulate choroid plexus function. eNeuro 6. https://doi.org/10.1523/ENEURO.0051-19.2019

Lee HW et al (2018) E-cigarette smoke damages DNA and reduces repair activity in mouse lung, heart, and bladder as well as in human lung and bladder cells. Proc Natl Acad Sci U S A 115: E1560–E1569. https://doi.org/10.1073/pnas.1718185115

Lewis AS, van Schalkwyk GI, Bloch MH (2017) Alpha-7 nicotinic agonists for cognitive deficits in neuropsychiatric disorders: a translational meta-analysis of rodent and human studies. Progress Neuro-psychopharmacol Biol Psychiat 75:45–53. https://doi.org/10.1016/j.pnpbp.2017.01.001

Lombardo S, Maskos U (2015) Role of the nicotinic acetylcholine receptor in Alzheimer's disease pathology and treatment. Neuropharmacology 96:255–262. https://doi.org/10.1016/j.neuropharm.2014.11.018

Lotfipour S et al (2013) Targeted deletion of the mouse alpha2 nicotinic acetylcholine receptor subunit gene (Chrna2) potentiates nicotine-modulated behaviors. J Neurosci 33:7728–7741

Lotfipour S et al (2017) alpha2∗ nicotinic acetylcholine receptors influence hippocampus-dependent learning and memory in adolescent mice. Learn Mem 24:231–244. https://doi.org/10.1101/lm.045369.117
Lyukmanova EN et al (2011) NMR structure and action on nicotinic acetylcholine receptors of water-soluble domain of human LYNX1. J Biol Chem 286:10618–10627. https://doi.org/10.1074/jbc.M110.189100
Ma X et al (2014) alpha5 nicotinic acetylcholine receptor mediates nicotine-induced HIF-1alpha and VEGF expression in non-small cell lung cancer. Toxicol Appl Pharmacol 278:172–179. https://doi.org/10.1016/j.taap.2014.04.023
Mameli-Engvall M, Evrard A, Pons S, Maskos U, Svensson TH, Changeux JP, Faure P (2006) Hierarchical control of dopamine neuron-firing patterns by nicotinic receptors. Neuron 50:911–921
Mansvelder HD, McGehee DS (2002) Cellular and synaptic mechanisms of nicotine addiction. J Neurobiol 53:606–617. https://doi.org/10.1002/neu.10148
Mansvelder HD, Keath JR, McGehee DS (2002) Synaptic mechanisms underlie nicotine-induced excitability of brain reward areas. Neuron 33:905–919
Markou A, Paterson NE (2001) The nicotinic antagonist methyllycaconitine has differential effects on nicotine self-administration and nicotine withdrawal in the rat. Nicotin Tob Res 3:361–373
Marks MJ, Burch JB, Collins AC (1983) Effects of chronic nicotine infusion on tolerance development and nicotinic receptors. J Pharm Exp Ther 226:817–825
Marks MJ, Pauly JR, Gross SD, Deneris ES, Hermans-Borgmeyer I, Heinemann SF, Collins AC (1992) Nicotine binding and nicotinic receptor subunit RNA after chronic nicotine treatment. J Neurosci 12:2765–2784
Maskos U et al (2005) Nicotine reinforcement and cognition restored by targeted expression of nicotinic receptors. Nature 436:103–107. https://doi.org/10.1038/nature03694
McGehee DS, Heath MJ, Gelber S, Devay P, Role LW (1995) Nicotine enhancement of fast excitatory synaptic transmission in CNS by presynaptic receptors. Science 269:1692–1696
McGranahan TM, Patzlaff NE, Grady SR, Heinemann SF, Booker TK (2011) Alpha4beta2 nicotinic acetylcholine receptors on dopaminergic neurons mediate nicotine reward and anxiety relief The Journal of neuroscience: the official journal of the Society for. Neuroscience 31:10891–10902. https://doi.org/10.1523/JNEUROSCI.0937-11.2011
Miech R, Johnston L, O'Malley PM, Bachman JG, Patrick ME (2019) Adolescent Vaping and nicotine use in 2017-2018 - U.S. National Estimates. N Engl J Med 380:192–193. https://doi.org/10.1056/NEJMc1814130
Mifsud JC, Hernandez L, Hoebel BG (1989) Nicotine infused into the nucleus accumbens increases synaptic dopamine as measured by in vivo microdialysis. Brain Res 478:365–367
Miller PS, Smart TG (2010) Binding, activation and modulation of Cys-loop receptors. Trends Pharmacol Sci 31:161–174. https://doi.org/10.1016/j.tips.2009.12.005
Miller MB, Wilson RS, Lam TT, Nairn AC, Picciotto MR (2018) Evaluation of the Phosphoproteome of mouse alpha 4/Beta 2-containing nicotinic acetylcholine receptors in vitro and in vivo. Proteomes 6:42. https://doi.org/10.3390/proteomes6040042
Miwa JM, Ibanez-Tallon I, Crabtree GW, Sanchez R, Sali A, Role LW, Heintz N (1999) Lynx1, an endogenous toxin-like modulator of nicotinic acetylcholine receptors in the mammalian CNS. Neuron 23:105–114
Mucchietto V, Crespi A, Fasoli F, Clementi F, Gotti C (2016) Neuronal acetylcholine nicotinic receptors as new targets for lung cancer treatment. Curr Pharm Des 22:2160–2169
Nakada T et al (2012) Lung tumorigenesis promoted by anti-apoptotic effects of cotinine, a nicotine metabolite through activation of PI3K/Akt pathway. J Toxicol Sci 37:555–563
Nichols WA, Henderson BJ, Yu C, Parker RL, Richards CI, Lester HA, Miwa JM (2014) Lynx1 shifts alpha4beta2 nicotinic receptor subunit stoichiometry by affecting assembly in the endoplasmic reticulum. J Biol Chem 289:31423–31432. https://doi.org/10.1074/jbc.M114.573667

Nisell M, Nomikos GG, Svensson TH (1994) Systemic nicotine-induced dopamine release in the rat nucleus accumbens is regulated by nicotinic receptors in the ventral tegmental area. Synapse 16:36–44. https://doi.org/10.1002/syn.890160105

Noda M, Kobayashi AI (2017) Nicotine inhibits activation of microglial proton currents via interactions with alpha7 acetylcholine receptors. J Physiol Sci 67:235–245. https://doi.org/10.1007/s12576-016-0460-5

Orejarena MJ, Herrera-Solis A, Pons S, Maskos U, Maldonado R, Robledo P (2012) Selective re-expression of beta2 nicotinic acetylcholine receptor subunits in the ventral tegmental area of the mouse restores intravenous nicotine self-administration. Neuropharmacology 63:235–241. https://doi.org/10.1016/j.neuropharm.2012.03.011

Pang X, Liu L, Ngolab J, Zhao-Shea R, McIntosh JM, Gardner PD, Tapper AR (2016) Habenula cholinergic neurons regulate anxiety during nicotine withdrawal via nicotinic acetylcholine receptors. Neuropharmacology 107:294–304. https://doi.org/10.1016/j.neuropharm.2016.03.039

Papke RL, Porter Papke JK (2002) Comparative pharmacology of rat and human alpha7 nAChR conducted with net charge analysis. Br J Pharm 137:49–61. https://doi.org/10.1038/sj.bjp.0704833

Paradiso KG, Steinbach JH (2003) Nicotine is highly effective at producing desensitization of rat alpha4beta2 neuronal nicotinic receptors. J Physiol 553:857–871. https://doi.org/10.1113/jphysiol.2003.053447

Patrick J, Stallcup B (1977) Alpha-Bungarotoxin binding and cholinergic receptor function on a rat sympathetic nerve line. J Biol Chem 252:8629–8633

Perry DC, Davila-Garcia MI, Stockmeier CA, Kellar KJ (1999) Increased nicotinic receptors in brains from smokers: membrane binding and autoradiography studies. J Pharm Exp Ther 289:1545–1552

Picciotto MR et al (1995) Abnormal avoidance learning in mice lacking functional high-affinity nicotine receptor in the brain. Nature 374:65–67. https://doi.org/10.1038/374065a0

Picciotto MR et al (1998) Acetylcholine receptors containing the beta2 subunit are involved in the reinforcing properties of nicotine. Nature 391:173–177. https://doi.org/10.1038/34413

Picciotto MR, Addy NA, Mineur YS, Brunzell DH (2008) It is not "either/or": activation and desensitization of nicotinic acetylcholine receptors both contribute to behaviors related to nicotine addiction and mood. Prog Neurobiol 84:329–342. https://doi.org/10.1016/j.pneurobio.2007.12.005

Pons S et al (2008) Crucial role of alpha4 and alpha6 nicotinic acetylcholine receptor subunits from ventral tegmental area in systemic nicotine self-administration. J Neurosci 28:12318–12327. https://doi.org/10.1523/JNEUROSCI.3918-08.2008

Pontieri FE, Tanda G, Orzi F, Di Chiara G (1996) Effects of nicotine on the nucleus accumbens and similarity to those of addictive drugs. Nature 382:255–257

Portugal GS, Kenney JW, Gould TJ (2008) Beta2 subunit containing acetylcholine receptors mediate nicotine withdrawal deficits in the acquisition of contextual fear conditioning. Neurobiol Learn Mem 89:106–113. https://doi.org/10.1016/j.nlm.2007.05.002

Potasiewicz A, Golebiowska J, Popik P, Nikiforuk A (2018) Procognitive effects of varenicline in the animal model of schizophrenia depend on alpha4beta2- and alpha 7-nicotinic acetylcholine receptors. J Psychopharmacol 33:62–73. https://doi.org/10.1177/0269881118812097

Quick MW, Lester RA (2002) Desensitization of neuronal nicotinic receptors. J Neurobiol 53:457–478. https://doi.org/10.1002/neu.10109

Ramirez-Latorre J, Yu CR, Qu X, Perin F, Karlin A, Role L (1996) Functional contributions of alpha5 subunit to neuronal acetylcholine receptor channels. Nature 380:347–351

Rennie LJ, Bazillier-Bruneau C, Rouesse J (2016) Harm reduction or harm introduction? Prevalence and correlates of E-cigarette use among French adolescents. J Adolesc Health 58:440–445. https://doi.org/10.1016/j.jadohealth.2015.12.013

Reuben M, Boye S, Clarke PB (2000) Nicotinic receptors modulating somatodendritic and terminal dopamine release differ pharmacologically. Eur J Pharmacol 393:39–49

Rice ME, Cragg SJ (2004) Nicotine amplifies reward-related dopamine signals in striatum. Nat Neurosci 7:583–584
Saccone SF et al (2007) Cholinergic nicotinic receptor genes implicated in a nicotine dependence association study targeting 348 candidate genes with 3713 SNPs. Hum Mol Genet 16:36–49
Saccone NL et al (2009) Multiple distinct risk loci for nicotine dependence identified by dense coverage of the complete family of nicotinic receptor subunit (CHRN) genes. Am J Med Genet Part B Neuropsychiatr Genet 150B:453–466
Salas R, Pieri F, De Biasi M (2004) Decreased signs of nicotine withdrawal in mice null for the beta4 nicotinic acetylcholine receptor subunit. J Neurosci 24:10035–10039
Salas R, Main A, Gangitano D, De Biasi M (2007) Decreased withdrawal symptoms but normal tolerance to nicotine in mice null for the alpha7 nicotinic acetylcholine receptor subunit. Neuropharmacology 53:863–869
Salas R, Sturm R, Boulter J, De Biasi M (2009) Nicotinic receptors in the habenulo-interpeduncular system are necessary for nicotine withdrawal in mice. J Neurosci 29:3014–3018
Selya AS, Dierker L, Rose JS, Hedeker D, Mermelstein RJ (2018) The role of nicotine dependence in E-cigarettes' potential for smoking reduction. Nicotin Tob Res 20:1272–1277. https://doi.org/10.1093/ntr/ntx160
Shariff M et al (2016) Neuronal nicotinic acetylcholine receptor modulators reduce sugar intake. PloS One 11:e0150270. https://doi.org/10.1371/journal.pone.0150270
Shen JX, Yakel JL (2012) Functional alpha7 nicotinic ACh receptors on astrocytes in rat hippocampal CA1 slices. J Mol Neurosci 48:14–21. https://doi.org/10.1007/s12031-012-9719-3
Shim JC et al (2012) Adjunctive varenicline treatment with antipsychotic medications for cognitive impairments in people with schizophrenia: a randomized double-blind placebo-controlled trial. Neuropsychopharmacology 37:660–668. https://doi.org/10.1038/npp.2011.238
Shinohara F, Kihara Y, Ide S, Minami M, Kaneda K (2014) Critical role of cholinergic transmission from the laterodorsal tegmental nucleus to the ventral tegmental area in cocaine-induced place preference. Neuropharmacology 79:573–579. https://doi.org/10.1016/j.neuropharm.2014.01.019
Shoaib M, Gommans J, Morley A, Stolerman IP, Grailhe R, Changeux JP (2002) The role of nicotinic receptor beta-2 subunits in nicotine discrimination and conditioned taste aversion. Neuropharmacology 42:530–539
Shytle RD et al (2004) Cholinergic modulation of microglial activation by alpha 7 nicotinic receptors. J Neurochem 89:337–343. https://doi.org/10.1046/j.1471-4159.2004.02347.x
Sifat AE, Vaidya B, Kaisar MA, Cucullo L, Abbruscato TJ (2018) Nicotine and electronic cigarette (E-Cig) exposure decreases brain glucose utilization in ischemic stroke. J Neurochem 147:204–221. https://doi.org/10.1111/jnc.14561
Sine SM et al (2002) Naturally occurring mutations at the acetylcholine receptor binding site independently alter ACh binding and channel gating. J Gen Physiol 120:483–496. https://doi.org/10.1085/jgp.20028568
Sottile SY, Ling L, Cox BC, Caspary DM (2017) Impact of ageing on postsynaptic neuronal nicotinic neurotransmission in auditory thalamus. J Physiol 595:5375–5385. https://doi.org/10.1113/JP274467
Srinivasan R, Pantoja R, Moss FJ, Mackey ED, Son CD, Miwa J, Lester HA (2011) Nicotine up-regulates alpha4beta2 nicotinic receptors and ER exit sites via stoichiometry-dependent chaperoning. J Gen Physiol 137:59–79. https://doi.org/10.1085/jgp.201010532
St Helen G, Havel C, Dempsey DA, Jacob P 3rd, Benowitz NL (2016) Nicotine delivery, retention and pharmacokinetics from various electronic cigarettes. Addiction 111:535–544. https://doi.org/10.1111/add.13183
Stoker AK, Olivier B, Markou A (2012) Role of alpha7- and beta4-containing nicotinic acetylcholine receptors in the affective and somatic aspects of nicotine withdrawal: studies in knockout mice. Behav Genet 42:423–436
Stolerman IP, Jarvis MJ (1995) The scientific case that nicotine is addictive. Psychopharmacology 117:2–10; discussion 14-20

Strasser AA, Souprountchouk V, Kaufmann A, Blazekovic S, Leone F, Benowitz NL, Schnoll RA (2016) Nicotine replacement, topography, and smoking phenotypes of E-cigarettes tobacco regulatory. Science 2:352–362. https://doi.org/10.18001/TRS.2.4.7
Sun H, Ma X (2015) alpha5-nAChR modulates nicotine-induced cell migration and invasion in A549 lung cancer cells. Exp Toxicol Pathol 67:477–482. https://doi.org/10.1016/j.etp.2015.07.001
Taoka N, Kamiizawa R, Wada S, Minami M, Kaneda K (2016) Chronic cocaine exposure induces noradrenergic modulation of inhibitory synaptic transmission to cholinergic neurons of the laterodorsal tegmental nucleus. Eur J Neurosci 44:3035–3045. https://doi.org/10.1111/ejn.13405
Tapper AR et al (2004) Nicotine activation of alpha4∗ receptors: sufficient for reward, tolerance, and sensitization. Science 306:1029–1032. https://doi.org/10.1126/science.1099420
Tuesta LM et al (2017) GLP-1 acts on habenular avoidance circuits to control nicotine intake. Nat Neurosci 20:708–716. https://doi.org/10.1038/nn.4540
Villani L, Contestabile A, Niso R (1983) Ultrastructural features and acetylcholinesterase histochemistry of the rat habenular complex. Acta Anat (Basel) 117:112–120
Walters CL, Brown S, Changeux JP, Martin B, Damaj MI (2006) The beta2 but not alpha7 subunit of the nicotinic acetylcholine receptor is required for nicotine-conditioned place preference in mice. Psychopharmacology 184:339–344. https://doi.org/10.1007/s00213-005-0295-x
Wanamaker CP, Green WN (2007) Endoplasmic reticulum chaperones stabilize nicotinic receptor subunits and regulate receptor assembly. J Biol Chem 282:31113–31123. https://doi.org/10.1074/jbc.M705369200
Wang F, Gerzanich V, Wells GB, Anand R, Peng X, Keyser K, Lindstrom J (1996) Assembly of human neuronal nicotinic receptor alpha5 subunits with alpha3, beta2, and beta4 subunits. J Biol Chem 271:17656–17665
Wang JC et al (2009) Genetic variation in the CHRNA5 gene affects mRNA levels and is associated with risk for alcohol dependence. Mol Psychiatry 14:501–510
Wang X, Lippi G, Carlson DM, Berg DK (2013) Activation of alpha7-containing nicotinic receptors on astrocytes triggers AMPA receptor recruitment to glutamatergic synapses. J Neurochem 127:632–643. https://doi.org/10.1111/jnc.12436
Wang TW, Gentzke A, Sharapova S, Cullen KA, Ambrose BK, Jamal A (2018) Tobacco product use among middle and high school students - United States, 2011-2017. MMWR Morb Mortal Wkly Rep 67:629–633. https://doi.org/10.15585/mmwr.mm6722a3
Watkins SS, Epping-Jordan MP, Koob GF, Markou A (1999) Blockade of nicotine self-administration with nicotinic antagonists in rats. Pharmacol Biochem Behav 62:743–751
Wonnacott S (1997) Presynaptic nicotinic ACh receptors. Trends Neurosci 20:92–98
Wonnacott S, Barik J, Dickinson J, Jones IW (2006) Nicotinic receptors modulate transmitter cross talk in the CNS: nicotinic modulation of transmitters. J Mol Neurosci 30:137–140. https://doi.org/10.1385/JMN:30:1:137
Yan Y et al (2018) Nicotinic cholinergic receptors in VTA glutamate neurons modulate excitatory transmission. Cell Rep 23:2236–2244. https://doi.org/10.1016/j.celrep.2018.04.062
Yoshikawa H, Hellstrom-Lindahl E, Grill V (2005) Evidence for functional nicotinic receptors on pancreatic beta cells. Metabolism 54:247–254. https://doi.org/10.1016/j.metabol.2004.08.020
Zia S, Ndoye A, Nguyen VT, Grando SA (1997) Nicotine enhances expression of the alpha 3, alpha 4, alpha 5, and alpha 7 nicotinic receptors modulating calcium metabolism and regulating adhesion and motility of respiratory epithelial cells. Res Commun Mol Pathol Pharmacol 97:243–262
Zoli M, Pistillo F, Gotti C (2015) Diversity of native nicotinic receptor subtypes in mammalian brain. Neuropharmacology 96:302–311. https://doi.org/10.1016/j.neuropharm.2014.11.003

Cholinergic Receptors and Addiction

Roger L. Papke, Darlene H. Brunzell, and Mariella De Biasi

Contents

Abstract Human behavior can be controlled by physical or psychological dependencies associated with addiction. One of the most insidious addictions in our society is the use of tobacco products which contain nicotine. This addiction can be associated with specific receptors in the brain that respond to the natural neurotransmitter acetylcholine. These nicotinic acetylcholine receptors (nAChR) are ligand-gated ion channels formed by the assembly of one or multiple types of nAChR receptor subunits. In this paper, we review the structure and diversity of nAChR subunits and our understanding for how different nAChR subtypes play specific roles in the phenomenon of nicotine addiction. We focus on receptors containing β2 and/or α6 subunits and the special significance of α5-containing receptors. These subtypes all have roles in regulating dopamine-mediated

R. L. Papke (✉)
Department of Pharmacology and Therapeutics, University of Florida, Gainesville, FL, USA
e-mail: rlpapke@ufl.edu

D. H. Brunzell
Department of Pharmacology and Toxicology, Virginia Commonwealth University School of Medicine, Richmond, VA, USA

M. De Biasi
Department of Psychiatry and Neuroscience, University of Pennsylvania School of Medicine, Philadelphia, PA, USA

Curr Topics Behav Neurosci (2020) 45: 123–152
https://doi.org/10.1007/7854_2020_139
Published Online: 26 May 2020

neurotransmission in the mesolimbic reward pathways of the brain. We also discuss the unique roles of homomeric α7 nAChR in behavioral responses to nicotine and how our knowledge of nAChR functional diversity may help guide pharmacotherapeutic approaches for treating nicotine addiction. While nicotine addiction is a truly global problem, the use of areca nut (betel) products is also a serious addiction associated with public health issues across most of South Asia, impacting as many as 600 million people. We discuss how cholinergic receptors of the brain are also involved with areca addiction and the unique challenges for dealing with addiction to this substance.

Keywords Addiction · Areca · Betel nut · Cholinergic receptors · Partial agonists · Tobacco

1 Addiction

The DSM-5 criteria for the identification of a substance use disorder (Hasin et al. 2013) (i.e., addiction) have aspects that are very value-laden, implying that the use of an addictive drug may not only be associated with a social impairment but, in most cases, may also produce intoxication. Although the DSM-5 specifically indicates that intoxication is not a criterion that applies to tobacco, most addictions are associated with mood lability and impaired judgment. Maintaining that a key component of any meaningful definition of addiction was the concept of intoxication, it was possible for tobacco industry representatives to assert that "science and common-sense support the view that nicotine is not addictive" and rather that smoking could be more accurately labeled as a habit rather than an addiction (Robinson and Pritchard 1992). Indeed, while the subjective experiences related by cigarette smokers are subtle and variable, the effects are not overtly intoxicating or immediately debilitating, as compared to a drug like alcohol. However, the abstemious cigarette smoker satisfies the criteria for cravings and withdrawal and will continue the use of the drug, with full awareness of the deleterious effects of the drug on his/her long-term health. With the current restriction on indoor smoking, smokers can also experience significant social inconvenience. So, while the immediate effects of smoking seem mild, the long-term health liability is enormous. The insidious aspects of nicotine addiction make it the greatest preventable cause of death in the world, accounting for 18.1% of all deaths in the United States in 2000 (Mokdad et al. 2004).

Appreciation for the long-term health liabilities of smoking, second-hand smoke in particular, has led to the ban of smoking in most public areas. Ironically, these bans can increase the likelihood that individuals meet criteria for a diagnosis of "tobacco use disorder," such as Criterion 5, "failure to fulfill major role obligations at work, school, or home"; Criterion 6, "continued tobacco use despite having persistent or recurrent social or interpersonal problems caused or exacerbated by the effects of tobacco (e.g., arguments with others about tobacco use)"; and Criterion 7, "important social, occupational, or recreational activities are given up or reduced because of tobacco use." For some smokers, these factors doubtlessly have helped motivate cessation attempts. The majority of smokers express a desire to quit, and as a testament to the addictive properties of tobacco, most have made numerous unsuccessful attempts to do so (Picciotto and Mineur 2014).

There are two principle aspects to drug-taking behavior that together can promote addiction: immediate reinforcement and, ultimately, dependence. We gauge the reinforcing qualities of a drug in animal studies by determining if the drug will be self-administered. In some cases, the drug is not immediately found to be reinforcing, since there may be adverse effects on naïve users. However, it is often the case that with repeated exposure to the drug, the aversive effects diminish, and drug use increases. It is at this point that tolerance and dependence become more important factors promoting the continued use of the drug. Tolerance takes different forms. "Metabolic tolerance," a term that can be related to medications as well as addictive drugs, simply refers to an upregulation of the enzymes that determine the half-life of a drug, so that over time higher dosages of the drug are required to achieve the same concentrations in blood or brain.

Arguably, most relevant to addiction and dependence is "functional tolerance," the condition where the brain has adapted to the continued presence of the drug and higher levels are required to obtain the same subjective reinforcing experience. When drug concentrations fall, the brain's functional adaptations result in the symptoms of abstinence associated with withdrawal. In some cases, for example, in the case of alcoholics, the symptoms of withdrawal can be quite serious, or even life-threatening, compelling the addicted individual to seek out more of the drug. In the case of tobacco, the overt symptoms of withdrawal listed in the DSM-5 are irritability, frustration, or anger, anxiety, difficulty concentrating, increased appetite, restlessness, depressed mood, and insomnia. While these conditions are admittedly unpleasant, on their own, they do not appear to be sufficiently compelling to overcome the desire to quit in the face of the well-known health concerns related to tobacco use. Rather, smoking creates a "psychological dependence" that produces irresistible cravings, especially in social settings that provide the cues normally associated with the smoking behavior, effectively enhancing the salience of other cues (Perkins et al. 2017; Robinson and Berridge 2003). These cues become associated with the reinforcing effects of the drug and become reinforcers themselves.

2 Neuronal Nicotinic Receptors

The influence on the brain's mesolimbic reward system is a feature common to numerous self-administered drugs, including nicotine (Corrigall et al. 1992, 1994; Mansvelder et al. 2002). While central nervous system stimulants, such as amphetamine and cocaine, work directly on dopamine release and reuptake (Hall et al. 2004; Jones et al. 1999; Uhl et al. 2002), nicotine's effects appear to rely on the presence of specific nicotinic receptor subtypes on the cell body and terminals of the dopamine neurons projecting from the ventral tegmentum to the nucleus accumbens (NAc) (De Biasi and Dani 2011; Gotti et al. 2006; Millar and Gotti 2009). Systemic injection of nicotine causes dopamine to rise in the NAc (Corrigall et al. 1994). Ex vivo studies of isolated dopamine-containing synaptic terminals and brain slices have suggested that the activation of presynaptic receptors might be sufficient to promote dopamine release (Grady et al. 2007, 2010; Threlfell et al. 2012; Wang et al. 2014) independent of dopamine neuron firing. While in vivo microdialysis studies have indicated that the delivery of nAChR antagonists to the ventral tegmental area (VTA) is more effective at reducing dopamine release from systemic injections than delivery of antagonists directly to the nucleus accumbens (Rahman et al. 2007), other studies have shown that local antagonism of both VTA and NAc neuronal nicotinic acetylcholine receptors (nAChRs) can impact nicotine self-administration and reward (Brunzell et al. 2010; Brunzell and McIntosh 2012; Corrigall et al. 1994; Gotti et al. 2010; Pons et al. 2008; Sanjakdar et al. 2015). Together these data suggest that whereas there is a prominent role of VTA nAChRs in dopaminergic neuron activity-dependent dopamine release, nAChR on the NAc terminals as well as in the VTA can support both dopamine release and behavior that is relevant to the development and maintenance of nicotine use. Studies are beginning to reveal that a diversity of neuronal nAChR subtypes differentially regulates behaviors that support nicotine addiction. One principle effect in vivo is likely to be the increased firing of dopaminergic neurons rather than evoked release mediated by the presynaptic receptors. However, on a behavioral level, although VTA activity can supersede antagonism in the NAc, it has been shown that blockade of nAChR in the NAc shell (terminals) impacts nicotine self-administration and reward (Brunzell et al. 2010; Brunzell and McIntosh 2012; Sanjakdar et al. 2015).

Brain nAChRs are members of the superfamily of Cys-loop ligand-gated ion channels (Dent 2010; Jaiteh et al. 2016; Ortells and Lunt 1995), which includes receptors for the inhibitory neurotransmitters GABA and glycine. In the peripheral nervous system, nAChRs mediate synaptic transmission at neuromuscular junctions and through autonomic ganglia (De Biasi 2002). Early studies of the nAChR of the neuromuscular junction and the related receptor of the electric eel *Torpedo* ultimately led to the isolation and cloning of the neuronal nAChR subunit genes (reviewed in Papke 2014). All nAChRs function as pentameric assemblies of subunits arrayed around a central ion pore, the opening of which is controlled by the binding of acetylcholine (ACh) or other ligands, including nicotine. Muscle-type receptors exist in only two configurations, with the receptors at mature neuromuscular junctions containing α1, β1, ε, and δ subunits, with two α1 subunits in each pentamer. In embryonic or denervated muscle, a γ subunit takes the place of ε. Early work identified α1 as the most critical subunit for ACh-binding and channel activation, both of which could be blocked by the snake toxin, α-bungarotoxin (α-BTX).

Early site-directed mutagenesis studies identified three key subdomain domains (designated A, B, and C) in alpha subunits (Corringer et al. 2000) and provided evidence that the complete binding site was located at the interface between subunits, with a complementary surface on the adjacent subunit containing three additional subdomains (designated D, E, and F). The subsequent isolation and crystallization of an ACh-binding protein with homology to a homopentamer of nAChR alpha subunits (Smit et al. 2001) confirmed the importance of these subdomains for agonist binding. The alpha subunit C domain takes the form of a loop that can close down on the agonist when it is accommodated in a hydrophobic pocket formed by converged subdomains. This C-loop has a unique pair of disulfide-linked vicinal cysteines at the apex of the loop. This feature is the signature structure present in all subunits subsequently designated as α subunits (α2–α10, although α8 has been identified only in chicken). The non-alpha subunits of neuronal tissues were designated beta subunits (β2–β4) since they could replace the β1 subunit when co-expressed with other muscle subunits.

Autoradiographic studies of tritiated ACh and nicotine showed a wide distribution of high-affinity binding sites throughout the rodent brain (Clarke et al. 1984). Interestingly, the neuromuscular blocker α-BTX identified a second, largely nonoverlapping, population of an alternative nAChR subtype that lacked high affinity to ACh and nicotine (Clarke et al. 1985). The cloning of neuronal nAChR subunits (Heinemann et al. 1990) and studies of their expression patterns in brain (Wada et al. 1989) suggested a primary association between the high-affinity nicotine receptors and the overlapping co-expression of α4 and β2 subunits. The α2 and α3 subunits, expressed at lower levels in the brain than α4, also appear to overlap with β2 expression. Heterologous pair-wise expression of β2 with α2, α3, or α4 in *Xenopus* oocytes demonstrated that each pair could form functional receptors with unique properties (Boulter et al. 1987; Luetje and Patrick 1991). It was subsequently shown that the alternative beta subunit β4 could also form functional receptors when co-expressed with α2, α3, or α4 (Duvoisin et al. 1989). While the expression of β4 is largely limited to the medial habenula and interpeduncular nucleus (Duvoisin et al. 1989), it is strongly expressed in the adrenal gland and autonomic ganglia (Skok 2002). The α5, α6, and β3 subunits did not readily form receptors in pairwise co-expression studies and were for a time considered "orphans." It was subsequently shown that α5 and β3 subunits could co-assemble with other subunits but did not contribute agonist binding sites (Cui et al. 2003; Gerzanich et al. 1998). These subunits, along with α6, however, all have great relevance to nicotine addiction, as will be discussed.

Likewise, the identity of the brain's α-BTX binding sites remained a mystery until it was shown that they could be associated with the expression of α7 subunits (Seguela et al. 1993). The α7 nAChR subunits form functional receptors without the necessary co-expression of non-alpha subunits, and indeed, co-expression of α7 with β2 may negatively impact the formation of functional receptors (Murray et al. 2012). Although α7 nAChRs have five potential agonist binding sites (Palma et al. 1996), channel activation requires binding to only a small fraction of these sites and, in fact, occurs with very low probability (Uteshev et al. 2002; Williams et al. 2012a, 2011). α7 nAChRs do not convert to forms with high affinity for agonist, a condition associated with the desensitized, i.e., closed and inactive, form of heteromeric receptor, such as those containing α4 and β2 subunits (Papke 2014). Although α7 receptors do not adopt high affinity for agonist, they do desensitize rapidly in the continued presence of agonist (Uteshev et al. 2002), a property that limits their probability of activating, even in the presence of a strong, rapidly applied stimulus (Williams et al. 2012a). Several lines of evidence suggest that the five putative binding sites at the subunit interfaces of α7 are not equivalent (Gulsevin et al. 2019; Helekar et al. 1994; Rakhilin et al. 1999), and indeed the limited activation of the α7 channel is best promoted by a low level of binding site occupancy (Papke and Papke 2002; Williams et al. 2011).

Although not as well studied as α7, α-BTX-sensitive receptors of the inner ear and peripheral tissues also can form as homopentamers of α9 subunits, although sometimes co-assembled with α10 subunits (Elgoyhen et al. 1994, 2001).

3 Special Significance of β2-Containing nAChRs

As noted above, the primary high-affinity nicotine receptors of rodent brain are heteromeric pentamers containing α4 and β2 subunits, while in primates, α2-containing receptors may play similar roles (Han et al. 2000). In a heterologous expression system, the expression of two subunits would presumably permit the formation of receptors with differing subunit stoichiometry and arrangement unless intrinsic factors existed that excluded or limited certain combinations or arrangements, at least in regard to functional detection. Based on what is known about muscle-type receptors, it is reasonable to hypothesize that receptors with fewer than two beta or two alpha subunits might assemble but not function efficiently. Likewise, it is reasonable to assume that in functional receptors, there would be two alpha-beta dimers forming the ACh-binding sites. Would there then be in a heterologous expression system constraints on the fifth subunit, or could it be either an α or a β? The very first study of the single-channel currents of receptors formed from the pairwise expression of cloned subunits (Papke et al. 1989) reported that each alpha-beta pair (α2β2, α3β2, and α4β2) generated two types of single-channel currents distinguished by conductances and open times. To test the hypothesis that these arose from receptors with different subunit stoichiometry, cells were injected with RNAs for each subunit at different ratios. When α2 was co-expressed with a ninefold excess of β2, only channels with the lower single-channel conductance were observed, suggesting that they corresponded to a configuration with two α2 and three β2 subunits (Papke et al. 1989). The significance of this observation went unappreciated for a number of years, although it was well known in the author's (RLP) laboratory that consistent concentration-response data for, for example, α3β2 receptors relied on the quality of both α and β RNAs. If the β2 RNA was not recently synthesized, the ACh concentration-response curves shifted to the right (unpublished). Similar data were later published with α4β2 receptors (Moroni et al. 2006), and it is now well accepted that α4β2 receptors have two functional configurations: α4(2)β2(3) and α4(3)β2(2) (Fig. 1). The receptor populations can be

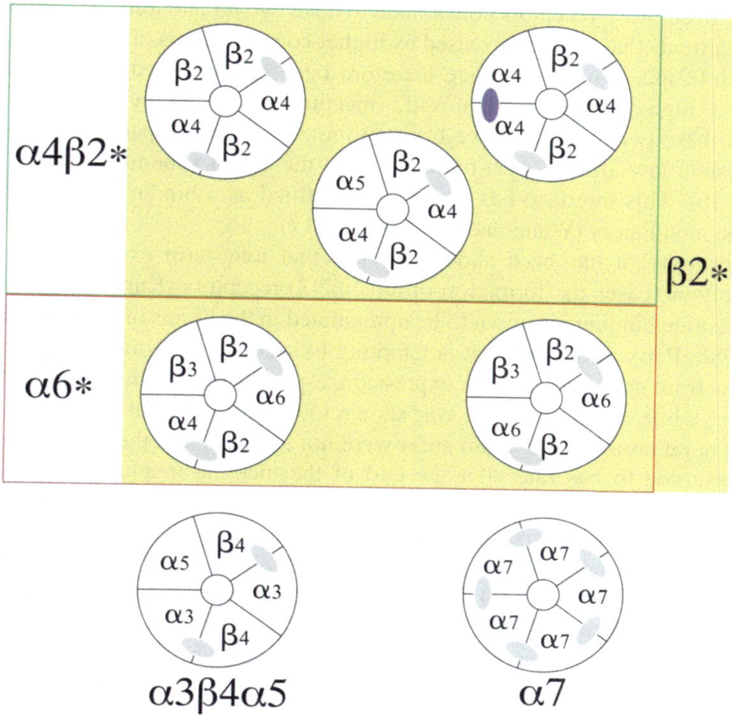

Fig. 1 Subunit composition of nAChR subtypes associated with nicotine addiction. The box in yellow shows the subunit composition of the β2-containing receptors implicated in release of dopamine in the mesolimbic reward system. These may be subdivided based on the absence (upper three) or presence of α6 subunits (lower two). It should be noted that these represent only a most likely subset of all the subunit configurations potentially involved for rodents and that in primates α2-containing nAChR may also be involved. Shown below are two other nAChR subtypes discussed in the text. The gray ovals represent the locations of binding sites for ACh or nicotine. The purple oval represents the putative low-affinity binding site unique to the LSα4β2 (α4(3)β2(2)) subtype

shifted through mass action by overexpressing one subunit relative to the other (Nelson et al. 2003) or, more elegantly, through the use of linked subunits (concatamers) of β2 and α4 that each configure an ACh-binding dimer. The fifth position in the functional pentamer can then be regulated by co-expressing the concatamer with monomeric α4 or β2. Similar results can be obtained with β2-α2 as well as β4-α3 concatamers (Papke et al. 2013). Concatamers have also proven useful for assessing the effects of α5 subunits (Kuryatov et al. 2011) and α6-containing receptors (Kuryatov and Lindstrom 2011) (see below).

Comparisons of the two stoichiometries of receptors containing just α4 and β2 subunits show that receptors constrained to have three α4 and two β2 subunits express more rapidly and give larger currents with a higher EC_{50} for ACh and nicotine. In contrast, receptors constrained to have two α4 and three β2 subunits give smaller currents that are not increased by higher concentrations of ACh. The α4(3)β2(2) and α4(2)β2(3) receptors have therefore been characterized as low-sensitivity (LS) and high-sensitivity (HS) α4β2 receptors, respectively. The differences between these two isoforms have been attributed, at least in part, to the existence of a putative low-affinity ACh-binding site at the α4-α4 subunit interface (Lucero et al. 2016). This interface has also been identified as a binding site for selective allosteric modulators (Wang and Lindstrom 2017).

Interestingly, it has been shown in vitro that long-term exposure to nicotine selectively increases the formation of α4(2)β2(3) receptors (Kuryatov et al. 2005). Since nicotine binding is known to be upregulated in the brains of smokers (Benwell et al. 1988; Perry et al. 1999), it is tempting to speculate that nicotine dependence may arise from increased receptor expression, especially that of the HSα4β2 isoform. However, while chronic nicotine was shown to cause a transient increase in the β2/α4 ratio in rat cortex, other brain areas were not affected, and the β2/α4 ratio in the cortex returned to baseline after the end of the nicotine treatments (Fasoli et al. 2016).

One approach for understanding the importance of specific nAChR subtypes has been to delete (knock out) specific nAChR subunit genes (Picciotto et al. 1998) (Cordero-Erausquin et al. 2000; Marubio and Changeux 2000) and evaluate changes in nicotine reinforcement/self-administration. One of the earliest studies established the essential role for β2-containing receptors for these behaviors (Picciotto et al. 1998). Knockouts of α4 were less effective at reducing the reinforcing effects of nicotine since it appeared that α6-containing receptors could compensate (Exley et al. 2011; Peng et al. 2017; Pons et al. 2008), but studies using α4 subunit "gain-of-function" transgenic mice have revealed that subthreshold doses of nicotine can support nicotine conditioned place preference in these mice, suggesting that stimulation of α4β2-containing receptors is sufficient to support nicotine reward (Tapper et al. 2004). Lentiviral studies that rescue nicotinic receptors on a knockout background have also demonstrated a critical role for VTA α4β2- and α6β2-containing receptors in nicotine reward (Maskos et al. 2005; Pons et al. 2008). Interestingly, the α4α6β2 receptor subtype has been shown in the VTA to be the most sensitive to nicotine and the only subtype to be preferentially activated, rather than desensitized, in response to physiological levels of nicotine (Liu et al. 2012).

While much work on animal models of nicotine addiction/dependence has focused on the α4β2 receptor subtypes, recent work has pointed toward the importance of other subtypes as well. One other important subclass of β2 subunit-containing receptors (Fig. 1) is those containing α6 subunits, especially regarding nicotine's effects in the mesolimbic reward pathway. Receptors containing α6 subunits were difficult to express in oocytes or other systems except in chimeras (Papke et al. 2008), until the development of α6 concatamers (Kuryatov and Lindstrom 2011) that additionally incorporated β3 subunits. Unlike α4β2 nAChR, which are ubiquitously expressed throughout the brain, the α6 gene is more selectively expressed in brain with a high expression in dopaminergic neurons (Charpantier et al. 1998; Salas et al. 2003) along with the accessory subunit β3 (Cui et al. 2003).

Multiple forms of α6-containing receptors are known to exist in the mesolimbic dopaminergic pathway. A combination of selective pharmacology and molecular studies has revealed that activation of α6β2 nAChR supports the acquisition, maintenance, and motivation for systemic nicotine self-administration (Brunzell et al. 2010; Gotti et al. 2010; Pons et al. 2008). Studies using intraventricular administration of conotoxin antagonists suggest that α6β2* nAChRs in brain are also critical for nicotine withdrawal (Jackson et al. 2009), but given that these compounds can target brain α3β2* nAChRs, further work needs to clarify which conotoxin-sensitive nAChRs support withdrawal behavior. It is presumed that these behaviors are largely influenced by dopamine release. It has been shown that α6β3* receptors are important for the in vitro effects of nicotine for evoking dopamine release from striatal synaptosomes (Grady et al. 2007) and the primary receptors on terminals to stimulate dopamine release (Exley et al. 2008), although α4β2 receptors are also important and, in fact, more abundant than α6-containing receptors (Grady et al. 2007). Ex vivo slice studies using α6 subunit "gain-of-function" transgenic mice suggest that stimulation of α6β2* nAChR is sufficient for both ACh- and nicotine-stimulated activation of dopaminergic neurons, and behavioral studies in these mice show that α6β2* nAChRs are critical for nicotine-stimulated locomotor activity (Drenan et al. 2010; Wang et al. 2014).

4 Special Significance of α5-Containing Receptors

The α5 subunit is encoded by CHRNA5, a gene expressed in chromosome 15 in humans and chromosome 9 in the mouse (Chini et al. 1992; Eng et al. 1991). CHRNA5 is part of the CHRNA3/A5/B4 gene cluster and shares several regulatory elements with the two genes that encode the α3 and β4 nAChR subunits (Boulter et al. 1990; Corriveau and Berg 1993; Sivilotti et al. 1997). α5 forms receptor complexes with either the α3 and β4 (Sivilotti et al. 1997; Vailati et al. 2003) or the α4 and β2 subunits (Gerzanich et al. 1998), and such receptors are expressed in both peripheral and central nervous system neurons as well as in nonneuronal cells (De Biasi 2002; Zoli et al. 2018). α5 co-expression enhances the sensitivity of α3β2 nAChRs to nicotine and ACh (Wang et al. 1996) and increases the calcium permeability of α3β2 and α3β4 nAChRs in both cell lines and ganglionic neurons (Gerzanich et al. 1998; Ramirez-Latorre et al. 1996). ACh-evoked currents from *Xenopus* oocytes expressing α4, α5, and β2 show a prominent rightward shift of the concentration-response curves for acetylcholine, while the single-channel conductance of α4α5β2 nAChR is almost double that of α4β2 receptors (Ramirez-Latorre et al. 1996). Furthermore, α4β2 nAChRs that include α5 show faster rates of recovery from desensitization than α4β2 nAChRs without α5 (Grady et al. 2012). Another study demonstrated an important effect of α5 on nicotine-evoked dopamine release. Synaptosomal preparations obtained from mice lacking the α5 subunit display a lower sensitivity to ACh and release less dopamine in response to nicotine than synaptosomes from wild-type mice (Grady et al. 2010).

The interest in the functional roles of α5 increased significantly about a decade ago, when a series of large-scale genetic studies pointed to a role of gene variation in the CHRNA3/A5/B4 cluster in smoking-related phenotypes – such as number of cigarettes/day and smoking heaviness – as well as lung cancer, chronic obstructive pulmonary disease (COPD), alcoholism, and peripheral arterial disease (Amos et al. 2008; Berrettini et al. 2008; Bierut et al. 2008; Grucza et al. 2008; Pillai et al. 2009; Schlaepfer et al. 2008; Stevens et al. 2008). The most notable effect, replicated by many studies in several populations (Li et al. 2010; Saccone et al. 2009), is that of a non-synonymous single nucleotide polymorphism (SNP), rs16969968. This polymorphism changes an aspartic acid residue into asparagine at position 398 (D398N) in the second intracellular loop of α5. Individuals heterozygous for the risk allele have a 1.3-fold increased risk for nicotine dependence, while homozygous individuals have almost a 2-fold increase in risk. In addition, the presence of this SNP influences the biophysical properties of α5∗ nAChRs. When expressed in HEK 293T cells, α4β2α5N398 nAChRs exhibit reduced maximal response to nicotine compared to α4β2α5D398 nAChRs. This phenomenon is not due to an intrinsic inability of the mutant α subunit to form functional nAChRs nor to an altered Ca^{2+} permeability, but likely to changes in intracellular modulation (Bierut et al. 2008; Sciaccaluga et al. 2015). Incorporation of α5D398 into α3β4 nAChRs also decreases the maximal response to agonist, especially when extracellular calcium is high, a phenomenon that may lead to distinct downstream cellular signaling (Tammimaki et al. 2012).

5 α5 Knockout and Knock-In

The creation of mutant mice that either lack the α5 subunit or express the rs16969948 polymorphism has provided significant insight into the functional roles of α5. Overall, the lack of α5 does not have major effects on development or baseline behaviors (De Biasi 2002; Gangitano et al. 2009) besides leading to reduced anxiety in the elevated plus maze in female mice (Gangitano et al. 2009) and reducing attention under highly demanding conditions (Bailey et al. 2011). In contrast, the null mutation has a profound influence on the effects of nicotine. It confers resistance to nicotine-induced hypolocomotion and nicotine-induced seizure (De Biasi 2002), reduces the physical manifestations of nicotine withdrawal (Salas et al. 2009), and markedly increases nicotine self-administration (Fowler et al. 2011; Morel et al. 2014). Normal nicotine self-administration can be “rescued” by re-expression of α5D398 in either the medial habenula (Fowler et al. 2011) or VTA neurons (Morel et al. 2014). Interestingly, when α5N398 is re-expressed in VTA dopaminergic neurons, mice continue to self-administer larger nicotine volumes, suggesting that the rs16969948 polymorphism, similar to the null mutation, alters dopaminergic function and creates a hypodopaminergic state that could promote nicotine abuse (Morel et al. 2014). Mice expressing the α5 SNP exhibit neurocognitive behavioral deficits in social interaction and sensorimotor gating tasks due to increased inhibitory drive over layer II/III pyramidal neurons (Koukouli et al. 2017). The decreased cortical activity observed in the mutant mice, which resembles the hypofrontality, a state of decreased cerebral blood flow (CBF) in the prefrontal cortex of the brain, reported in patients with schizophrenia or drug addiction, was reversed by chronic nicotine treatment. Such findings might provide a physiological basis for the tendency of patients with schizophrenia to self-medicate with the nicotine contained in tobacco products (Mallet et al. 2017). It might also explain why subjects carrying the risk allele for rs16969948 have the most success at quitting smoking when treated with nicotine replacement therapy, and such treatment would help to reduce cravings and withdrawal symptoms as well as potential hypofrontality (Bergen et al. 2013). Finally, recent work has also shown that rs16969948 has important transgenerational effects. Offspring born to females carrying the α5N398 variant consumed the most nicotine at the highest concentration presented compared to offspring born to females expressing α5D398, which drank the least amount of nicotine at all concentrations tested (O'Neill et al. 2018). Overall, the α5 subunit, once considered to have little impact on nAChR function, clearly has an important role in normal nAChR function and the mechanisms of nicotine addiction.

6 Unique Contributions of α7-Containing Receptors

With a preponderance of the evidence suggesting that stimulation of heteromeric receptors supported nicotine use and addiction behavior, a failure to observe similar nicotine-associated phenotypes in α7 knockout mice initially led to the conclusion that α7 nAChRs were not critically involved in tobacco use phenotypes (Pons et al. 2008; Walters et al. 2006). More recent work supports a unique role for α7 nicotinic receptors in tobacco dependence (Brunzell et al. 2014). Unlike β2 nAChRs, where a loss of function reduces nicotine reward and reinforcement, selective antagonism of α7 nicotinic receptors has been shown to significantly increase motivation to self-administer nicotine in rats maintained on a progressive ratio schedule of reinforcement (Brunzell and McIntosh 2012). Expanded dose-response curves further revealed that knockout of the α7 subunit significantly reduced threshold doses required to achieve nicotine conditioned place preference (Harenza et al. 2014), demonstrating an increased sensitivity to nicotine reward. To the contrary, selective stimulation of α7 nicotinic receptors greatly reduced how hard rats were willing to work for a single infusion of nicotine and blocked nicotine reward as measured using nicotine conditioned place preference in mice (Brunzell and McIntosh 2012; Harenza et al. 2014). Mice engineered to have an α7 nAChR gain-of-function mutation completely failed to establish nicotine place preference at any dose tested (Harenza et al. 2014). Although it is worth noting that one study has reported that nicotine self-administration maintained on a fixed ratio 1 schedule of reinforcement was reduced by systemic administration of the α7 nAChR antagonist methyllycaconitine (MLA) (Markou and Paterson 2001), it is not clear if the higher doses of MLA in this study may have antagonized α6β2 nicotinic receptors as well as α7 (Mogg et al. 2002). Together, these findings support the hypothesis that stimulation of α7 nAChR supports nicotine satiety. Hence, it would appear that α7 nAChRs work in opposition to β2 nAChRs to curb nicotine use. Stimulation of α7 nAChRs has been shown to inhibit β2 nAChR function on VTA dopamine neurons, presumably via stimulation of PPARα (Melis et al. 2013). In mice, a PPARα antagonist blocked the ability of a selective α7 nAChR agonist to attenuate nicotine place conditioning, suggesting that stimulation of α7 nAChRs attenuates nicotine reward via a PPARα mechanism (Jackson et al. 2017) and suggesting that PPARα, as well as α7 nAChR, may be effective pharmacological targets for tobacco cessation.

7 Special Considerations Related to α7 nAChR

The basic considerations related to the reinforcing effects of nicotine discussed above focused on the mesolimbic dopaminergic reward systems. Genetic studies suggest that CHRNA7 polymorphisms may confer vulnerability to tobacco use (De Luca et al. 2004); however, there are factors in addition to nicotine reinforcement that encourage smoking among the mentally ill. The prevalence of smoking in psychiatric outpatients is significantly higher than in the general population and is especially high (88%) in patients diagnosed with schizophrenia (Hughes et al. 1986). Individuals with schizophrenia typically show impaired sensory gating, manifested as an inability to adapt to repetitive or unexpected stimuli. This has been shown to be normalized by nicotine (Adler et al. 1992), so that smoking in this population has been proposed to be a form of self-medication (Kumari and Postma 2005; Leonard et al. 2007). The defect in sensory gating has been associated with a reduction of α7 nAChR expression (Freedman et al. 2000; Guan et al. 1999), and α7 receptors have been proposed as a specific target for managing schizophrenia (Freedman et al. 2000; Guan et al. 1999). This has led to small-scale trials of α7-selective agonists and allosteric modulators for the treatment of schizophrenia (Gee et al. 2017; Kem et al. 2017; Tregellas et al. 2010; Walling et al. 2016). An ideal smoking cessation therapy that would be applicable to both the general population and psychiatric outpatients should address this desire to self-medicate. In this regard, the fact that varenicline, while a partial agonist for β2-containing receptors, is a full agonist for α7 nAChRs (Mihalak et al. 2006) supports its suitability for both populations (Liu et al. 2011; Pachas et al. 2012; Williams et al. 2012b), although at least one case of adverse effects has also been reported (Freedman 2007). A Phase IIb clinical trial utilizing an α7 nAChR partial agonist with nicotine patch did not improve cognition or cessation (Schuster et al. 2018), perhaps because a partial agonist can inhibit α7 nAChRs in the presence of full agonist. A Phase IIa clinical trial utilizing an α7-positive allosteric modulator also did not deter smoking in individuals with schizophrenia or otherwise healthy smokers, but only one dose was used in this short-term, double-blind crossover study (Perkins et al. 2017). A small but significant finding reported that PPARα polymorphism was associated with tobacco dependence in females with schizophrenia diagnosis (Nadalin et al. 2016), suggesting that PPAR may be a novel target to support tobacco cessation in individuals with schizophrenia diagnosis. Early findings show that a PPARγ agonist effectively reduces craving in smokers (Jones et al. 2017), but these studies need to be expanded and repeated in smokers with schizophrenia.

8 Pharmacotherapies for Smoking Cessations

While there are no truly effective therapies to manage nicotine addiction/dependence, a few approaches have been developed, largely based on the hypothesis that addressing the functional tolerance to nicotine with drugs that act upon the receptors will alleviate withdrawal and cravings. Arguably, the most straightforward approach to achieve this has been with nicotine replacement (Benowitz 2009), typically with formulations such as patches or lozenges that would be intrinsically less rewarding (Atzori et al. 2008; Rose et al. 1990). These approaches will of course target all of the same receptors as the nicotine delivered in cigarettes, but a slower delivery of nicotine to the brain is less likely to promote the receptor activation required to stimulate dopamine release.

These nicotine replacement therapies have the goal of ultimately enabling smokers to quit and should not be confused with the use of e-cigarettes, which promote continued "smoking" with surrogate nicotine delivery systems that theoretically reduce the harmful effects of combusted nicotine (Farsalinos and Niaura 2019; Notley et al. 2018; Smith et al. 2018).

An alternative approach for aiding smoking cessation has been to utilize partial agonists for β2-containing nAChR (Hogg and Bertrand 2007). This approach was inspired by the identification of cytisine, a *Laburnum* alkaloid, as such an agent, able to produce only weak activation of β2-containing receptors and to reduce the activation of these receptors by full agonists such as ACh (Papke and Heinemann 1994). While cytisine itself is used as a cessation aid in Europe, (Etter et al. 2008), in the United States, it inspired the development of the related compound varenicline (Coe et al. 2005).

The basic concept for the use of β2 receptor partial agonists (Papke et al. 2011) is that when present at relatively low concentrations, they can suppress the phasic activation of the α4β2- and α6β2-containing receptors on dopaminergic neurons that is produced by puffs of nicotine (Fig. 2), reducing the reinforcing effects while simultaneously producing low levels of tonic activation that could diminish cravings.

Another approved pharmacotherapy for smoking cessation is the antidepressant bupropion (Benowitz 2009; Cryan et al. 2003). Although roughly as effective as varenicline, its mechanism of action is less clear. It has been shown in vitro to have some activity as nAChR antagonist (Slemmer et al. 2000) and to reduce nicotine-evoked dopamine release in brain slice experiments (Miller et al. 2002). However,

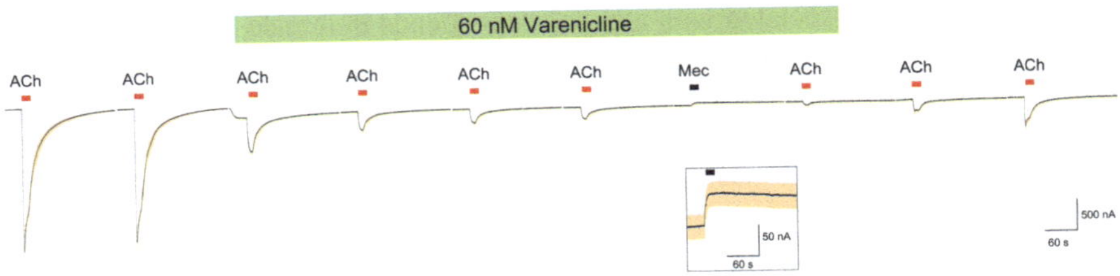

Fig. 2 Modulation of HSα4β2 (α4(2)β2(3)) receptor phasic activation by varenicline. Two-electrode voltage-clamp recordings were made as previously described (Papke and Stokes 2010). Oocytes expressing the β2-α4 concatamer and monomeric β2 were repeatedly stimulated with applications of 10 μM ACh to mimic the repeated effects of nicotine pulses. After the second ACh application, 60 nM varenicline was added to the bath. After the sixth ACh application, 100 μM of the nonselective antagonist mecamylamine was delivered to the cells to demonstrate the inhibition of the tonically active current produced by the bath-applied varenicline. The insert shows this inhibition at a tenfold increased scale. Shown are the averaged responses (solid lines) of seven cells ±SEM (tan area), normalized to the average of the two initial ACh responses prior to the addition of varenicline to the bath. Each trace of 10,322 points (30 s preapplication baseline, 6 s drug application, and 170 s of a 241 s washout in Ringer's solution) is 206.44 s long

bupropion is also known to block dopamine reuptake, a feature common to the CNS stimulants cocaine and amphetamine (Banks et al. 2016). Additionally, as an antidepressant, bupropion may also help manage the depression that frequently occurs during nicotine withdrawal.

8.1 *Areca: Another Cholinergic-Based Addiction?*

The previous sections focused on addiction associated with nAChRs, which function as ligand-gated ion channels. The second major class of acetylcholine receptors contains metabotropic receptors that couple through G-proteins with ion channels among numerous possible downstream mediators of their effects. Just as the ion channel-coupled receptors for acetylcholine are associated with activation by the plant toxin nicotine, the G-protein-coupled acetylcholine receptors are identified by their sensitivity to muscarine, a toxin first isolated from the mushroom *Amanita muscaria*. The structures of acetylcholine and these eponymous agonists are shown in Fig. 3a, along with another plant alkaloid that is also associated with addiction, arecoline, which is well known to be a muscarinic agonist. There are five subtypes of muscarinic receptors, and the M1 and M3 subtypes, which are widely expressed in the brain, produce increases in intracellular calcium upon activation. This allows for the downstream activation of calcium-dependent channels for potassium or chloride to be used as reporters of muscarinic receptor activation. As shown in Fig. 3b, when the M1 and M3 muscarinic receptor subtypes are expressed in *Xenopus* oocytes, the application of ACh, muscarine, or arecoline, but not nicotine, produces strong activation of calcium-dependent chloride currents in voltage-clamp experiments.

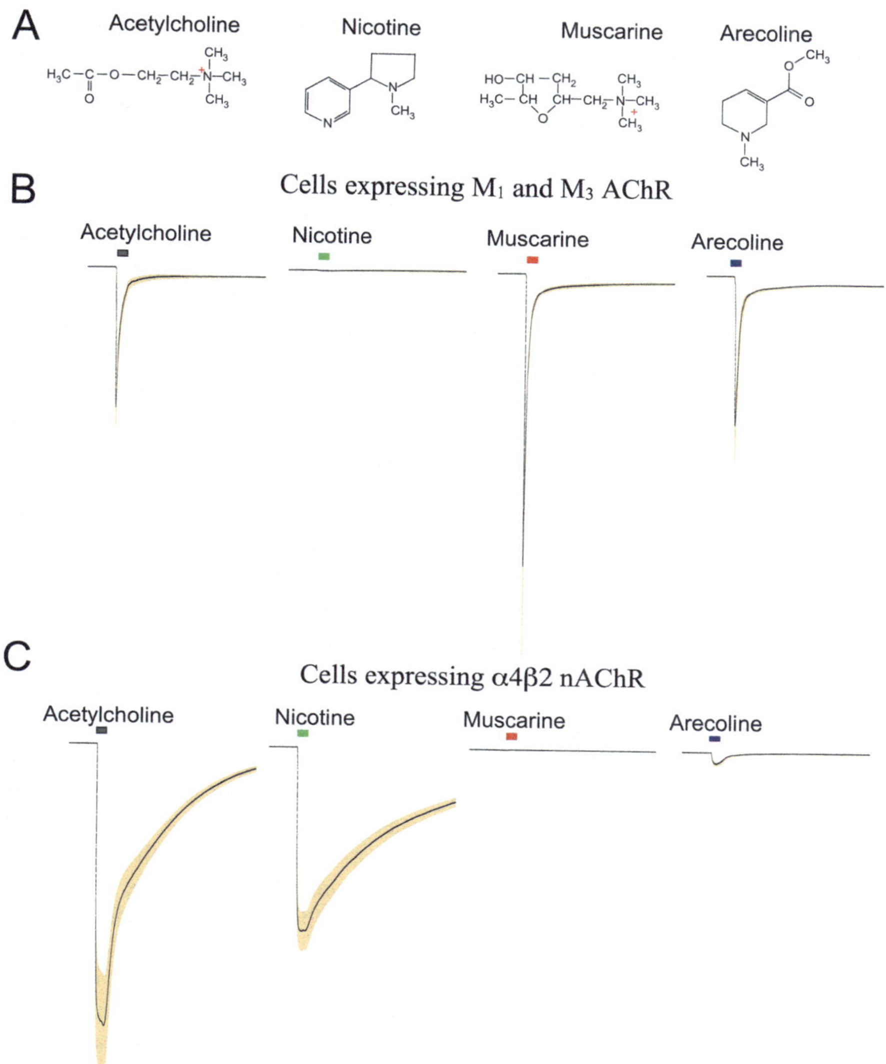

Fig. 3 (**a**) Structures of acetylcholine and the alkaloids nicotine, muscarine, and arecoline. Note that although nicotine and arecoline are tertiary amines, they are likely to be charged at physiological pH. (**b**) Shown are the averaged responses (solid lines) of eight cells to each of the drugs shown in A, plus and minus the standard error of the mean calculated at each of the of 10,322 points in each trace (tan area). Drugs were applied at 100 μM. In these experiments, the currents recorded were associated with the downstream activation of calcium-dependent chloride currents that desensitize after a single activation, so the different drugs were applied to separate sets of cells from the same batch of oocytes injected with RNA coding for M1 and M3 muscarinic receptors. Each trace is 206.44 s long. (**c**) Responses of oocytes injected with RNA for α4 and β2 nAChR subunits to applications of the drugs shown in A at 100 μM ($n \geq 5$ for each trace). Prior to averaging, data from each cell were normalized to the average of two initial 30 μM ACh responses from the same cells prior to the experimental drugs. The solid lines are the averaged normalized data, and the tan areas are the standard errors of the means calculated at each point. Each trace is 206.44 s long

When these same ligands are applied to oocytes expressing the human α4 and β2 nAChR subunits, there is, of course, strong activation by ACh or nicotine and no response to muscarine (Fig. 3c). However, arecoline also produces a small activation of these receptors, which are strongly associated with nicotine addiction (Papke et al. 2015).

Arecoline is one of the four major alkaloids present in areca, along with guvacine, guvacoline, and arecaidine (Jain et al. 2017). With an estimated 600 million users, mostly in South Asia, areca is the fourth most commonly used addictive substance in the world, after alcohol, caffeine, and tobacco (Gupta and Warnakulasuriya 2002). Areca users have a twofold higher chance of oral cancer than the general population and a 50–100-fold higher chance of precancerous disease such as oral submucosal fibrosis (Auluck et al. 2009), making it a serious health problem in countries where areca use is high. The chewing of preparations of areca nut (most commonly known as betel quid or betel nut) is an ancient custom throughout much of South Asia, with physical evidence for areca use on human dental remains predating written history (Oxenham et al. 2002). There are references to the use and cultural importance of betel in the most ancient Indian sutras (Raghavan and Baruah 1958). It was a custom enjoyed by all strata of society, with nobility possessing elegant accessories for the storage and processing of areca nut. Europeans visiting royal courts, as far back as the time of Marco Polo, would be invited to share betel with their noble hosts as signs of acceptance and respect. In some cases, the guests were even offered quids that had first been partially chewed by their hosts (Rooney 1993). How could they refuse?

The essential ingredient of all forms of "betel" is the fruit of the palm, *Areca catechu*, usually referred to as a nut, although actually a drupe (Fig. 4a). Areca is prepared in various ways by different cultures, and these preparations have many different names (Patidar et al. 2015). As noted above, one of the most common formulations is the "betel quid," the preparation of which begins with a leaf of the vine *Piper betle*, believed to be the source of the somewhat misleading association between areca and "betel nut." To prepare a quid, the betle leaf is first spread with a paste of slaked lime (calcium hydroxide); then pieces of areca nut (Fig. 4b), often with other spices, seeds, and flavorants, are added prior to folding the leaf into small packet (Fig. 4c). This "quid" is then inserted into the cheek to be chewed and sucked on for up to an hour or more. While some cultures use dry, ripe areca nuts, as popular in India (Fig. 4b), others use fresh, often immature nuts (Fig. 4a), sometimes including the husks (Lin et al. 2017), and chewed without betle leaf. As is often the case with drug use habits, areca users often change their preferences over time and with their age, starting with sweetly flavored quids, such as meetha paan masala. The spread of tobacco products through Asia in recent centuries has also affected betel quid use, so presently, on average, about half of adult betel chewers add tobacco to their quid (IARC 2004). Accommodating modern mass marketing, areca is also available in foil packets premixed with other ingredients (Gupta and Warnakulasuriya 2002). These are sold as "pan masala" when formulated without tobacco and as "gutka" when processed smokeless tobacco is included. Asian

Fig. 4 (**a**) Fresh areca nuts from Vietnam, sourced from eBay and shipped to Florida. One fruit is intact and the other split to reveal the nut inside the husk. (**b**) Dried areca nuts being sold in bulk at the wholesale produce market in Yangon, Myanmar, variously whole, cut in half, or chopped into pieces. (**c**) A betel quid being assembled by a street vendor in Yangon, Myanmar. The betle leaf has been partially coated with white lime, and the vendor is adding sliced areca to a quid already containing a small amount of tobacco. Also shown are prepared quids, ready for sale. (**d**) Pavement in Yangon, Myanmar, decorated with red expectorate from a betel quid chewer. The muscarinic action of the areca alkaloid arecoline stimulates copious amounts of saliva that acquire red pigments from components in the nut. All photos were taken by the author (RLP)

specialty stores in the United States commonly carry raw areca products and these prepared packets (Bachman 2013; Changrani et al. 2006).

The areca alkaloid arecoline, first isolated in 1891, has been known to be a muscarinic agonist since the time of the earliest studies of autonomic pharmacology (Epstein 1932). Since the muscarinic activation of salivary gland is the most overt indication of betel quid use (Fig. 4c), it has been hypothesized that the muscarinic activity of arecoline in the brain is sufficient to account for the short-term reinforcing effects of areca. Although the muscarinic antagonist scopolamine is considered a deliriant hallucinogen (Graff 1969), muscarinic agonists have no history of being drugs of abuse, aside from the unsubstantiated supposition that arecoline must account for areca addiction (Nelson and Heischober 1999). As is the case with tobacco, it is difficult, even for users, to identify the psychoactive effects of areca that are likely to account for its use-promoting reward. Anecdotal accounts sometimes liken areca's effects to a strong cup of coffee. In a survey of 370 addicted betel quid users in Karachi (Khan et al. 2013), 47% of the subjects identified betel as a central nervous system (CNS) stimulant. However, 14% identified it as a CNS depressant, and 38% believed it had no CNS effects. Nonetheless, in the same study, 74% of the respondents believed that chewing betel could cause cancers of the mouth or throat. Similar results were obtained in a recent study of betel quid users in Yangon, Myanmar (Papke et al. 2019), although in that study more subjects (55.5%) reported that the effects were like that of alcohol. The Yangon study also included a survey of quid vendors and confirmed that 75% added tobacco to the quids in equal amount to the areca.

Seriously understudied, areca use is an orphan addiction (Little and Papke 2015). Although overlooked in the official DSM-5, a recent large-scale study of six endemic Asian populations (8922 participants) applied DSM-5 type criteria to identify Betel-Quid Use Disorder (BUD) (Lee et al. 2018). They found that among current users of betel quid, 86.0% met the criteria for BUD (mild BUD, 15.5%; moderate BUD, 20.6%; and severe BUD, 50.0%). Similar results were reported for betel quid users in Guam (Herzog et al. 2014).

While the effects of coffee and alcohol are easily appreciated by the consumer of these beverages, the subjective CNS effects of areca, like tobacco, are relatively subtle (Khan et al. 2013). Profuse salivation, the most obvious effect of chewing betel quid, is certainly a cue that the betel user will associate with their drug, but not being a CNS effect, it is unlikely to relate to the actual addiction. As noted earlier, addictions typically have two components, the first being a short-term reinforcing effect that targets the brain's reward mechanism associated with neurotransmitters like dopamine (Stolerman and Shoaib 1991). The second component of a serious addiction is the physical dependence produced by the chronic use of the drug that leads to withdrawal symptoms should the user try to quit. The nicotinic activity of arecoline has been proposed to relate to this second component, as well as to contribute to the predilection for adult betel users to add tobacco to their quid (Papke et al. 2015).

As western societies have in recent decades been trying to deal with tobacco use, Asian countries are now beginning to enact new public policies to deal with the public health cost of betel use (Garg et al. 2014; Gupta and Warnakulasuriya 2002; Mehrtash et al. 2017). However, on the level of the individual, awareness of a health risk is generally not sufficient motivation for a person to quit an addiction (Herzog et al. 2014; Little et al. 2014), providing a reason for developing targeted cessation therapies. Although, as noted above, the essential reinforcing properties of areca remain in question, it is clear that the large fraction of users that mix tobacco with their quid is likely to develop a nicotine dependence similar to users of smokeless tobacco alone. Any cessation therapy for this population would need to address this aspect of the dependence in addition to unique aspects associated with areca, such as the muscarinic activity of arecoline that produces an association between drug use and the copious production of saliva. Salivation would also then become a cue associated with the delivery of nicotine, and nicotine itself would augment the salience of such a cue (Perkins et al. 2017). This suggests that an areca cessation therapy targeted to users of betel quid with tobacco might combine a nicotine replacement in the form of a gum with a clinically approved muscarinic agonist like pilocarpine.

Betel quid was once considered by Europeans as a quaint and somewhat curious custom and by South Asians as a societal norm. Nowadays, it is largely unknown by Americans, outside of the substantial Asian-American communities. Based on data from the US census, there are a total of approximately nine million first-generation immigrants in the United States from countries where areca product use is high, including India, the Philippines, Vietnam, Pakistan, Cambodia, Taiwan, and Bangladesh. Additionally, areca use is prevalent in the US territory of Guam (population of 168,000). As with other cultural traditions, this use may be passed on to second, third, and even further generations, especially in tight social communities, such as those of the Indian-Gujarati immigrants. It is reasonable to estimate the total number of people in the United States who have a cultural tradition of areca product use to be close to 20 million, and based on the limited survey data available (Changrani et al. 2006; Murphy and Herzog 2015), more than one million people in the United States may be current users, with elevated risks for oral disease and cancers (Auluck et al. 2009). In some cases, the use of areca products may spread to other populations. For example, in Hawaii, where Micronesian students of a betel-using culture mix with the general student population, 2% of the high school students identified themselves as current users of betel nut (Pobutsky and Neri 2012), although Micronesians constitute only 1% of the total population. Furthermore, as noted above, approximately 50% of the time, the use of areca products leads to the addition of smokeless tobacco or the switch from pan masala to gutka, which also greatly increases the likelihood of oral cancers. The sum impact of areca product use is probably hundreds of preventable cancers in the United States every year and much more precancerous oral disease. The path to prevention of areca-related disease is the management of the addiction, and with these new insights into the root causes of areca addiction, we can hope to find that path.

Acknowledgments RLP is supported by NIH RO1 GM57481, DHB is supported by NIH RO1 DA042749, and MDB is supported by NIH RO1 DA044205 and UO1 AA025931.

References

Adler LE, Hoffer LJ, Griffith J, Waldo MC, Freedman R (1992) Normalization by nicotine of deficient auditory sensory gating in the relatives of schizophrenics. Biol Psychiatry 32 (7):607–616

Amos CI, Wu X, Broderick P, Gorlov IP, Gu J, Eisen T, Dong Q, Zhang Q, Gu X, Vijayakrishnan J, Sullivan K, Matakidou A, Wang Y, Mills G, Doheny K, Tsai YY, Chen WV, Shete S, Spitz MR, Houlston RS (2008) Genome-wide association scan of tag SNPs identifies a susceptibility locus for lung cancer at 15q25.1. Nat Genet 40(5):616–622

Atzori G, Lemmonds CA, Kotler ML, Durcan MJ, Boyle J (2008) Efficacy of a nicotine (4 mg)-containing lozenge on the cognitive impairment of nicotine withdrawal. J Clin Psychopharmacol 28(6):667–674

Auluck A, Hislop G, Poh C, Zhang L, Rosin MP (2009) Areca nut and betel quid chewing among South Asian immigrants to Western countries and its implications for oral cancer screening. Rural Remote Health 9(2):1118

Bachman SA (2013) Betel nut product characteristics and availability in King County, Washington: a secret shopper study. In: Global health. University of Washington, Seattle
Bailey CD, De Biasi M, Fletcher PJ, Lambe EK (2011) The nicotinic acetylcholine receptor alpha5 subunit plays a key role in attention circuitry and accuracy. J Neurosci 30(27):9241–9252
Banks ML, Smith DA, Blough BE (2016) Methamphetamine-like discriminative stimulus effects of bupropion and its two hydroxy metabolites in male rhesus monkeys. Behav Pharmacol 27(2–3 Spec Issue):196–203
Benowitz NL (2009) Pharmacology of nicotine: addiction, smoking-induced disease, and therapeutics. Annu Rev Pharmacol Toxicol 49:57–71
Benwell ME, Balfour DJ, Anderson JM (1988) Evidence that tobacco smoking increases the density of (-)-[3H]nicotine binding sites in human brain. J Neurochem 50(4):1243–1247
Bergen AW, Javitz HS, Krasnow R, Nishita D, Michel M, Conti DV, Liu J, Lee W, Edlund CK, Hall S, Kwok PY, Benowitz NL, Baker TB, Tyndale RF, Lerman C, Swan GE (2013) Nicotinic acetylcholine receptor variation and response to smoking cessation therapies. Pharmacogenet Genomics 23(2):94–103
Berrettini W, Yuan X, Tozzi F, Song K, Francks C, Chilcoat H, Waterworth D, Muglia P, Mooser V (2008) Alpha-5/alpha-3 nicotinic receptor subunit alleles increase risk for heavy smoking. Mol Psychiatry 13(4):368–373
Bierut LJ, Stitzel JA, Wang JC, Hinrichs AL, Grucza RA, Xuei X, Saccone NL, Saccone SF, Bertelsen S, Fox L, Horton WJ, Breslau N, Budde J, Cloninger CR, Dick DM, Foroud T, Hatsukami D, Hesselbrock V, Johnson EO, Kramer J, Kuperman S, Madden PA, Mayo K, Nurnberger J Jr, Pomerleau O, Porjesz B, Reyes O, Schuckit M, Swan G, Tischfield JA, Edenberg HJ, Rice JP, Goate AM (2008) Variants in nicotinic receptors and risk for nicotine dependence. Am J Psychiatry 165(9):1163–1171
Boulter J, Connolly J, Deneris E, Goldman D, Heinemann S, Patrick J (1987) Functional expression of two neural nicotinic acetylcholine receptors from cDNA clones identifies a gene family. Proc Natl Acad Sci U S A 84:7763–7767
Boulter J, O'Shea-Greenfield A, Duvoisin RM, Connolly JG, Wada E, Jensen A, Gardner PD, Ballivet M, Deneris ES, McKinnon D et al (1990) Alpha 3, alpha 5, and beta 4: three members of the rat neuronal nicotinic acetylcholine receptor-related gene family form a gene cluster. J Biol Chem 265(8):4472–4482
Brunzell DH, McIntosh JM (2012) Alpha7 nicotinic acetylcholine receptors modulate motivation to self-administer nicotine: implications for smoking and schizophrenia. Neuropsychopharmacology 37(5):1134–1143
Brunzell DH, Boschen KE, Hendrick ES, Beardsley PM, McIntosh JM (2010) Alpha-conotoxin MII-sensitive nicotinic acetylcholine receptors in the nucleus accumbens shell regulate progressive ratio responding maintained by nicotine. Neuropsychopharmacology 35(3):665–673
Brunzell DH, McIntosh JM, Papke RL (2014) Diverse strategies targeting alpha7 homomeric and alpha6beta2∗ heteromeric nicotinic acetylcholine receptors for smoking cessation. Ann N Y Acad Sci 1327:27–45
Changrani J, Gany FM, Cruz G, Kerr R, Katz R (2006) Paan and gutka use in the United States: a pilot study in Bangladeshi and Indian-Gujarati immigrants in New York City. J Immigr Refug Stud 4(1):99–110
Charpantier E, Barneoud P, Moser P, Besnard F, Sgard F (1998) Nicotinic acetylcholine subunit mRNA expression in dopaminergic neurons of the rat substantia nigra and ventral tegmental area. Neuroreport 9(13):3097–3101
Chini B, Clementi F, Hukovic N, Sher E (1992) Neuronal-type alpha-bungarotoxin receptors and the alpha 5-nicotinic receptor subunit gene are expressed in neuronal and nonneuronal human cell lines. Proc Natl Acad Sci U S A 89(5):1572–1576
Clarke PBS, Pert CB, Pert A (1984) Autoradiographic distribution of nicotinic receptors in rat brain. Brain Res 323:390–395

Clarke PBS, Schwartz RD, Paul SM, Pert CB, Pert A (1985) Nicotinic binding in rat brain: autoradiographic comparison of [^{3}H] acetylcholine [$_3$H] nicotine and [^{125}I]-alpha-bungarotoxin. J Neurosci 5:1307–1315
Coe JW, Brooks PR, Vetelino MG, Wirtz MC, Arnold EP, Huang J, Sands SB, Davis TI, Lebel LA, Fox CB, Shrikhande A, Heym JH, Schaeffer E, Rollema H, Lu Y, Mansbach RS, Chambers LK, Rovetti CC, Schulz DW, Tingley FD 3rd, O'Neill BT (2005) Varenicline: an alpha4beta2 nicotinic receptor partial agonist for smoking cessation. J Med Chem 48(10):3474–3477
Cordero-Erausquin M, Marubio LM, Klink R, Changeux JP (2000) Nicotinic receptor function: new perspectives from knockout mice. Trends Pharmacol Sci 21(6):211–217
Corrigall WA, Franklin KB, Coen KM, Clarke PB (1992) The mesolimbic dopaminergic system is implicated in the reinforcing effects of nicotine. Psychopharmacology (Berl) 107(2-3):285–289
Corrigall WA, Coen KM, Adamson KL (1994) Self-administered nicotine activates the mesolimbic dopamine system through the ventral tegmental area. Brain Res 653(1–2):278–284
Corringer PJ, Le Novere N, Changeux JP (2000) Nicotinic receptors at the amino acid level. Annu Rev Pharmacol Toxicol 40:431–458
Corriveau RA, Berg DK (1993) Coexpression of multiple acetylcholine receptor genes in neurons: quantification of transcripts during development. J Neurosci 13(6):2662–2671
Cryan JF, Gasparini F, van Heeke G, Markou A (2003) Non-nicotinic neuropharmacological strategies for nicotine dependence: beyond bupropion. Drug Discov Today 8(22):1025–1034
Cui C, Booker TK, Allen RS, Grady SR, Whiteaker P, Marks MJ, Salminen O, Tritto T, Butt CM, Allen WR, Stitzel JA, McIntosh JM, Boulter J, Collins AC, Heinemann SF (2003) The beta3 nicotinic receptor subunit: a component of alpha-conotoxin MII-binding nicotinic acetylcholine receptors that modulate dopamine release and related behaviors. J Neurosci 23 (35):11045–11053
De Biasi M (2002) Nicotinic mechanisms in the autonomic control of organ systems. J Neurobiol 53 (4):568–579
De Biasi M, Dani JA (2011) Reward, addiction, withdrawal to nicotine. Annu Rev Neurosci 34:105–130
De Luca V, Wong AH, Muller DJ, Wong GW, Tyndale RF, Kennedy JL (2004) Evidence of association between smoking and alpha7 nicotinic receptor subunit gene in schizophrenia patients. Neuropsychopharmacology 29(8):1522–1526
Dent JA (2010) The evolution of pentameric ligand-gated ion channels. Adv Exp Med Biol 683:11–23
Drenan RM, Grady SR, Steele AD, McKinney S, Patzlaff NE, McIntosh JM, Marks MJ, Miwa JM, Lester HA (2010) Cholinergic modulation of locomotion and striatal dopamine release is mediated by alpha6alpha4* nicotinic acetylcholine receptors. J Neurosci 30(29):9877–9889
Duvoisin RM, Deneris E, Patrick J, Heinemann S (1989) The functional diversity of the neuronal acetylcholine receptors is increased by a novel subunit: b4. Neuron 3:487–496
Elgoyhen AB, Johnson DS, Boulter J, Vetter DE, Heinemann S (1994) a9: an acetylcholine receptor with novel pharmacological properties expressed in rat cochlear hair cells. Cell 79:705–715
Elgoyhen AB, Vetter DE, Katz E, Rothlin CV, Heinemann SF, Boulter J (2001) alpha10: a determinant of nicotinic cholinergic receptor function in mammalian vestibular and cochlear mechanosensory hair cells. Proc Natl Acad Sci U S A 98(6):3501–3506
Eng CM, Kozak CA, Beaudet AL, Zoghbi HY (1991) Mapping of multiple subunits of the neuronal nicotinic acetylcholine receptor to chromosome 15 in man and chromosome 9 in mouse. Genomics 9(2):278–282
Epstein D (1932) The responses of the batrachian alimentary canal to autonomic drugs. Rana and Bufo arecoline. J Physiol 75(1):99–111
Etter JF, Lukas RJ, Benowitz NL, West R, Dresler CM (2008) Cytisine for smoking cessation: a research agenda. Drug Alcohol Depend 92(1-3):3–8
Exley R, Clements MA, Hartung H, McIntosh JM, Cragg SJ (2008) Alpha6-containing nicotinic acetylcholine receptors dominate the nicotine control of dopamine neurotransmission in nucleus accumbens. Neuropsychopharmacology 33(9):2158–2166

Exley R, Maubourguet N, David V, Eddine R, Evrard A, Pons S, Marti F, Threlfell S, Cazala P, McIntosh JM, Changeux JP, Maskos U, Cragg SJ, Faure P (2011) Distinct contributions of nicotinic acetylcholine receptor subunit alpha4 and subunit alpha6 to the reinforcing effects of nicotine. Proc Natl Acad Sci U S A 108(18):7577–7582
Farsalinos K, Niaura R (2019) E-cigarettes and smoking cessation in the United States according to frequency of e-cigarette use and quitting duration: analysis of the 2016 and 2017 National Health Interview Surveys. Nicotine Tob Res 22(5):655–662
Fasoli F, Moretti M, Zoli M, Pistillo F, Crespi A, Clementi F, Mc Clure-Begley T, Marks MJ, Gotti C (2016) In vivo chronic nicotine exposure differentially and reversibly affects upregulation and stoichiometry of alpha4beta2 nicotinic receptors in cortex and thalamus. Neuropharmacology 108:324–331
Fowler CD, Lu Q, Johnson PM, Marks MJ, Kenny PJ (2011) Habenular alpha5 nicotinic receptor subunit signalling controls nicotine intake. Nature 471(7340):597–601
Freedman R (2007) Exacerbation of schizophrenia by varenicline. Am J Psychiatry 164(8):1269
Freedman R, Adams CE, Leonard S (2000) The alpha7-nicotinic acetylcholine receptor and the pathology of hippocampal interneurons in schizophrenia. J Chem Neuroanat 20(3-4):299–306
Gangitano D, Salas R, Teng Y, Perez E, De Biasi M (2009) Progesterone modulation of alpha5 nAChR subunits influences anxiety-related behavior during estrus cycle. Genes Brain Behav 8 (4):398–406
Garg A, Chaturvedi P, Gupta PC (2014) A review of the systemic adverse effects of areca nut or betel nut. Indian J Med Paediatr Oncol 35(1):3–9
Gee KW, Olincy A, Kanner R, Johnson L, Hogenkamp D, Harris J, Tran M, Edmonds SA, Sauer W, Yoshimura R, Johnstone T, Freedman R (2017) First in human trial of a type I positive allosteric modulator of alpha7-nicotinic acetylcholine receptors: pharmacokinetics, safety, and evidence for neurocognitive effect of AVL-3288. J Psychopharmacol 31(4):434–441
Gerzanich V, Wang F, Kuryatov A, Lindstrom J (1998) Alpha5 Subunit alters desensitization, pharmacology, Ca++ permeability and Ca++ modulation of human neuronal alpha 3 nicotinic receptors. J Pharmacol Exp Ther 286(1):311–320
Gotti C, Zoli M, Clementi F (2006) Brain nicotinic acetylcholine receptors: native subtypes and their relevance. Trends Pharmacol Sci 27(9):482–491
Gotti C, Guiducci S, Tedesco V, Corbioli S, Zanetti L, Moretti M, Zanardi A, Rimondini R, Mugnaini M, Clementi F, Chiamulera C, Zoli M (2010) Nicotinic acetylcholine receptors in the mesolimbic pathway: primary role of ventral tegmental area alpha6beta2* receptors in mediating systemic nicotine effects on dopamine release, locomotion, and reinforcement. J Neurosci 30(15):5311–5325
Grady SR, Salminen O, Laverty DC, Whiteaker P, McIntosh JM, Collins AC, Marks MJ (2007) The subtypes of nicotinic acetylcholine receptors on dopaminergic terminals of mouse striatum. Biochem Pharmacol 74(8):1235–1246
Grady SR, Salminen O, McIntosh JM, Marks MJ, Collins AC (2010) Mouse striatal dopamine nerve terminals express alpha4alpha5beta2 and two stoichiometric forms of alpha4beta2-*-nicotinic acetylcholine receptors. J Mol Neurosci 40(1–2):91–95
Grady SR, Wageman CR, Patzlaff NE, Marks MJ (2012) Low concentrations of nicotine differentially desensitize nicotinic acetylcholine receptors that include alpha5 or alpha6 subunits and that mediate synaptosomal neurotransmitter release. Neuropharmacology 62(5–6):1935–1943
Graff H (1969) Marihuana and scopolamine "High". Am J Psychiatry 125(9):1258–1259
Grucza RA, Wang JC, Stitzel JA, Hinrichs AL, Saccone SF, Saccone NL, Bucholz KK, Cloninger CR, Neuman RJ, Budde JP, Fox L, Bertelsen S, Kramer J, Hesselbrock V, Tischfield J, Nurnberger JI Jr, Almasy L, Porjesz B, Kuperman S, Schuckit MA, Edenberg HJ, Rice JP, Goate AM, Bierut LJ (2008) A risk allele for nicotine dependence in CHRNA5 is a protective allele for cocaine dependence. Biol Psychiatry 64(11):922–929
Guan ZZ, Zhang X, Blennow K, Nordberg A (1999) Decreased protein level of nicotinic receptor alpha7 subunit in the frontal cortex from schizophrenic brain. Neuroreport 10(8):1779–1782

Gulsevin A, Papke RL, Stokes C, Garai S, Thakur GA, Quadri M, Horenstein N (2019) Allosteric agonism of alpha7 nicotinic acetylcholine receptors. Mol Pharmacol 95(6):604–614
Gupta PC, Warnakulasuriya S (2002) Global epidemiology of areca nut usage. Addict Biol 7 (1):77–83
Hall FS, Sora I, Drgonova J, Li XF, Goeb M, Uhl GR (2004) Molecular mechanisms underlying the rewarding effects of cocaine. Ann N Y Acad Sci 1025:47–56
Han ZY, Le Novere N, Zoli M, Hill JA Jr, Champtiaux N, Changeux JP (2000) Localization of nAChR subunit mRNAs in the brain of Macaca mulatta. Eur J Neurosci 12(10):3664–3674
Harenza JL, Muldoon PP, De Biasi M, Damaj MI, Miles MF (2014) Genetic variation within the Chrna7 gene modulates nicotine reward-like phenotypes in mice. Genes Brain Behav 13 (2):213–225
Hasin DS, O'Brien CP, Auriacombe M, Borges G, Bucholz K, Budney A, Compton WM, Crowley T, Ling W, Petry NM, Schuckit M, Grant BF (2013) DSM-5 criteria for substance use disorders: recommendations and rationale. Am J Psychiatry 170(8):834–851
Heinemann S, Boulter J, Deneris E, Conolly J, Duvoisin R, Papke R, Patrick J (1990) The brain nicotinic acetylcholine receptor gene family. Prog Brain Res 86:195–203
Helekar SA, Char D, Neff S, Patrick J (1994) Prolyl isomerase requirement for the expression of functional homo-oligomeric ligand-gated ion channels. Neuron 12(1):179–189
Herzog TA, Murphy KL, Little MA, Suguitan GS, Pokhrel P, Kawamoto CT (2014) The Betel Quid Dependence Scale: replication and extension in a Guamanian sample. Drug Alcohol Depend 138:154–160
Hogg RC, Bertrand D (2007) Partial agonists as therapeutic agents at neuronal nicotinic acetylcholine receptors. Biochem Pharmacol 73(4):459–468
Hughes JR, Hatsukami DK, Mitchell JE, Dahlgren LA (1986) Prevalence of smoking among psychiatric outpatients. Am J Psychiatry 143(8):993–997
IARC (2004) Betel-quid and areca-nut chewing and some areca-nut derived nitrosamines. IARC Monogr Eval Carcinog Risks Hum 85:1–334. PMID: 15635762
Jackson KJ, McIntosh JM, Brunzell DH, Sanjakdar SS, Damaj MI (2009) The role of alpha6-containing nicotinic acetylcholine receptors in nicotine reward and withdrawal. J Pharmacol Exp Ther 331(2):547–554
Jackson A, Bagdas D, Muldoon PP, Lichtman AH, Carroll FI, Greenwald M, Miles MF, Damaj MI (2017) In vivo interactions between alpha7 nicotinic acetylcholine receptor and nuclear peroxisome proliferator-activated receptor-alpha: Implication for nicotine dependence. Neuropharmacology 118:38–45
Jain V, Garg A, Parascandola M, Chaturvedi P, Khariwala SS, Stepanov I (2017) Analysis of alkaloids in areca nut-containing products by liquid chromatography-Tandem mass spectrometry. J Agric Food Chem 65(9):1977–1983
Jaiteh M, Taly A, Henin J (2016) Evolution of pentameric ligand-gated ion channels: pro-loop receptors. PLoS One 11(3):e0151934
Jones SR, Joseph JD, Barak LS, Caron MG, Wightman RM (1999) Dopamine neuronal transport kinetics and effects of amphetamine. J Neurochem 73(6):2406–2414
Jones JD, Comer SD, Metz VE, Manubay JM, Mogali S, Ciccocioppo R, Martinez S, Mumtaz M, Bisaga A (2017) Pioglitazone, a PPARgamma agonist, reduces nicotine craving in humans, with marginal effects on abuse potential. Pharmacol Biochem Behav 163:90–100
Kem WR, Olincy A, Johnson L, Harris J, Wagner BD, Buchanan RW, Christians U, Freedman R (2017) Pharmacokinetic limitations on effects of an alpha7 nicotinic receptor agonist in schizophrenia: randomized trial with an extended release formulation. Neuropsychopharmacology 43 (3):583–589
Khan MS, Bawany FI, Shah SR, Hussain M, Arshad MH, Nisar N (2013) Comparison of knowledge, attitude and practices of betelnut users in two socio-economic areas of Karachi. J Pak Med Assoc 63(10):1319–1325

Koukouli F, Rooy M, Tziotis D, Sailor KA, O'Neill HC, Levenga J, Witte M, Nilges M, Changeux JP, Hoeffer CA, Stitzel JA, Gutkin BS, DiGregorio DA, Maskos U (2017) Nicotine reverses hypofrontality in animal models of addiction and schizophrenia. Nat Med 23(3):347–354

Kumari V, Postma P (2005) Nicotine use in schizophrenia: the self medication hypotheses. Neurosci Biobehav Rev 29(6):1021–1034

Kuryatov A, Lindstrom J (2011) Expression of functional human alpha6beta2beta3∗ acetylcholine receptors in Xenopus laevis oocytes achieved through subunit chimeras and concatamers. Mol Pharmacol 79(1):126–140

Kuryatov A, Luo J, Cooper J, Lindstrom J (2005) Nicotine acts as a pharmacological chaperone to up-regulate human alpha4beta2 acetylcholine receptors. Mol Pharmacol 68(6):1839–1851

Kuryatov A, Berrettini W, Lindstrom J (2011) Acetylcholine receptor (AChR) alpha5 subunit variant associated with risk for nicotine dependence and lung cancer reduces (alpha4beta2) alpha5 AChR function. Mol Pharmacol 79(1):119–125

Lee CH, Ko AM, Yang FM, Hung CC, Warnakulasuriya S, Ibrahim SO, Zain RB, Ko YC (2018) Association of DSM-5 Betel-quid use disorder with oral potentially malignant disorder in 6 Betel-quid endemic Asian populations. JAMA Psychiat 75(3):261–269

Leonard S, Mexal S, Freedman R (2007) Smoking, genetics and schizophrenia: evidence for self medication. J Dual Diagn 3(3-4):43–59

Li MD, Yoon D, Lee JY, Han BG, Niu T, Payne TJ, Ma JZ, Park T (2010) Associations of variants in CHRNA5/A3/B4 gene cluster with smoking behaviors in a Korean population. PLoS One 5 (8):e12183

Lin CC, Tami-Maury I, Ma WF, Lam C, Tsai MH, Lin MT, Li CI, Liu CS, Li TC, Chiu CF, Lu IY, Gritz ER (2017) Social and cultural context of Betel quid consumption in Taiwan and implications for prevention and cessation interventions. Subst Use Misuse 52(5):646–655

Little MA, Papke RL (2015) Betel, the orphan addiction. J Addiction Res Ther 6:130–132

Little MA, Pokhrel P, Murphy KL, Kawamoto CT, Suguitan GS, Herzog TA (2014) Intention to quit betel quid: a comparison of betel quid chewers and cigarette smokers. Oral Health Dental Manag 13(2):512–518

Liu ME, Tsai SJ, Jeang SY, Peng SL, Wu SL, Chen MC, Tsai YL, Yang ST (2011) Varenicline prevents affective and cognitive exacerbation during smoking abstinence in male patients with schizophrenia. Psychiatry Res 190(1):79–84

Liu L, Zhao-Shea R, McIntosh JM, Gardner PD, Tapper AR (2012) Nicotine persistently activates ventral tegmental area dopaminergic neurons via nicotinic acetylcholine receptors containing alpha4 and alpha6 subunits. Mol Pharmacol 81(4):541–548

Lucero LM, Weltzin MM, Eaton JB, Cooper JF, Lindstrom JM, Lukas RJ, Whiteaker P (2016) Differential alpha4(+)/(-)beta2 agonist-binding site contributions to alpha4beta2 nicotinic acetylcholine receptor function within and between isoforms. J Biol Chem 291(5):2444–2459

Luetje CW, Patrick J (1991) Both a- and b-subunits contribute to the agonist sensitivity of neuronal nicotinic acetylcholine receptors. J Neurosci 11(3):837–845

Mallet J, Le Strat Y, Schurhoff F, Mazer N, Portalier C, Andrianarisoa M, Aouizerate B, Berna F, Brunel L, Capdevielle D, Chereau I, D'Amato T, Denizot H, Dubreucq J, Faget C, Gabayet F, Lancon C, Llorca PM, Misdrahi D, Rey R, Roux P, Schandrin A, Urbach M, Vidailhet P, Fond G, Dubertret C, FACE-SZ (FondaMental Academic Centers of Expertise for Schizophrenia) Group (2017) Cigarette smoking and schizophrenia: a specific clinical and therapeutic profile? Results from the FACE-Schizophrenia cohort. Prog Neuropsychopharmacol Biol Psychiatry 79(Pt B):332–339

Mansvelder HD, Keath JR, McGehee DS (2002) Synaptic mechanisms underlie nicotine-induced excitability of brain reward areas. Neuron 33(6):905–919

Markou A, Paterson NE (2001) The nicotinic antagonist methyllycaconitine has differential effects on nicotine self-administration and nicotine withdrawal in the rat. Nicotine Tob Res 3 (4):361–373

Marubio LM, Changeux J (2000) Nicotinic acetylcholine receptor knockout mice as animal models for studying receptor function. Eur J Pharmacol 393(1–3):113–121

Maskos U, Molles BE, Pons S, Besson M, Guiard BP, Guilloux JP, Evrard A, Cazala P, Cormier A, Mameli-Engvall M, Dufour N, Cloez-Tayarani I, Bemelmans AP, Mallet J, Gardier AM, David V, Faure P, Granon S, Changeux JP (2005) Nicotine reinforcement and cognition restored by targeted expression of nicotinic receptors. Nature 436(7047):103–107

Mehrtash H, Duncan K, Parascandola M, David A, Gritz ER, Gupta PC, Mehrotra R, Amer Nordin AS, Pearlman PC, Warnakulasuriya S, Wen CP, Zain RB, Trimble EL (2017) Defining a global research and policy agenda for betel quid and areca nut. Lancet Oncol 18(12):e767–e775

Melis M, Scheggi S, Carta G, Madeddu C, Lecca S, Luchicchi A, Cadeddu F, Frau R, Fattore L, Fadda P, Ennas MG, Castelli MP, Fratta W, Schilstrom B, Banni S, De Montis MG, Pistis M (2013) PPARalpha regulates cholinergic-driven activity of midbrain dopamine neurons via a novel mechanism involving alpha7 nicotinic acetylcholine receptors. J Neurosci 33 (14):6203–6211

Mihalak KB, Carroll FI, Luetje CW (2006) Varenicline is a partial agonist at alpha4beta2 and a full agonist at alpha7 neuronal nicotinic receptors. Mol Pharmacol 70(3):801–805

Millar NS, Gotti C (2009) Diversity of vertebrate nicotinic acetylcholine receptors. Neuropharmacology 56(1):237–246

Miller DK, Sumithran SP, Dwoskin LP (2002) Bupropion inhibits nicotine-evoked [(3)H]overflow from rat striatal slices preloaded with [(3)H]dopamine and from rat hippocampal slices preloaded with [(3)H]norepinephrine. J Pharmacol Exp Ther 302(3):1113–1122

Mogg AJ, Whiteaker P, McIntosh JM, Marks M, Collins AC, Wonnacott S (2002) Methyllycaconitine is a potent antagonist of alpha-conotoxin-MII-sensitive presynaptic nicotinic acetylcholine receptors in rat striatum. J Pharmacol Exp Ther 302(1):197–204

Mokdad AH, Marks JS, Stroup DF, Gerberding JL (2004) Actual causes of death in the United States, 2000. JAMA 291(10):1238–1245

Morel C, Fattore L, Pons S, Hay YA, Marti F, Lambolez B, De Biasi M, Lathrop M, Fratta W, Maskos U, Faure P (2014) Nicotine consumption is regulated by a human polymorphism in dopamine neurons. Mol Psychiatry 19(8):930–936

Moroni M, Zwart R, Sher E, Cassels BK, Bermudez I (2006) alpha4beta2 nicotinic receptors with high and low acetylcholine sensitivity: pharmacology, stoichiometry, and sensitivity to long-term exposure to nicotine. Mol Pharmacol 70(2):755–768

Murphy KL, Herzog TA (2015) Sociocultural factors that affect chewing behaviors among betel nut chewers and ex-chewers on Guam. Hawai'i J Med Publ Health 74(12):406–411

Murray TA, Bertrand D, Papke RL, George AA, Pantoja R, Srinivasan R, Liu Q, Wu J, Whiteaker P, Lester HA, Lukas RJ (2012) alpha7beta2 nicotinic acetylcholine receptors assemble, function, and are activated primarily via their alpha7-alpha7 interfaces. Mol Pharmacol 81 (2):175–188

Nadalin S, Buretic-Tomljanovic A, Rebic J, Plesa I, Sendula Jengic V (2016) An association between the PPARalpha-L162V polymorphism and nicotine dependency among patients with schizophrenia. Compr Psychiatry 70:118–124

Nelson BS, Heischober B (1999) Betel nut: a common drug used by naturalized citizens from India, Far East Asia, and the South Pacific Islands. Ann Emerg Med 34(2):238–243

Nelson ME, Kuryatov A, Choi CH, Zhou Y, Lindstrom J (2003) Alternate stoichiometries of alpha4beta2 nicotinic acetylcholine receptors. Mol Pharmacol 63(2):332–341

Notley C, Ward E, Dawkins L, Holland R (2018) The unique contribution of e-cigarettes for tobacco harm reduction in supporting smoking relapse prevention. Harm Reduct J 15(1):31

O'Neill HC, Wageman CR, Sherman SE, Grady SR, Marks MJ, Stitzel JA (2018) The interaction of the Chrna5 D398N variant with developmental nicotine exposure. Genes Brain Behav 17(7): e12474

Ortells MO, Lunt GG (1995) Evolutionary history of the ligand-gated ion-channel superfamily of receptors. Trends Neurosci 18(3):121–127

Oxenham MF, Locher C, Nguyen LC, Nguyen KT (2002) Identification of Areca catechu (betel nut) residues on the dentitions of bronze age inhabitants of Nui Nap, Northern Vietnam. J Archeol Sci 29:909–915

Pachas GN, Cather C, Pratt SA, Hoeppner B, Nino J, Carlini SV, Achtyes ED, Lando H, Mueser KT, Rigotti NA, Goff DC, Evins AE (2012) Varenicline for smoking cessation in Schizophrenia: safety and effectiveness in a 12-week, open-label trial. J Dual Diagn 8(2):117–125
Palma E, Bertrand S, Binzoni T, Bertrand D (1996) Neuronal nicotinic alpha 7 receptor expressed in Xenopus oocytes presents five putative binding sites for methyllycaconitine. J Physiol 491:151–161
Papke RL (2014) Merging old and new perspectives on nicotinic acetylcholine receptors. Biochem Pharmacol 89(1):1–11
Papke RL, Heinemann SF (1994) The partial agonist properties of cytisine on neuronal nicotinic receptors containing the beta2 subunit. Mol Pharm 45:142–149
Papke RL, Papke JKP (2002) Comparative pharmacology of rat and human alpha7 nAChR conducted with net charge analysis. Br J Pharmacol 137(1):49–61
Papke RL, Stokes C (2010) Working with OpusXpress: methods for high volume oocyte experiments. Methods 51(1):121–133
Papke RL, Boulter J, Patrick J, Heinemann S (1989) Single-channel currents of rat neuronal nicotinic acetylcholine receptors expressed in Xenopus laevis oocytes. Neuron 3:589–596
Papke RL, Dwoskin LP, Crooks PA, Zheng G, Zhang Z, McIntosh JM, Stokes C (2008) Extending the analysis of nicotinic receptor antagonists with the study of alpha6 nicotinic receptor subunit chimeras. Neuropharmacology 54(8):1189–1200
Papke RL, Trocme-Thibierge C, Guendisch D, Abbas Al Rubaiy SA, Bloom SA (2011) Electrophysiological perspectives on the therapeutic use of nicotinic acetylcholine receptor partial agonists. J Pharmacol Exp Ther 337(2):367–379
Papke RL, Stokes C, Muldoon P, Imad Damaj M (2013) Similar activity of mecamylamine stereoisomers in vitro and in vivo. Eur J Pharmacol 720(1-3):264–275
Papke RL, Horenstein NA, Stokes C (2015) Nicotinic activity of arecoline, the psychoactive element of "Betel Nuts", suggests a basis for habitual use and anti-inflammatory activity. PLoS One 10(10):e0140907
Papke RL, Bhattacharyya I, Hatsukami DK, Moe I, Glatman S (2019) Betel nut (areca) and smokeless tobacco use in Myanmar. Subst Use Misuse 54(10):1–10
Patidar KA, Parwani R, Wanjari SP, Patidar AP (2015) Various terminologies associated with areca nut and tobacco chewing: a review. J Oral Maxillofac Pathol 19(1):69–76
Peng C, Engle SE, Yan Y, Weera MM, Berry JN, Arvin MC, Zhao G, McIntosh JM, Chester JA, Drenan RM (2017) Altered nicotine reward-associated behavior following alpha4 nAChR subunit deletion in ventral midbrain. PLoS One 12(7):e0182142
Perkins KA, Karelitz JL, Boldry MC (2017) Nicotine acutely enhances reinforcement from non-drug rewards in humans. Front Psych 8:65
Perry DC, Davila-Garcia MI, Stockmeier CA, Kellar KJ (1999) Increased nicotinic receptors in brains from smokers: membrane binding and autoradiography studies. J Pharmacol Exp Ther 289(3):1545–1552
Picciotto MR, Mineur YS (2014) Molecules and circuits involved in nicotine addiction: the many faces of smoking. Neuropharmacology 76(Pt B):545–553
Picciotto M, Zoli M, Rimondini R, Lena C, Marubio L, Pich E, Fuxe K, Changeux J (1998) Acetylcholine receptors containing the beta2 subunit are involved in the reinforcing properties of nicotine. Nature 391:173–177
Pillai SG, Ge D, Zhu G, Kong X, Shianna KV, Need AC, Feng S, Hersh CP, Bakke P, Gulsvik A, Ruppert A, Lodrup Carlsen KC, Roses A, Anderson W, Rennard SI, Lomas DA, Silverman EK, Goldstein DB, Investigators I (2009) A genome-wide association study in chronic obstructive pulmonary disease (COPD): identification of two major susceptibility loci. PLoS Genet 5(3): e1000421
Pobutsky AM, Neri EI (2012) Betel nut chewing in Hawai'i: is it becoming a public health problem? Historical and socio-cultural considerations. Hawai'i J Med Publ Health 71(1):23–26

Pons S, Fattore L, Cossu G, Tolu S, Porcu E, McIntosh JM, Changeux JP, Maskos U, Fratta W (2008) Crucial role of alpha4 and alpha6 nicotinic acetylcholine receptor subunits from ventral tegmental area in systemic nicotine self-administration. J Neurosci 28(47):12318–12327

Raghavan V, Baruah HK (1958) Arecanut: India's popular masticatory history, chemistry and utilization. Econ Bot 12(4):315–345

Rahman S, Zhang Z, Papke RL, Crooks PA, Dwoskin LP, Bardo MT (2007) Region-specific effects of N,N'-dodecane-1,12-diyl-bis-3-picolinium dibromide on nicotine-induced increase in extracellular dopamine in vivo. Br J Pharmacol 153(4):792–804

Rakhilin S, Drisdel RC, Sagher D, McGehee DS, Vallejo Y, Green WN (1999) alpha-bungarotoxin receptors contain alpha7 subunits in two different disulfide-bonded conformations. J Cell Biol 146(1):203–218

Ramirez-Latorre J, Yu CR, Qu X, Perin F, Karlin A, Role L (1996) Functional contributions of alpha5 subunit to neuronal acetylcholine receptor channels. Nature 380(6572):347–351

Robinson TE, Berridge KC (2003) Addiction. Annu Rev Psychol 54:25–53

Robinson JH, Pritchard WS (1992) The role of nicotine in tobacco use. Psychopharmacology (Berl) 108(4):397–407

Rooney DF (1993) Betel chewing traditions in South-East Asia. Oxford University Press, Kuala Lumpur

Rose JE, Levin ED, Behm FM, Adivi C, Schur C (1990) Transdermal nicotine facilitates smoking cessation. Clin Pharmacol Ther 47(3):323–330

Saccone NL, Wang JC, Breslau N, Johnson EO, Hatsukami D, Saccone SF, Grucza RA, Sun L, Duan W, Budde J, Culverhouse RC, Fox L, Hinrichs AL, Steinbach JH, Wu M, Rice JP, Goate AM, Bierut LJ (2009) The CHRNA5-CHRNA3-CHRNB4 nicotinic receptor subunit gene cluster affects risk for nicotine dependence in African-Americans and in European-Americans. Cancer Res 69(17):6848–6856

Salas R, Orr-Urtreger A, Broide RS, Beaudet A, Paylor R, De Biasi M (2003) The nicotinic acetylcholine receptor subunit alpha 5 mediates short-term effects of nicotine in vivo. Mol Pharmacol 63(5):1059–1066

Salas R, Sturm R, Boulter J, De Biasi M (2009) Nicotinic receptors in the habenulo-interpeduncular system are necessary for nicotine withdrawal in mice. J Neurosci 29(10):3014–3018

Sanjakdar SS, Maldoon PP, Marks MJ, Brunzell DH, Maskos U, McIntosh JM, Bowers MS, Damaj MI (2015) Differential roles of alpha6beta2* and alpha4beta2* neuronal nicotinic receptors in nicotine- and cocaine-conditioned reward in mice. Neuropsychopharmacology 40(2):350–360

Schlaepfer IR, Hoft NR, Collins AC, Corley RP, Hewitt JK, Hopfer CJ, Lessem JM, McQueen MB, Rhee SH, Ehringer MA (2008) The CHRNA5/A3/B4 gene cluster variability as an important determinant of early alcohol and tobacco initiation in young adults. Biol Psychiatry 63 (11):1039–1046

Schuster RM, Pachas GN, Stoeckel L, Cather C, Nadal M, Mischoulon D, Schoenfeld DA, Zhang H, Ulysse C, Dodds EB, Sobolewski S, Hudziak V, Hanly A, Fava M, Evins AE (2018) Phase IIb trial of an alpha7 nicotinic receptor partial agonist with and without nicotine patch for withdrawal-associated cognitive deficits and tobacco abstinence. J Clin Psychopharmacol 38(4):307–316

Sciaccaluga M, Moriconi C, Martinello K, Catalano M, Bermudez I, Stitzel JA, Maskos U, Fucile S (2015) Crucial role of nicotinic alpha5 subunit variants for Ca2+ fluxes in ventral midbrain neurons. FASEB J 29(8):3389–3398

Seguela P, Wadiche J, Dinely-Miller K, Dani JA, Patrick JW (1993) Molecular cloning, functional properties and distribution of rat brain alpha 7: a nicotinic cation channel highly permeable to calcium. J Neurosci 13(2):596–604

Sivilotti LG, McNeil DK, Lewis TM, Nassar MA, Schoepfer R, Colquhoun D (1997) Recombinant nicotinic receptors, expressed in Xenopus oocytes, do not resemble native rat sympathetic ganglion receptors in single-channel behaviour. J Physiol 500(Pt 1):123–138

Skok VI (2002) Nicotinic acetylcholine receptors in autonomic ganglia. Auton Neurosci 97(1):1–11

Slemmer JE, Martin BR, Damaj MI (2000) Bupropion is a nicotinic antagonist. J Pharmacol Exp Ther 295(1):321–327
Smit AB, Syed NI, Schaap D, van Minnen J, Klumperman J, Kits KS, Lodder H, van der Schors RC, van Elk R, Sorgedrager B, Brejc K, Sixma TK, Geraerts WP (2001) A glia-derived acetylcholine-binding protein that modulates synaptic transmission. Nature 411(6835):261–268
Smith TT, Hatsukami DK, Benowitz NL, Colby SM, McClernon FJ, Strasser AA, Tidey JW, White CM, Donny EC (2018) Whether to push or pull? Nicotine reduction and non-combusted alternatives – two strategies for reducing smoking and improving public health. Prev Med 117:8–14
Stevens VL, Bierut LJ, Talbot JT, Wang JC, Sun J, Hinrichs AL, Thun MJ, Goate A, Calle EE (2008) Nicotinic receptor gene variants influence susceptibility to heavy smoking. Cancer Epidemiol Biomarkers Prev 17(12):3517–3525
Stolerman IP, Shoaib M (1991) The neurobiology of tobacco addiction. Trends Pharmacol Sci 12:467–473
Tammimaki A, Herder P, Li P, Esch C, Laughlin JR, Akk G, Stitzel JA (2012) Impact of human D398N single nucleotide polymorphism on intracellular calcium response mediated by alpha3beta4alpha5 nicotinic acetylcholine receptors. Neuropharmacology 63(6):1002–1011
Tapper AR, McKinney SL, Nashmi R, Schwarz J, Deshpande P, Labarca C, Whiteaker P, Marks MJ, Collins AC, Lester HA (2004) Nicotine activation of alpha4∗ receptors: sufficient for reward, tolerance, and sensitization. Science 306(5698):1029–1032
Threlfell S, Lalic T, Platt NJ, Jennings KA, Deisseroth K, Cragg SJ (2012) Striatal dopamine release is triggered by synchronized activity in cholinergic interneurons. Neuron 75(1):58–64
Tregellas JR, Olincy A, Johnson L, Tanabe J, Shatti S, Martin LF, Singel D, Du YP, Soti F, Kem WR, Freedman R (2010) Functional magnetic resonance imaging of effects of a nicotinic agonist in Schizophrenia. Neuropsychopharmacology 35(4):938–942
Uhl GR, Hall FS, Sora I (2002) Cocaine, reward, movement and monoamine transporters. Mol Psychiatry 7(1):21–26
Uteshev VV, Meyer EM, Papke RL (2002) Activation and inhibition of native neuronal alpha-bungarotoxin-sensitive nicotinic ACh receptors. Brain Res 948(1-2):33–46
Vailati S, Moretti M, Longhi R, Rovati GE, Clementi F, Gotti C (2003) Developmental expression of heteromeric nicotinic receptor subtypes in chick retina. Mol Pharmacol 63(6):1329–1337
Wada E, Wada K, Boulter J, Deneris E, Heinemann S, Patrick J, Swanson LW (1989) Distribution of alpha2, alpha3, alpha4, and beta2 neuronal nicotinic receptor subunit mRNAs in the central nervous system: a hybridization histochemical study in the rat. J Comp Neurol 284:314–335
Walling D, Marder SR, Kane J, Fleischhacker WW, Keefe RS, Hosford DA, Dvergsten C, Segreti AC, Beaver JS, Toler SM, Jett JE, Dunbar GC (2016) Phase 2 trial of an alpha-7 nicotinic receptor agonist (TC-5619) in negative and cognitive symptoms of Schizophrenia. Schizophr Bull 42(2):335–343
Walters CL, Brown S, Changeux JP, Martin B, Damaj MI (2006) The beta2 but not alpha7 subunit of the nicotinic acetylcholine receptor is required for nicotine-conditioned place preference in mice. Psychopharmacology (Berl) 184(3–4):339–344
Wang J, Lindstrom J (2017) Orthosteric and allosteric potentiation of heteromeric neuronal nicotinic acetylcholine receptors. Br J Pharmacol 175(11):1805–1821
Wang F, Gerzanich V, Wells GB, Anand R, Peng X, Keyser K, Lindstrom J (1996) Assembly of human neuronal nicotinic receptor alpha5 subunits with alpha3, beta2, and beta4 subunits. J Biol Chem 271(30):17656–17665
Wang Y, Lee JW, Oh G, Grady SR, McIntosh JM, Brunzell DH, Cannon JR, Drenan RM (2014) Enhanced synthesis and release of dopamine in transgenic mice with gain-of-function alpha6∗ nAChRs. J Neurochem 129(2):315–327
Williams DK, Stokes C, Horenstein NA, Papke RL (2011) The effective opening of nicotinic acetylcholine receptors with single agonist binding sites. J Gen Physiol 137(4):369–384
Williams DK, Peng C, Kimbrell MR, Papke RL (2012a) The intrinsically low open probability of alpha7 nAChR can be overcome by positive allosteric modulation and serum factors leading to

the generation of excitotoxic currents at physiological temperatures. Mol Pharmacol 82 (4):746–759

Williams JM, Anthenelli RM, Morris CD, Treadow J, Thompson JR, Yunis C, George TP (2012b) A randomized, double-blind, placebo-controlled study evaluating the safety and efficacy of varenicline for smoking cessation in patients with schizophrenia or schizoaffective disorder. J Clin Psychiatry 73(5):654–660

Zoli M, Pucci S, Vilella A, Gotti C (2018) Neuronal and extraneuronal nicotinic acetylcholine receptors. Curr Neuropharmacol 16(4):338–349

Behavioral and Molecular Basis of Cholinergic Modulation of Pain: Focus on Nicotinic Acetylcholine Receptors

Wisam Toma, Esad Ulker, Mashael Alqasem, Shakir D. AlSharari, J. Michael McIntosh, and M. Imad Damaj

Contents

Abstract Nicotinic acetylcholine receptors (nAChRs) have emerged as a novel therapeutic strategy for pain and inflammatory disorders. In particular, α4β2*, α7, and α9α10 nAChR subtypes have been investigated as potential targets to treat pain. The nAChRs are distributed on the pain transmission pathways, including central and peripheral nervous systems and immune cells as well. Several agonists for α4β2* nAChR subtypes have been investigated in multiple animal pain models with promising results. However, studies in human indicated a narrow therapeutic window for α4β2* agonists. Furthermore, animal studies suggest that using agonists

W. Toma, E. Ulker, and M. I. Damaj (✉)
Department of Pharmacology and Toxicology, Virginia Commonwealth University, Richmond, VA, USA
e-mail: m.damaj@vcuhealth.org

M. Alqasem
Department of Pharmacology and Toxicology, College of Pharmacy, King Saud University, Riyadh, Saudi Arabia

S. D. AlSharari
Department of Pharmacology and Toxicology, Virginia Commonwealth University, Richmond, VA, USA

Department of Pharmacology and Toxicology, College of Pharmacy, King Saud University, Riyadh, Saudi Arabia

J. M. McIntosh
Departments of Psychiatry and Biology, University of Utah, Salt Lake City, UT, USA

Curr Topics Behav Neurosci (2020) 45: 153–166
https://doi.org/10.1007/7854_2020_135
Published Online: 29 May 2020

for α7 nAChR subtype and antagonists for α9α10 nAChR subtypes are potential novel therapies for chronic pain management, including inflammatory and neuropathic pain. More recently, alternative nAChRs ligands such as positive allosteric modulators and silent agonists have shown potential to develop into new treatments for chronic pain.

Keywords Inflammatory pain · Neuropathic pain · Nicotinic acetylcholine receptors · α4β2 nAChRs · α7 nAChR · α9α10 nAChRs

1 Introduction

Effective and safe treatment of pain remains one of the most significant challenges in medicine. Pain affects more than 100 million Americans (Institute of Medicine (US) Committee on Advancing Pain Research, Care, and Education 2011) at the cost of greater than $635 billion a year (Gaskin and Richard 2012). The current opioid crisis, partly fueled by prescription opioids for pain control, has highlighted the enormous challenge to the medical community of managing chronic pain. Relatively few options are available for therapy, including nonsteroidal anti-inflammatory drugs (NSAIDs), steroids, opioids, gabapentinoids, reuptake inhibitors, and local anesthetic agents. Each of these classes of drugs possesses varying degrees of clinical efficacy in specific patient groups, and all are associated with untoward side effects that undermine utility. There is an urgent need to discover new drugs to treat acute and chronic pain that are effective and also safe. While over the last several years, there has been a major push in the identification of new targets for pain relief, this review will focus on the recent progress in our understanding of the cholinergic system as a target for pain modulation. More specifically, we will expand on the new advances in our understanding of the cholinergic nicotinic system modulation of pain at the level of neuronal and nonneuronal systems. Acetylcholine (ACh) is one of the most abundant neurotransmitters in both the central and peripheral nervous system. It acts primarily on two types of receptors: muscarinic acetylcholine receptors (mAChRs) and nicotinic acetylcholine receptors (nAChRs). mAChRs are G-protein-coupled receptors, and there are five different subtypes: M1, M2, M3, M4, and M5. M1, M3, and M5 act through Gq increasing intracellular calcium and M2 and M4 through Gi, inhibiting the formation of cAMP. Several pieces of evidence support the notion that mAChRs play an important role in pain modulation. These receptors can be found throughout the peripheral and central nervous system. The role of ACh and mAChRs in pain transmission was reviewed recently in an excellent paper by (Naser and Kuner 2018).

2 nAChRs in Pain Modulation

Nicotinic receptors are members of the pentameric ligand-gated ion channels Cys-loop superfamily consisting of five subunits surrounding an ion pore. nAChRs are engaged throughout the peripheral and central nervous systems in signal transduction of ACh-mediated signals. Furthermore, nAChRs play a significant role in pain mediation and several reactions evoked by nicotine (Decker et al. 2004). Multiple nAChRs were classified, including homomeric and heteromeric combinations of α2–α10 and β2–β4 subunits forming different combinations of α and β subunits or homomeric forms expressing only α subunits (Gotti et al. 2006). The most common nAChR subtypes are the heteromeric α4β2∗ and homomeric α7 receptors. We will concentrate in this review on the role of subtypes α4β2∗, α7, and α9 α10 in pain modulation.

3 The α4β2∗ nAChRs in Pain Modulation

Although nicotine shows reinforcing properties, it has been reported that nicotine also displays an important pain-relieving properties both in human and rodents studies as well, especially through the α4β2∗ nAChRs (Damaj et al. 2014). The lack of α4 or β2 nAChR subunit has been shown to reduce sensitivity to the antinociceptive properties of nicotinic compounds in acute pain testing in genetically mutated animals (Marubio et al. 1999). In addition, nicotine has been shown to alleviate the mechanical hypersensitivity in chronic constriction injury (CCI), a mouse model of peripheral neuropathy (Bagdas et al. 2018). Although nicotine short-half life and side effects restrict its implication for pain treatment, it brings to the attention that the antinociceptive properties of nicotine shadow a light into underlying mechanisms of α4β2∗ nAChRs in pain and promote developing better α4β2∗ therapeutic ligands. The α4β2∗ nAChRs are heteromeric and formed through the combination of α4 and β2 subunits or occasionally additional subunits of either α or β subunits to form functional receptors (represented as α4β2∗ in which the asterisk denotes that another nAChR subunits either α or β could be part of the receptor pentamer). Additionally, the α4β2 receptors are known to show different sensitivity to agonist, i.e., high sensitive (HS), where two α and three β subunits respond to agonist application while in case of presence of three α and two β subunits are considered low sensitive (LS) to agonist application (for further details, see review by Hendrickson et al. 2013). The α4β2 nAChRs are expressed on neuronal cells centrally (spinal and supraspinal) (Nashmi and Lester 2006; Posadas et al. 2013; Zoli et al. 2018) and on nonneuronal cells such as macrophages and microglia as well (Saika et al. 2015). The α4β2 nAChR subtype could be modulated either through binding by an agonist or by an allosteric modulator(s) to produce analgesia/antinociception. For example, several agonists have been tested for potential pain relief in animal models. Metanicotine, a selective agonist for α4β2 nAChRs has been reported to induce antinociception in heat and pressure tests in a postoperative mouse model (Rowley et al. 2008). Additionally, it has been reported that epibatidine, a nonselective agonist for α4β2 nAChRs produced antinociception in several acute tonic, inflammatory, and neuropathic pain assays in rodents (Badio and Daly 1994; Damaj et al. 1998; Curzon et al. 1998; Kesingland et al. 2000). Similarly, ABT-594, a moderately selective α4β2 nAChRs agonist, has been shown to have antinociceptive properties in acute (hot plate and tail flick) and inflammatory pain assays in rats (Boyce et al. 2000; Kesingland et al. 2000) as well as in multiple neuropathic pain models in rodents. For example, Bannon et al. (1998) showed that ABT-594 produced a reversal of mechanical hypersensitivity and tactile hyperalgesia in sciatic nerve ligation and diabetic neuropathic pain models in rats. Likewise, Lynch et al. (2005) described that ABT-594 produced a reversal of mechanical hypersensitivity in the rat model of chemotherapy-induced peripheral neuropathy. Also, administration of 5-Iodo-A-85380, an agonist for α4β2 nAChRs, showed antinociception in the Hargreaves test in rats (Rueter et al. 2000). Recently, Kiguchi et al. (2018) reported that the systemic administration of TC-2559, an agonist for α4β2 nAChRs, was shown to decrease the upregulated IL-1β level in the sciatic nerve following partial sciatic nerve injury in rats and lowered the elevated IL-1β level in cultured macrophages following incubation with lipopolysaccharide. Additionally, perineural administration of TC-2559 alleviated mechanical hypersensitivity during early (0–3 days) and late (21–24 days) post partial sciatic nerve ligation injury. For example, sazetidine-A, varenicline, attenuated licking pain-like behavior of the formalin test in mice (AlSharari et al. 2012; Bagdas et al. 2015a) and NS3956 in rats (Rode et al. 2012). A growing body of evidence suggests that positive allosteric modulators are capable of enhancing the pharmacologic effect of the agonists for the α4β2 nAChRs. The low sensitivity, selective positive allosteric modulator NS9283 produced an antinociceptive effect in the formalin test in rats when co-administered with NS3956, a partial agonist for the α4β2 nAChRs. Additionally, NS9283 has been shown to alleviate carrageenan-induced thermal hyperalgesia, reverses the mechanical hypersensitivity in a rat paw skin incision model of postoperative pain, and reduces monoiodo-acetate induced knee joint pain when co-administered with ABT-594 (Zhu et al. 2011). Likewise, NS9283 potentiated the attenuation of mechanical hypersensitivity of ABT-594 in spinal nerve ligation, a model of neuropathic pain in rats (Lee et al. 2011). Desformylflustrabromine, another α4β2 low sensitivity PAM, potentiated nicotine's mechanical antihypersensitivity in a mouse model of peripheral neuropathy (Bagdas et al. 2018). Similarly, desformylflustrabromine alleviated pain-like behavior in CD-1 mice in both formalin and acetic acid-induced writhing response tests (Weggel and Pandya 2019).

To date, the clinical utility of α4β2∗ agonists as analgesics has been limited. ABT-594, the relatively selective α4β2∗ agonist, have shown efficacy in treating diabetic neuropathic pain in a phase II trial but was plagued by side effects including dizziness, nausea, vomiting, and unpleasant dreams (Rowbotham et al. 2009). However, when ABT-894, a highly selective α4β2∗ agonist, was tested in patients with diabetic peripheral neuropathic pain, it was well tolerated but failed to show significant analgesic efficacy (Rowbotham et al. 2012). These results suggest that the direct activation of the α4β2* subtype nicotinic receptor may modulate pain transmission, but the identity of the α4β2* nAChR subtype(s) mediating the analgesic properties of nicotinic agonists is still unclear. In that regard, α5 and α6 nicotinic subunits are accessory subunits that can form functional receptors when co-expressed with α4β2 to form α4β2α5 or α4β2α6 subtypes (Brown et al. 2007). A possible role for the α5 nAChR subunit in the processing of nociceptive information in neuropathic pain has been suggested (Vincler and Eisenach 2004), and nicotine's antinociceptive properties are reduced or absent in α5 subunit knockout mice (Bagdas et al. 2015a). In addition, the expression levels of CHRNA6, which encodes the α6 nicotinic subunit, were highly associated with mechanical allodynia in a chronic neuropathic pain mouse model. Furthermore, mechanical allodynia associated with neuropathic and inflammatory injuries is significantly altered in α6* null mutants and that α6* but not α4* nicotinic receptors are absolutely required for peripheral and/or spinal nicotine antinociception in these models (Wieskopf et al. 2015).

4 The α7 nAChRs in Pain Modulation

The α7 nAChR subtype has a number of unique physiological and pharmacological properties that distinguish it from other nicotinic subtypes, including a high permeability to calcium, rapid and reversible desensitization, and pronounced inward rectification (Séguéla et al. 1993). α7 nAChR subtype is activated by ACh and also selectively activated by choline, therefore ideally suited to respond to manifestly different kinds of signals: targeted transmission, localized tissue damage, and paracrine signals. α7 nAChR subtype is expressed in supraspinal and spinal pain transmission pathways (Wada et al. 1989; Cordero-Erausquin et al. 2004). Autoradiographic analyses showed that α7 nAChR binding sites were diverse within the substantia gelatinosa in the rat (Hunt and Schmidt 1978) and human (Gillberg and Aquilonius 1985) spinal cord, and these sites were decreased following dorsal rhizotomy (Gillberg and Wiksten 1986). α7 nAChR subtype is also expressed on macrophages and other types of immune cells (Tracey 2002; Wang et al. 2005). It has been suggested that activation of α7 nAChR subtype causes a downregulation of proinflammatory cytokine synthesis and prevention of tissue damage (Tracey 2002; De Rosa et al. 2009) representing a "cholinergic anti-inflammatory pathway" for modulation of the immune system. Supporting this concept, several preclinical studies have confirmed the therapeutic potential of targeting α7 nAChRs-mediated anti-inflammatory effects (Tracey 2002; de Jonge et al. 2007; De Rosa et al. 2009). For example, α7 nAChR knockout mice showed a significant increase in the incidence and severity of arthritis (van Maanen et al. 2010). Also, a marked increase in edema, hyperalgesia, and allodynia associated with intraplantar complete Freund's adjuvant injection was observed in α7 knockout mice compared with wild-type littermates (AlSharari et al. 2013). Neuronal and dorsal root ganglia expression of α7 nAChR has been found to be significantly lowered in experimental models of inflammatory or neuropathic pain (Hoffmeister et al. 2011; Di Cesare et al. 2014b). In contrast, expression of α7 nAChR on macrophages was upregulated (Albuquerque et al. 2009; Hoffmeister et al. 2011; Khan et al. 2012). In addition, selective α7 nAChR agonists such as PHA-543613, JN403, and AR-R17779 were shown to be active in tonic and chronic inflammatory and neuropathic pain models in rodents (Damaj et al. 2000; Medhurst et al. 2008; Feuerbach et al. 2009; van Maanen et al. 2010; Marrero et al. 2011; Loram et al. 2012; Freitas et al. 2013b). However, the rapid desensitization of α7 nAChR, a process that occurs in milliseconds, raises concerns about the long-term administration of this class of compounds. In addition, other limitations to the development α7 nAChR agonists for human use such as receptor selectivity issues (cross-reactivity with 5-HT_3 receptors, which have high homology with α7 nAChRs binding sites and possibly related to adverse effects seen in clinical trials with α7 agonists), overactivation and rapid desensitization of the receptor persist. Finally, α7 agonists seem to possess a narrow window of antinociceptive effect in vivo, as reflected by a U-shaped dose-effect curve in the formalin test in mice (Freitas et al. 2013a). These limitations led to the development of compounds that modulate α7 nAChR function by binding to allosteric sites instead of the orthosteric site that binds agonists and antagonists. α7 nAChR-positive allosteric modulators (PAMs) are compounds that can potentiate α7 currents in the presence of an endogenous agonist such as acetylcholine and choline. These PAMs were classified as type I and type II based on their electrophysiological properties. Type I PAMs increase agonist response with little or no effect on desensitization of α7 nAChRs, whereas type II PAMs increase agonist response and slow down the apparent desensitization profile of the agonist response (Hurst et al. 2005). Both PAM types have been tested in vivo for their efficacy in animal models of inflammatory and neuropathic pain. For example, it has been shown that PNU-120596, an α7 selective type II PAM, reversed mechanical allodynia in rat and mouse chronic inflammatory and neuropathic pain models (Munro et al. 2012; Freitas et al. 2013b). Further, PNU-120596, TQS, 3-furan-2-yl-*N*-p-tolyl-acrylamide, and 2,4,2′,5′-tetrahydroxychalcone, all type II α7 PAMs, were shown to possess antinociceptive and anti-inflammatory effects in tonic and chronic inflammatory pain models in rodents (Munro et al. 2012; Bagdas et al. 2015b; Abbas and Rahman 2016; Balsera et al. 2018). Interestingly, GAT107, a new α7 selective dual allosteric agonist-PAM, showed anti-inflammatory and antinociceptive properties in several chronic pain models in mice (Bagdas et al. 2016).

As mentioned above, the conformation of α7 nAChR subtype is regulated by the binding of agonists and allosteric modulators (Williams et al. 2011, 2012). An additional class of ligands identified as "silent agonists" was recently described (Chojnacka et al. 2013; Papke et al. 2014). Silent agonists are relatively inefficient ligands at inducing the open-channel state of α7 nAChRs and are far more effective at stabilizing agonist-dependent non-conducting states, traditionally referred to as "desensitized." These agonists do not activate the expressed α7 ion channel under normal conditions. For instance, the silent agonist NS6740 stimulated no detectable currents upon application in oocytes expressing α7 nAChRs but reduced time spent paw licking in the formalin test and attenuated mechanical allodynia in a neuropathic pain model after systemic administration via an unknown signaling mechanism (Papke et al. 2015). In addition, the novel α7-selective ligand PMP-072 (also called R-47) (Clark et al. 2014) suppressed inflammation in asthma and arthritis models (van Maanen et al. 2015) and prevented chemotherapy-induced peripheral neuropathy in mice (Toma et al. 2019). Similar results have been reported for the new silent agonist CF3-diEPP, which was found to be active in relieving neuropathic and inflammatory pain (Quadri et al. 2018). Furthermore, the silent agonist NS6740 effectively reduced acetic acid-induced conditioned place aversion (Papke et al. 2015). Collectively, this new class of α7 nAChR ligands challenges the traditional notion that receptors in non-conducting states are functionally unimportant (Williams et al. 2011). These ligand-bound non-conducting states of α7 nAChR likely regulate intracellular signal transduction processes in both neuronal and nonneuronal cells.

CHRFAM7A is a uniquely human partial gene duplication of CHRNA7 (the gene for α7 nAChRs) that is missing the first 4 exons, which code for binding loops A and D, beta-strands 1–4, the main immunogenic region, and the important loop 2 close to the membrane (Sinkus et al. 2015). Co-expression of this gene with full-length α7 has dominant-negative effects on channel function, although interestingly these can be partially reversed by the PAM PNU-120596 (Araud et al. 2011), suggesting that CHRFAM7A promotes the stabilization of a desensitized conformational state. There are deletion polymorphisms in this gene associated with neurological disorders including inflammation, and it has also been implicated in cholinergic control of immune function (de Lucas-Cerrillo et al. 2011). It may also contribute to the unique pharmacology of α7-mediated control of inflammation (Costantini et al. 2019).

Traditionally, it is known that α7 nAChR subtype composed of five identical α7 subunits to form a functional homomeric α7 nAChR subtype. However, it has been recently reported that α7 subunit can also associate with β2 subunit to form α7β2 heteromeric nAChRs (Wu et al. 2016). However, little is known about the role of heteromeric α7β2 nAChRs in pain regulation.

5 The α9/α9α10 nAChRs in Pain Modulation

It has been reported that α9 nAChR subunit can be combined into a homopentamer with other α9 subunits, as well as a heteropentamer with α10 nAChR subunits, to form functional nAChRs. Furthermore, it has been demonstrated that α9 and α10 subunits are expressed in the peripheral nervous system (Gotti et al. 2006), but evidence for brain expression is lacking or controversial (Elgoyhen et al. 2001; Lykhmus et al. 2017; Morley et al. 2018). α9/α10-containing nAChRs are also present in immune cells where they may modulate neuroinflammation (Fujii et al. 2017; Liu et al. 2017; Grau et al. 2018). Also, reports have shown that acetylcholine activates α9α10 nAChRs; on the other hand, nicotine acts as an antagonist by inhibiting ACh-evoked currents (Elgoyhen et al. 1994, 2001).

It has been reported that α-conotoxin family is a class of short peptides derived from the venom of *Conus* marine snails and that some of the compounds isolated from this class of drugs are potent antagonists of the α9α10 nAChRs which could be used to mitigate pain and to reduce inflammation (Hone and McIntosh 2018; Abraham and Lewis 2018). The α-conotoxin RgIA, a selective mammalian α9α10 antagonist, has been shown to reverse mechanical allodynia and hyperalgesia in the chronic constriction injury model (Vincler et al. 2006; McIntosh et al. 2009). In addition, it has been demonstrated that treatment with RgIA decreased the loss of sciatic nerve fibers and myelin and reduced macrophage infiltration consistent with disease-modifying properties (Vincler et al. 2006; Di Cesare et al. 2014a). Furthermore, RgIA was effective at reducing oxaliplatin-induced neuropathic pain (Pacini et al. 2016). However, RgIA peptide is 300-fold less potent at human α9α10 nAChRs (Azam and McIntosh 2012) and was also reported to have agonist activity on GABAB receptors, proposed to account for its analgesic activity (Callaghan et al. 2008). Structure-activity studies led to the development of an RgIA analog, RgIA4, that has a high affinity for both the rat and human α9α10 nAChRs and lacks GABAB and opioid receptor activity. RgIA4 has been shown to prevent oxaliplatin-induced neuropathic pain in rats and wild-type but not α9 subunit null mice (Romero et al. 2017). Remarkably, the therapeutic effect of RgIA4 persists for at least 3 weeks after cessation of drug treatment consistent with a disease-modifying effect (Christensen et al. 2017). Vc1.1 and analogs which are another α-conotoxins have been found to produce analgesic activity in neuropathic and visceral pain models (Satkunanathan et al. 2005; Napier et al. 2012; Sadeghi et al. 2017; Castro et al. 2018). Also, it has been reported that Vc1.1 successfully passed Phase I human clinical trials, but did not continue past Phase 2A after in vitro data indicated that Vc1.1 is much less potent on and selective for the human α9α10 nAChR compared to the rat α9α10 nAChR (Azam and McIntosh 2012).

GeXIVA, an αO-conotoxin which is an α9α10 antagonist peptide, structurally unrelated to the above conotoxins was also found to be analgesic in chronic constriction injury and chemotherapy-induced neuropathic pain models (Luo et al. 2015; Wang et al. 2019). Furthermore, ZZ1-61c, a bis-azaaromatic quaternary ammonium analog of nicotine, has been demonstrated to produce prevention and reversal of chemotherapy-induced mechanical allodynia when administered in combination with or after vincristine (Wala et al. 2012). In addition, treatment with ZZ1-61c was found not transient, suggesting that acute administration of ZZ1-61c elicited lasting pharmacological changes. ZZ-204-G, a tetrakis-quaternary ammonium α9α10 antagonist was also analgesic in formalin and chronic constriction nerve injury models of neuropathic pain (Holtman et al. 2011).

6 Conclusions

The continuous research and characterization of nAChR modulators including but not limited to agonists, silent agonists, PAMs, and antagonists have opened new avenues to explore and to develop better therapies for pain management.

Acknowledgment This work was supported under NIH grants 1R01CA206028-01 and R01CA219637-01.

References

Abbas M, Rahman S (2016) Effects of alpha-7 nicotinic acetylcholine receptor positive allosteric modulator on lipopolysaccharide-induced neuroinflammatory pain in mice. Eur J Pharmacol 783:85–91. https://doi.org/10.1016/j.ejphar.2016.05.003

Abraham N, Lewis RJ (2018) Neuronal nicotinic acetylcholine receptor modulators from cone snails. Mar Drugs 16:208. https://doi.org/10.3390/md16060208

Albuquerque EX, Pereira EFR, Alkondon M, Rogers SW (2009) Mammalian nicotinic acetylcholine receptors: from structure to function. Physiol Rev 89:73–120. https://doi.org/10.1152/physrev.00015.2008

AlSharari SD, Carroll FI, McIntosh JM, Damaj MI (2012) The antinociceptive effects of nicotinic partial agonists varenicline and sazetidine-A in murine acute and tonic pain models. J Pharmacol Exp Ther 342:742–749. https://doi.org/10.1124/jpet.112.194506

AlSharari SD, Freitas K, Damaj MI (2013) Functional role of alpha7 nicotinic receptor in chronic neuropathic and inflammatory pain: studies in transgenic mice. Biochem Pharmacol 86:1201–1207. https://doi.org/10.1016/j.bcp.2013.06.018

Araud T, Graw S, Berger R et al (2011) The chimeric gene CHRFAM7A, a partial duplication of the CHRNA7 gene, is a dominant negative regulator of α7∗nAChR function. Biochem Pharmacol 82:904–914

Azam L, McIntosh JM (2012) Molecular basis for the differential sensitivity of rat and human α9α10 nAChRs to α-conotoxin RgIA. J Neurochem 122:1137–1144. https://doi.org/10.1111/j.1471-4159.2012.07867.x

Badio B, Daly JW (1994) Epibatidine, a potent analgetic and nicotinic agonist. Mol Pharmacol 45:563–569

Bagdas D, AlSharari SD, Freitas K et al (2015a) The role of alpha5 nicotinic acetylcholine receptors in mouse models of chronic inflammatory and neuropathic pain. Biochem Pharmacol 97:590–600. https://doi.org/10.1016/j.bcp.2015.04.013

Bagdas D, Targowska-Duda KM, López JJ et al (2015b) The antinociceptive and antiinflammatory properties of 3-furan-2-yl-N-p-tolyl-acrylamide, a positive allosteric modulator of α7 nicotinic acetylcholine receptors in mice. Anesth Analg 121:1369–1377. https://doi.org/10.1213/ANE.0000000000000902

Bagdas D, Wilkerson JL, Kulkarni A et al (2016) The α7 nicotinic receptor dual allosteric agonist and positive allosteric modulator GAT107 reverses nociception in mouse models of inflammatory and neuropathic pain. Br J Pharmacol 173:2506–2520. https://doi.org/10.1111/bph.13528

Bagdas D, Ergun D, Jackson A et al (2018) Allosteric modulation of α4β2∗ nicotinic acetylcholine receptors: Desformylflustrabromine potentiates antiallodynic response of nicotine in a mouse model of neuropathic pain. Eur J Pain 22:84–93. https://doi.org/10.1002/ejp.1092

Balsera B, Mulet J, Sala S et al (2018) Amino acid and peptide prodrugs of diphenylpropanones positive allosteric modulators of α7 nicotinic receptors with analgesic activity. Eur J Med Chem 143:157–165. https://doi.org/10.1016/j.ejmech.2017.10.083

Bannon AW, Decker MW, Kim DJB et al (1998) ABT-594, a novel cholinergic channel modulator, is efficacious in nerve ligation and diabetic neuropathy models of neuropathic pain. Brain Res. https://doi.org/10.1016/S0006-8993(98)00596-4

Boyce S, Webb JK, Shepheard SL et al (2000) Analgesic and toxic effects of ABT-594 resemble epibatidine and nicotine in rats. Pain 85:443–450. https://doi.org/10.1016/S0304-3959(99)00303-6

Brown RWB, Collins AC, Lindstrom JM, Whiteaker P (2007) Nicotinic alpha5 subunit deletion locally reduces high-affinity agonist activation without altering nicotinic receptor numbers. J Neurochem 103:204–215. https://doi.org/10.1111/j.1471-4159.2007.04700.x

Callaghan B, Haythornthwaite A, Berecki G et al (2008) Analgesic alpha-conotoxins Vc1.1 and Rg1A inhibit N-type calcium channels in rat sensory neurons via GABAB receptor activation. J Neurosci 28:10943–10951. https://doi.org/10.1523/JNEUROSCI.3594-08.2008

Castro J, Grundy L, Deiteren A et al (2018) Cyclic analogues of α-conotoxin Vc1.1 inhibit colonic nociceptors and provide analgesia in a mouse model of chronic abdominal pain. Br J Pharmacol 175:2384–2398. https://doi.org/10.1111/bph.14115

Chojnacka K, Papke RL, Horenstein NA (2013) Synthesis and evaluation of a conditionally-silent agonist for the α7 nicotinic acetylcholine receptor. Bioorg Med Chem Lett 23:4145–4149. https://doi.org/10.1016/j.bmcl.2013.05.039

Christensen SB, Hone AJ, Roux I et al (2017) RgIA4 potently blocks mouse α9α10 nAChRs and provides long lasting protection against Oxaliplatin-induced cold Allodynia. Front Cell Neurosci 11:219. https://doi.org/10.3389/fncel.2017.00219

Clark RB, Lamppu D, Libertine L et al (2014) Discovery of novel 2-((pyridin-3-yloxy)methyl) piperazines as α7 nicotinic acetylcholine receptor modulators for the treatment of inflammatory disorders. J Med Chem 57:3966–3983. https://doi.org/10.1021/jm5004599

Cordero-Erausquin M, Pons S, Faure P, Changeux JP (2004) Nicotine differentially activates inhibitory and excitatory neurons in the dorsal spinal cord. Pain 109:308–318. https://doi.org/10.1016/j.pain.2004.01.034

Costantini TW, Chan TW, Cohen O et al (2019) Uniquely human CHRFAM7A gene increases the hematopoietic stem cell reservoir in mice and amplifies their inflammatory response. Proc Natl Acad Sci U S A 116:7932–7940. https://doi.org/10.1073/pnas.1821853116

Curzon P, Nikkel AL, Bannon AW et al (1998) Differences between the antinociceptive effects of the cholinergic channel activators A-85380 and (+/−)-epibatidine in rats. J Pharmacol Exp Ther 287:847–853

Damaj MI, Fei-Yin M, Dukat M et al (1998) Antinociceptive responses to nicotinic acetylcholine receptor ligands after systemic and intrathecal administration in mice. J Pharmacol Exp Ther 284:1058–1065

Damaj MI, Meyer EM, Martin BR (2000) The antinociceptive effects of alpha7 nicotinic agonists in an acute pain model. Neuropharmacology 39:2785–2791. https://doi.org/10.1016/s0028-3908(00)00139-8

Damaj MI, Freitas K, Bagdas D, Flood P (2014) Nicotinic receptors as targets for novel analgesics and anti-inflammatory drugs. In: Lester RAJ (ed) Nicotinic receptors. Springer, New York, pp 239–254

de Jonge WJ, Ulloa L, De Jonge WJ, Ulloa L (2007) The alpha7 nicotinic acetylcholine receptor as a pharmacological target for inflammation. Br J Pharmacol 151:915–929. https://doi.org/10.1038/sj.bjp.0707264

de Lucas-Cerrillo AM, Maldifassi MC, Arnalich F et al (2011) Function of partially duplicated human α77 nicotinic receptor subunit CHRFAM7A gene: potential implications for the cholinergic anti-inflammatory response. J Biol Chem 286:594–606. https://doi.org/10.1074/jbc.M110.180067

De Rosa MJ, Dionisio L, Agriello E et al (2009) Alpha 7 nicotinic acetylcholine receptor modulates lymphocyte activation. Life Sci 85:444–449. https://doi.org/10.1016/j.lfs.2009.07.010

Decker MW, Rueter LE, Bitner RS (2004) Nicotinic acetylcholine receptor agonists: a potential new class of analgesics. Curr Top Med Chem 4:369–384

Di Cesare ML, Cinci L, Micheli L et al (2014a) α-Conotoxin RgIA protects against the development of nerve injury-induced chronic pain and prevents both neuronal and glial derangement. Pain 155:1986–1995. https://doi.org/10.1016/j.pain.2014.06.023

Di Cesare ML, Pacini A, Matera C et al (2014b) Involvement of α7 nAChR subtype in rat oxaliplatin-induced neuropathy: effects of selective activation. Neuropharmacology 79:37–48. https://doi.org/10.1016/j.neuropharm.2013.10.034

Elgoyhen AB, Johnson DS, Boulter J et al (1994) α9: an acetylcholine receptor with novel pharmacological properties expressed in rat cochlear hair cells. Cell 79:705–715. https://doi.org/10.1016/0092-8674(94)90555-x

Elgoyhen AB, Vetter DE, Katz E et al (2001) α10: a determinant of nicotinic cholinergic receptor function in mammalian vestibular and cochlear mechanosensory hair cells. Proc Natl Acad Sci U S A 98:3501–3506. https://doi.org/10.1073/pnas.051622798

Feuerbach D, Lingenhoehl K, Olpe H-R et al (2009) The selective nicotinic acetylcholine receptor alpha7 agonist JN403 is active in animal models of cognition, sensory gating, epilepsy and pain. Neuropharmacology 56:254–263. https://doi.org/10.1016/j.neuropharm.2008.08.025

Freitas K, Carroll FI, Damaj MI (2013a) The antinociceptive effects of nicotinic receptors α7-positive allosteric modulators in murine acute and tonic pain models. J Pharmacol Exp Ther 344:264–275. https://doi.org/10.1124/jpet.112.197871

Freitas K, Ghosh S, Ivy Carroll F et al (2013b) Effects of alpha 7 positive allosteric modulators in murine inflammatory and chronic neuropathic pain models. Neuropharmacology 65:156–164. https://doi.org/10.1016/j.neuropharm.2012.08.022

Fujii T, Mashimo M, Moriwaki Y et al (2017) Expression and function of the cholinergic system in immune cells. Front Immunol 8:1085. https://doi.org/10.3389/fimmu.2017.01085

Gaskin DJ, Richard P (2012) The economic costs of pain in the United States. J Pain 13:715–724. https://doi.org/10.1016/j.jpain.2012.03.009

Gillberg PG, Aquilonius SM (1985) Cholinergic, opioid and glycine receptor binding sites localized in human spinal cord by in vitro autoradiography. Changes in amyotrophic lateral sclerosis. Acta Neurol Scand 72:299–306. https://doi.org/10.1111/j.1600-0404.1985.tb00874.x

Gillberg PG, Wiksten B (1986) Effects of spinal cord lesions and rhizotomies on cholinergic and opiate receptor binding sites in rat spinal cord. Acta Physiol Scand 126:575–582. https://doi.org/10.1111/j.1748-1716.1986.tb07857.x

Gotti C, Zoli M, Clementi F (2006) Brain nicotinic acetylcholine receptors: native subtypes and their relevance. Trends Pharmacol Sci 27:482–491. https://doi.org/10.1016/j.tips.2006.07.004

Grau V, Richter K, Hone AJ, McIntosh JM (2018) Conopeptides [V11L;V16D]ArIB and RgIA4: powerful tools for the identification of novel nicotinic acetylcholine receptors in monocytes. Front Pharmacol 9:1499. https://doi.org/10.3389/fphar.2018.01499

Hendrickson LM, Guildford MJ, Tapper AR (2013) Neuronal nicotinic acetylcholine receptors: common molecular substrates of nicotine and alcohol dependence. Front Psych 4:1–16. https://doi.org/10.3389/fpsyt.2013.00029

Hoffmeister P-G, Donat CK, Schuhmann MU et al (2011) Traumatic brain injury elicits similar alterations in α7 nicotinic receptor density in two different experimental models. NeuroMolecular Med 13:44–53. https://doi.org/10.1007/s12017-010-8136-4

Holtman JR, Dwoskin LP, Dowell C et al (2011) The novel small molecule α9α10 nicotinic acetylcholine receptor antagonist ZZ-204G is analgesic. Eur J Pharmacol 670:500–508. https://doi.org/10.1016/j.ejphar.2011.08.053

Hone AJ, McIntosh JM (2018) Nicotinic acetylcholine receptors in neuropathic and inflammatory pain. FEBS Lett 592:1045–1062. https://doi.org/10.1002/1873-3468.12884

Hunt S, Schmidt J (1978) Some observations on the binding patterns of alpha-bungarotoxin in the central nervous system of the rat. Brain Res 157:213–232. https://doi.org/10.1016/0006-8993(78)90025-2

Hurst RS, Hajós M, Raggenbass M et al (2005) A novel positive allosteric modulator of the alpha7 neuronal nicotinic acetylcholine receptor: in vitro and in vivo characterization. J Neurosci 25:4396–4405. https://doi.org/10.1523/JNEUROSCI.5269-04.2005

Institute of Medicine (US) Committee on Advancing Pain Research, Care and E (2011) Relieving pain in America: a blueprint for transforming prevention, care, education, and research. National Academies Press (US), Washington

Kesingland AC, Gentry CT, Panesar MS et al (2000) Analgesic profile of the nicotinic acetylcholine receptor agonists, (+)-epibatidine and ABT-594 in models of persistent inflammatory and neuropathic pain. Pain 86:113–118. https://doi.org/10.1016/S0304-3959(00)00233-5

Khan MAS, Farkhondeh M, Crombie J et al (2012) Lipopolysaccharide upregulates α7 acetylcholine receptors. Shock 38:213–219. https://doi.org/10.1097/SHK.0b013e31825d628c

Kiguchi N, Kobayashi D, Saika F et al (2018) Inhibition of peripheral macrophages by nicotinic acetylcholine receptor agonists suppresses spinal microglial activation and neuropathic pain in mice with peripheral nerve injury. J Neuroinflammation 15:96. https://doi.org/10.1186/s12974-018-1133-5

Lee CH, Zhu C, Malysz J et al (2011) α4β2 neuronal nicotinic receptor positive allosteric modulation: an approach for improving the therapeutic index of α4β2 nAChR agonists in pain. Biochem Pharmacol 82:959–966. https://doi.org/10.1016/j.bcp.2011.06.044

Liu Q, Whiteaker P, Morley BJ et al (2017) Distinctive roles for α7∗- and α9∗-nicotinic acetylcholine receptors in inflammatory and autoimmune responses in the murine experimental autoimmune encephalomyelitis model of multiple sclerosis. Front Cell Neurosci 11:287. https://doi.org/10.3389/fncel.2017.00287

Loram LC, Taylor FR, Strand KA et al (2012) Systemic administration of an alpha-7 nicotinic acetylcholine agonist reverses neuropathic pain in male Sprague Dawley rats. J Pain 13:1162–1171. https://doi.org/10.1016/j.jpain.2012.08.009

Luo S, Zhangsun D, Harvey PJ et al (2015) Cloning, synthesis, and characterization of αO-conotoxin GeXIVA, a potent α9α10 nicotinic acetylcholine receptor antagonist. Proc Natl Acad Sci U S A 112:E4026–E4035. https://doi.org/10.1073/pnas.1503617112

Lykhmus O, Voytenko LP, Lips KS et al (2017) Nicotinic acetylcholine receptor α9 and α10 subunits are expressed in the brain of mice. Front Cell Neurosci 11:282. https://doi.org/10.3389/fncel.2017.00282

Lynch JJ, Wade CL, Mikusa JP et al (2005) ABT-594 (a nicotinic acetylcholine agonist): anti-allodynia in a rat chemotherapy-induced pain model. Eur J Pharmacol 509:43–48. https://doi.org/10.1016/j.ejphar.2004.12.034

Marrero MB, Bencherif M, Lippiello PM, Lucas R (2011) Application of alpha7 nicotinic acetylcholine receptor agonists in inflammatory diseases: an overview. Pharm Res 28:413–416. https://doi.org/10.1007/s11095-010-0283-7

Marubio LM, del Mar Arroyo-Jimenez M, Cordero-Erausquin M et al (1999) Reduced antinociception in mice lacking neuronal nicotinic receptor subunits. Nature 398:805–810. https://doi.org/10.1038/19756

McIntosh JM, Absalom N, Chebib M et al (2009) Alpha9 nicotinic acetylcholine receptors and the treatment of pain. Biochem Pharmacol 78:693–702. https://doi.org/10.1016/j.bcp.2009.05.020

Medhurst SJ, Hatcher JP, Hille CJ et al (2008) Activation of the α7-nicotinic acetylcholine receptor reverses complete Freund adjuvant-induced mechanical Hyperalgesia in the rat via a central site of action. J Pain 9:580–587. https://doi.org/10.1016/j.jpain.2008.01.336

Morley BJ, Whiteaker P, Elgoyhen AB (2018) Commentary: nicotinic acetylcholine receptor α9 and α10 subunits are expressed in the brain of mice. Front Cell Neurosci 12:104. https://doi.org/10.3389/fncel.2018.00104

Munro G, Hansen RR, Erichsen HK et al (2012) The alpha7 nicotinic ACh receptor agonist compound B and positive allosteric modulator PNU-120596 both alleviate inflammatory hyperalgesia and cytokine release in the rat. Br J Pharmacol 167:421–435. https://doi.org/10.1111/j.1476-5381.2012.02003.x

Napier IA, Klimis H, Rycroft BK et al (2012) Intrathecal α-conotoxins Vc1.1, AuIB and MII acting on distinct nicotinic receptor subtypes reverse signs of neuropathic pain. Neuropharmacology 62:2202–2207. https://doi.org/10.1016/j.neuropharm.2012.01.016

Naser PV, Kuner R (2018) Molecular, cellular and circuit basis of cholinergic modulation of pain. Neuroscience 387:135–148. https://doi.org/10.1016/j.neuroscience.2017.08.049

Nashmi R, Lester HA (2006) CNS localization of neuronal nicotinic receptors. J Mol Neurosci 30:181–184. https://doi.org/10.1385/JMN:30:1:181

Pacini A, Micheli L, Maresca M et al (2016) The α9α10 nicotinic receptor antagonist α-conotoxin RgIA prevents neuropathic pain induced by oxaliplatin treatment. Exp Neurol 282:37–48. https://doi.org/10.1016/j.expneurol.2016.04.022

Papke RL, Chojnacka K, Horenstein NA (2014) The minimal pharmacophore for silent agonism of the α7 nicotinic acetylcholine receptor. J Pharmacol Exp Ther 350:665–680. https://doi.org/10.1124/jpet.114.215236

Papke RL, Bagdas D, Kulkarni AR et al (2015) The analgesic-like properties of the alpha7 nAChR silent agonist NS6740 is associated with non-conducting conformations of the receptor. Neuropharmacology 91:34–42. https://doi.org/10.1016/j.neuropharm.2014.12.002

Posadas I, López-Hernández B, Ceña V (2013) Nicotinic receptors in neurodegeneration. Curr Neuropharmacol 11:298–314. https://doi.org/10.2174/1570159X11311030005

Quadri M, Bagdas D, Toma W et al (2018) The Antinociceptive and anti-inflammatory properties of the α 7 nAChR weak partial agonist p -CF 3 N, N-diethyl-N'-phenylpiperazine. J Pharmacol Exp Ther 367:203–214. https://doi.org/10.1124/jpet.118.249904

Rode F, Munro G, Holst D et al (2012) Positive allosteric modulation of α4β2 nAChR agonist induced behaviour. Brain Res 1458:67–75. https://doi.org/10.1016/j.brainres.2012.03.064

Romero HK, Christensen SB, Di Cesare ML et al (2017) Inhibition of α9α10 nicotinic acetylcholine receptors prevents chemotherapy-induced neuropathic pain. Proc Natl Acad Sci 114:E1825–E1832. https://doi.org/10.1073/pnas.1621433114

Rowbotham MC, Rachel Duan W, Thomas J et al (2009) A randomized, double-blind, placebo-controlled trial evaluating the efficacy and safety of ABT-594 in patients with diabetic peripheral neuropathic pain. Pain 146:245–252. https://doi.org/10.1016/j.pain.2009.06.013

Rowbotham MC, Arslanian A, Nothaft W et al (2012) Efficacy and safety of the α4β2 neuronal nicotinic receptor agonist ABT-894 in patients with diabetic peripheral neuropathic pain. Pain 153:862–868. https://doi.org/10.1016/j.pain.2012.01.009

Rowley TJ, Payappilly J, Lu J, Flood P (2008) The Antinociceptive response to nicotinic agonists in a mouse model of postoperative pain. Anesth Analg 107:1052–1057. https://doi.org/10.1213/ane.0b013e318165e0c0

Rueter LE, Meyer MD, Decker MW (2000) Spinal mechanisms underlying A-85380-induced effects on acute thermal pain. Brain Res 872:93–101

Sadeghi M, McArthur JR, Finol-Urdaneta RK, Adams DJ (2017) Analgesic conopeptides targeting G protein-coupled receptors reduce excitability of sensory neurons. Neuropharmacology 127:116–123. https://doi.org/10.1016/j.neuropharm.2017.05.020

Saika F, Kiguchi N, Kobayashi Y, Kishioka S (2015) Peripheral alpha4beta2 nicotinic acetylcholine receptor signalling attenuates tactile allodynia and thermal hyperalgesia after nerve injury in mice. Acta Physiol 213:462–471. https://doi.org/10.1111/apha.12437

Satkunanathan N, Livett B, Gayler K et al (2005) Alpha-conotoxin Vc1.1 alleviates neuropathic pain and accelerates functional recovery of injured neurones. Brain Res 1059:149–158. https://doi.org/10.1016/j.brainres.2005.08.009

Séguéla P, Wadiche J, Dineley-Miller K et al (1993) Molecular cloning, functional properties, and distribution of rat brain alpha 7: a nicotinic cation channel highly permeable to calcium. J Neurosci 13:596–604

Sinkus ML, Graw S, Freedman R et al (2015) The human CHRNA7 and CHRFAM7A genes: a review of the genetics, regulation, and function. Neuropharmacology 96:274–288. https://doi.org/10.1016/j.neuropharm.2015.02.006

Toma W, Kyte SL, Bagdas D et al (2019) The α7 nicotinic receptor silent agonist R-47 prevents and reverses paclitaxel-induced peripheral neuropathy in mice without tolerance or altering nicotine reward and withdrawal. Exp Neurol 320:113010. https://doi.org/10.1016/j.expneurol.2019.113010

Tracey KJ (2002) The inflammatory reflex. Nature 420:835–859. https://doi.org/10.1111/j.1365-2796.2004.01440.x
van Maanen MA, Stoof SP, LaRosa GJ et al (2010) Role of the cholinergic nervous system in rheumatoid arthritis: aggravation of arthritis in nicotinic acetylcholine receptor α7 subunit gene knockout mice. Ann Rheum Dis 69:1717–1723. https://doi.org/10.1136/ard.2009.118554
van Maanen MA, Papke RL, Koopman FA et al (2015) Two novel α7 nicotinic acetylcholine receptor ligands: in vitro properties and their efficacy in collagen-induced arthritis in mice. PLoS One 10:e0116227. https://doi.org/10.1371/journal.pone.0116227
Vincler M, Eisenach JC (2004) Plasticity of spinal nicotinic acetylcholine receptors following spinal nerve ligation. Neurosci Res 48:139–145
Vincler M, Wittenauer S, Parker R et al (2006) Molecular mechanism for analgesia involving specific antagonism of alpha9alpha10 nicotinic acetylcholine receptors. Proc Natl Acad Sci U S A 103:17880–17884. https://doi.org/10.1073/pnas.0608715103
Wada E, Wada K, Boulter J et al (1989) Distribution of alpha 2, alpha 3, alpha 4, and beta 2 neuronal nicotinic receptor subunit mRNAs in the central nervous system: a hybridization histochemical study in the rat. J Comp Neurol 284:314–335. https://doi.org/10.1002/cne.902840212
Wala EP, Crooks PA, McIntosh JM, Holtman JR (2012) Novel small molecule α9α10 nicotinic receptor antagonist prevents and reverses chemotherapy-evoked neuropathic pain in rats. Anesth Analg 115:713–720. https://doi.org/10.1213/ANE.0b013e31825a3c72
Wang Y, Su D-M, Wang R-H et al (2005) Antinociceptive effects of choline against acute and inflammatory pain. Neuroscience 132:49–56. https://doi.org/10.1016/j.neuroscience.2004.12.026
Wang H, Li X, Zhangsun D et al (2019) The α9α10 nicotinic acetylcholine receptor antagonist αO-conotoxin GeXIVA[1,2] alleviates and reverses chemotherapy-induced neuropathic pain. Mar Drugs 17:265. https://doi.org/10.3390/md17050265
Weggel LA, Pandya AA (2019) Acute administration of desformylflustrabromine relieves chemically induced pain in CD-1 mice. Molecules 24:944. https://doi.org/10.3390/molecules24050944
Wieskopf JS, Mathur J, Limapichat W et al (2015) The nicotinic 6 subunit gene determines variability in chronic pain sensitivity via cross-inhibition of P2X2/3 receptors. Sci Transl Med 7:287ra72. https://doi.org/10.1126/scitranslmed.3009986
Williams DK, Wang J, Papke RL (2011) Positive allosteric modulators as an approach to nicotinic acetylcholine receptor-targeted therapeutics: advantages and limitations. Biochem Pharmacol 82:915–930. https://doi.org/10.1016/j.bcp.2011.05.001
Williams DK, Peng C, Kimbrell MR, Papke RL (2012) Intrinsically low open probability of 7 nicotinic acetylcholine receptors can be overcome by positive allosteric modulation and serum factors leading to the generation of Excitotoxic currents at physiological temperatures. Mol Pharmacol 82:746–759. https://doi.org/10.1124/mol.112.080317
Wu J, Liu Q, Tang P et al (2016) Heteromeric α7β2 nicotinic acetylcholine receptors in the brain. Trends Pharmacol Sci 37:562–574. https://doi.org/10.1016/j.tips.2016.03.005
Zhu CZ, Chin C, Rustay NR et al (2011) Potentiation of analgesic efficacy but not side effects: co-administration of an α4β2 neuronal nicotinic acetylcholine receptor agonist and its positive allosteric modulator in experimental models of pain in rats. Biochem Pharmacol 82:967–976. https://doi.org/10.1016/j.bcp.2011.05.007
Zoli M, Pucci S, Vilella A, Gotti C (2018) Neuronal and extraneuronal nicotinic acetylcholine receptors. Curr Neuropharmacol 16:338–349. https://doi.org/10.2174/1570159X15666170912110450

An Evolving Therapeutic Rationale for Targeting the α_7 Nicotinic Acetylcholine Receptor in Autism Spectrum Disorder

Stephen I. Deutsch and Jessica A. Burket

Contents

Abstract Abnormalities of cholinergic nuclei, cholinergic projections, and cholinergic receptors, as well as abnormalities of growth factors involved in the maturation and maintenance of cholinergic neurons, have been described in postmortem brains of persons with autism spectrum disorder (ASD). Further, microdeletions of the 15q13.3 locus that encompasses *CHRNA7*, the gene coding the α_7 nicotinic acetylcholine receptor (α_7 nAChR), are associated with a spectrum of neurodevelopmental disorders, including ASD. The heterozygous 15q13.3 microdeletion syndrome suggests that diminished or impaired transduction of the acetylcholine (ACh) signal by the α_7 nAChR can be a pathogenic mechanism of ASD. The α_7 nAChR has a role in regulating the firing and function of parvalbumin (PV)-expressing GABAergic

S. I. Deutsch (✉)
Department of Psychiatry and Behavioral Sciences, Eastern Virginia Medical School, Norfolk, VA, USA
e-mail: deutscsi@evms.edu

J. A. Burket
Department of Molecular Biology and Chemistry, Christopher Newport University, Newport News, VA, USA

Curr Topics Behav Neurosci (2020) 45: 167–208
https://doi.org/10.1007/7854_2020_136
Published Online: 29 May 2020

projections, which synchronize the oscillatory output of assemblies of pyramidal neurons onto which they project. Synchronous oscillatory output is an electrophysiological substrate for higher executive functions, such as working memory, and functional connectivity between discrete anatomic areas of the brain. The α_7 nAChR regulates PV expression and works cooperatively with the co-expressed NMDA receptor in subpopulations of GABAergic interneurons in mouse models of ASD. An evolving literature supports therapeutic exploration of selectively targeted cholinergic interventions for the treatment of ASD, especially compounds that target the α_7 nAChR subtype. Importantly, development and availability of high-affinity, brain-penetrable, α_7 nAChR-selective agonists, partial agonists, allosteric agonists, and positive allosteric modulators (PAMs) should facilitate "proof-of-principle/concept" clinical trials. nAChRs are pentameric allosteric proteins that function as ligand-gated ion channel receptors constructed from five constituent polypeptide subunits, all of which share a common structural motif. Importantly, in addition to α_7 nAChR-gated Ca^{2+} conductance causing membrane depolarization, there are emerging data consistent with possible metabotropic functions of this ionotropic receptor. The ability of α_7-selective type II PAMs to "destabilize" the desensitized state and promote ion channel opening may afford them therapeutic advantages over orthosteric agonists. The current chapter reviews historic and recent literature supporting selective therapeutic targeting of the α_7 nAChR in persons affected with ASD.

Keywords 15q13.3 · Alpha7 nicotinic acetylcholine receptor · Autism spectrum disorder · GABA interneurons

1 Introduction

Autism spectrum disorder (ASD) is really a heterogeneous group of highly heritable and increasingly prevalent neurodevelopmental disorders, whose core diagnostic symptoms include impairment of social communication, restricted interests, and repetitive stereotypic behaviors. In addition to these core symptoms, irritability may be a prominent secondary symptom, and affected persons also commonly have neuropsychiatric comorbidities, such as intellectual disability, seizure disorders, attention deficit hyperactivity disorder, and obsessive-compulsive disorder, among other disorders. Although syndromic monogenic etiologies, due to DNA sequence variants, and copy number variants with altered gene dosing effects, due to microdeletions and microduplications, account for only a minority of the presentations of ASD, they are instructive with respect to identifying possible mechanisms of pathogenesis and molecular therapeutic targets. For example, ASD is a comorbidity associated with tuberous sclerosis, neurofibromatosis 1, and fragile X syndrome; the pathogenesis of these syndromic forms of ASD converges on disrupted regulation of mTOR signaling, which stimulated interest in exploring and developing neuronal cell-specific mTORC1 inhibition strategies (Brodkin 2008; Sharma et al. 2010; Ehninger and Silva 2011; Tsai et al. 2012; Sahin 2012; Burket et al. 2014, 2015b; Garg et al. 2015; Magdalon et al. 2017; Winden et al. 2018). Similarly, because microdeletions of the 15q13.3 locus that includes *CHRNA7*, the gene coding the α_7 nicotinic acetylcholine receptor (α_7 nAChR), are associated with a spectrum of neurodevelopmental disorders, including ASD, and the α_7 nAChR subtype is involved in normal processes of attention and cognition, a rationale for selective targeting of this receptor is emerging. Importantly, the development of selective agonists, partial agonists, and allosteric modulatory ligands has opened up the possibility of "therapeutically" interrogating this receptor in validated animal models (e.g., BTBR [BTBR T^+ *tf*/J]) (Oginsky et al. 2014; Wang et al. 2015; Yoshimura et al. 2017).

In addition to the preclinical challenges of developing highly selective, brain-penetrable molecules that are tolerated and devoid of toxicity when administered chronically to young children, adolescents, and adults, development of therapeutic cholinergic interventions for an ASD indication must resolve potentially confounding issues related to developmental age and time to initiate therapy and the variety of neuropsychiatric comorbidities that may limit their effectiveness and safety, such as intellectual disability and seizure disorders. Equally important, medication effectiveness will be very much influenced and dependent on the availability of individualized, interdisciplinary, multimodal treatment modalities, such as psychosocial interventions, special education, speech and language therapy, and occupational and physical therapy. In spite of these challenges, given the poor functional outcomes of many persons affected with ASD, there is a moral imperative to pursue the development of newer and more effective medication strategies. This chapter will focus predominantly on an emerging literature supporting a role of the α_7 nAChR subtype in the pathogenesis of, and as a therapeutic molecular target for, ASD.

2 Structure, Physiology, and Subtypes of Nicotinic Receptors

Acetylcholine (ACh), a neurotransmitter discovered at the turn of the twentieth century by Henry H. Dale and Otto Loewi, is transduced by families of ACh-selective metabotropic muscarinic and fast ionotropic cationic nAChRs, whose activations occur on timescales of milliseconds to seconds and micro- to submicroseconds, respectively (Albuquerque et al. 2009; Deutsch et al. 2015). ACh is utilized by a modulatory projection system located in the basal forebrain (Záborszky et al. 2018). Cholinergic projections from the basal forebrain to the "awake" cortex work in complex ways to enhance attention and encoding of new information, among other higher executive functions. In addition to direct cholinergic projections to neocortical pyramidal neurons, they also inhibit and disinhibit pyramidal cell activity indirectly by their synapsing onto specialized GABAergic interneurons that express PV, somatostatin and vasoactive intestinal polypeptide (VIP) within "canonical microcircuits" (Záborszky et al. 2018). Importantly, the basal forebrain cholinergic neurons themselves receive synaptic inputs from diverse regions of the brain, including, but not limited to, ventral and dorsal striatum, hypothalamus, amygdala, and brainstem tegmentum, that utilize a variety of neurotransmitters and neuropeptides. In addition to remarkable spatial selectivity, cholinergic signaling in prefrontal cortex enjoys temporal precision and the time-scales of its effects range from subseconds to seconds (e.g., cue detection and cue-triggered changes in goal-oriented behavior) to minutes (e.g., support of general arousal) (Záborszky et al. 2018). These realities encourage exploration of therapeutic interventions that are least likely to disrupt the spatial and temporal specificity of physiological cholinergic activation; thus, as discussed below, there is increasing interest in allosteric modulatory ligands and deliberate avoidance of binding to orthosteric (i.e., agonist) binding sites (Deutsch et al. 2008a, b, 2011, 2013, 2014, 2015, 2016). The cell bodies of basal forebrain cholinergic projection neurons originate in nuclei that lie medial and ventral to the basal ganglia at the core of the telencephalon (Bear et al. 2016). The *medial septal nuclei* project to the hippocampus and the *basal nucleus of Meynert* provides diffuse innervation to the neocortex. There is a second diffuse cholinergic projection system, referred to as the *pontomesencephalotegmental complex*, whose cell bodies originate in the pons and midbrain tegmentum and include prominent projections to the dorsal thalamus. The projections synapse onto muscarinic (i.e., metabotropic) and nicotinic (i.e., ionotropic) receptors to regulate excitability, arousal, sleep-wake cycle, learning and memory, and many higher executive functions.

There are five genetically distinct classes of muscarinic acetylcholine receptors (mAChRs) that are divided into two functional classes based on the coupling of these "seven transmembrane hydrophobic domain" receptors to either the G_q (i.e., M_1, M_3, and M_5) or G_i (i.e., M_2 and M_4) G-protein. These five muscarinic receptors influence metabolic events within the cell dependent on activation of phospholipase C and adenylyl cyclase, resulting in formation of inositol trisphosphate, diacylglycerol, and cAMP. Importantly, stimulation of the M_1 mAChR leads to activation of the NMDA receptor and excitation of pyramidal cells in medial prefrontal cortex. M_1 mAChR activation of the NMDA receptor has aroused interest in selective therapeutic targeting of this receptor subtype to improve cognition across a variety of neuropsychiatric disorders (Jones et al. 2012; Deutsch et al. 2014).

The nAChRs are pentameric allosteric proteins that function as ligand-gated ion channel receptors constructed from five constituent polypeptide subunits, all of which share a common structural motif. Thus, all 16 homologous subunits have a conserved extracellular large NH_2-terminal domain; three "conserved" transmembrane domains (TM1–TM3); a cytoplasmic domain of variable size and amino acid sequence located between the third (TM3) and fourth (TM4) transmembrane domain; and TM4 with a short and variable extracellular COOH-terminal sequence (Albuquerque et al. 2009) (Fig. 1). The subunits all share a "cysteine-loop

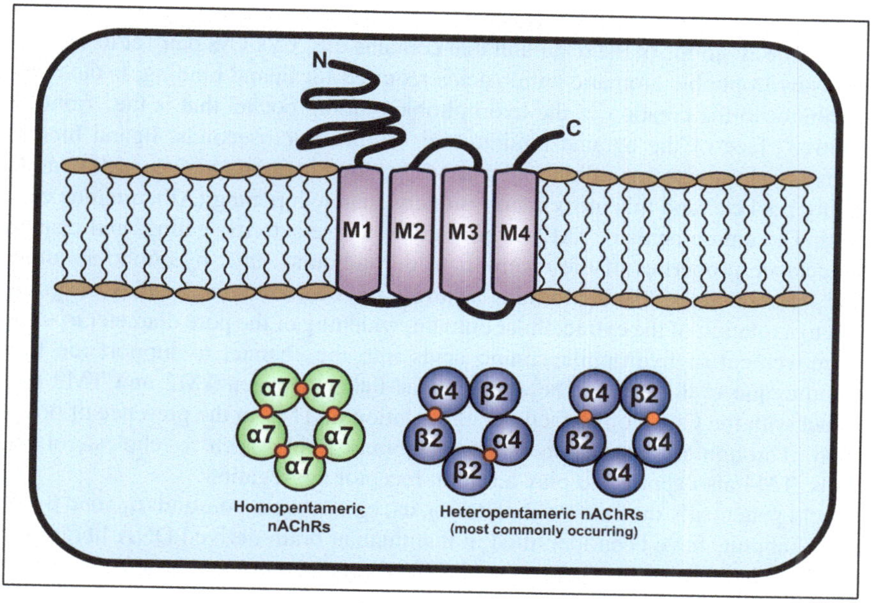

Fig. 1 Nicotinic acetylcholine receptors, a pentameric family of ligand-gated ion channel receptors. Nicotinic acetylcholine receptors (nAChRs) are constructed from five constituent polypeptide subunits that share a common structural motif, including four transmembrane hydrophobic domains (top panel). The M2 transmembrane domain from each of the five subunits create a channel, whose opening is dependent on the binding of the endogenous full agonists, acetylcholine (ACh) and choline (Ch), to the receptor's orthosteric binding site (also known as agonist recognition site) (bottom panel). The homopentameric α_7 nAChR binds five molecules of ACh/Ch, whereas heteropentameric receptors bind only two. A transmembrane site exists that binds allosteric agonists and positive allosteric modulatory ligands (PAMs). PAMs preserve the spatial and temporal specificity of endogenous ligands but lack intrinsic efficacy of their own; α_7 nAChR-selective PAMs act only where and when endogenous ACh or Ch is released to increase the likelihood the channel assumes an open configuration. α_7 nAChR-selective PAMs can also influence the deactivation kinetics of these ligand-gated ion channel receptors (see text for details)

(Cys-loop)," which consists of two disulfide-linked cystines separated by a fixed number of amino acids in the NH_2-terminal extracellular domain (Albuquerque et al. 2009). Also, the α subunits are characterized by a "Cys-Cys pair," which is required for agonist binding, located near the "entrance" to TM1. Electron microscopy revealed the structure of the heteropentameric nAChR expressed at very high density in the electric organ of the *Torpedo* fish. The receptor is shaped like a cone embedded in the lipid bilayer. The secondary structure of the large NH_2-extracellular domain consists largely of β-pleated sheets and assumes a configuration referred to as a β-barrel. TM1-4 assumes α-helical configurations that traverse the entire membrane and extend ~10 Å beyond the extracellular surface. The TM2 helices from each of the five polypeptide subunits align themselves to form and enclose the receptor's ion channel. Whereas TM1 and TM3 are situated close to the ion pore,

TM4 "stands" apart and has a special role in terms of interacting with the lipid milieu. The "C-loop" of the α subunit that contains the "Cys-Cys pair" at its apex, as well as hydrophobic aromatic amino acids required for ligand binding, is the major contributor to the creation of the hydrophobic binding pocket that is the "front" or "positive" face of the agonist binding site. Within microseconds, ligand binding results in rearrangements of hydrogen bonds "among invariant amino acids near the binding pocket," and this movement results in channel opening (Albuquerque et al. 2009). The amino acids in TM2 determine the ion selectivity, gating, and channel conductance properties of the receptor. In the unbound state, hydrophobic amino acids project into the channel forming a narrow ~3 Å constriction. Binding of agonist leads to a rotation of the extracellular domain, widening of the pore diameter to ~8 Å, and movement of hydrophilic amino acids into the channel to support ion flow (Albuquerque et al. 2009). The extracellular linker between TM2 and TM3 may interact with the Cys-loop to facilitate the rotation of TM2 in the presence of bound ligand. Through its interactions with membrane lipids, such as cholesterol and sterols, TM4 also appears to play a role in receptor aggregation.

Eight genetically discrete α subunits (α_2, α_3, α_4, α_5, α_6, α_7, α_9, and α_{10}) and β_2, β_3, and β_4 subunits have been identified in mammalian brain-derived DNA libraries or cloned from neuronal-like cells (e.g., PC12) (Albuquerque et al. 2009). However, two different stoichiometric combinations of pentameric nAChRs containing α_4 and β_2 subunits (i.e., two or three α_4 and three or two β_2 subunits, respectively) and homopentameric and heteropentameric α_7-containing nAChRs occur most commonly in the brain (Albuquerque et al. 2009; Deutsch et al. 2015) (Fig. 1). From an evolutionary perspective, the α_7 subunit is an ancestral one that is expressed by a variety of neuronal and non-neuronal cell types (including astrocytes, microglia, oligodendrocyte precursor cells and endothelial cells), some of whose functions in these non-neuronal cells are probably more related to such things as immunity, inflammation, and neuroprotection than to fast synaptic neurotransmission. Most nAChRs quickly desensitize, especially those containing the α_7 subunit, upon brief acute exposures to agonist; thus, elucidation of the kinetic properties of channel opening was dependent on technological developments affording fast delivery and removal of agonists (Albuquerque et al. 2009). Importantly, as noted, the α_7 subunit can form a fully functional ligand-gated homopentameric ion channel receptor with a unique type IA electrophysiological signature characterized by fast activation with a brief open-time (~100 μs), rapid desensitization, relatively low-affinity for agonists, and a ratio of higher permeability to Ca^{2+} than Na^{+} that is greater than the ratio for the NMDA receptor (Albuquerque et al. 2009). Also, consistent with the ancestral origin of the α_7 subunit, choline, the precursor and metabolic split product of ACh, is a full agonist. Choline itself may have an important role as an endogenous ligand, including during brain development, because expression of choline acetyltransferase (ChAT) lags behind expression of nAChRs in developing neurons and acetylcholinesterase (AChE) is a catalytically very efficient enzyme. Because AChE is enriched and associated with cholinergic synapses, in addition to the reuptake of choline that makes it available for synthesis of new ACh, incorporation into membrane phospholipids, and participation in one-carbon metabolism, the local generation of choline may influence the "sensitivity" of α_7 nAChRs. Importantly, although choline and ACh cause similar single-channel open-times and conductance changes at the α_7 nAChR, choline dissociates more rapidly from the receptor and causes a "less stable state of desensitization" than ACh (Albuquerque et al. 2009).

In addition to α_7 nAChR-gated Ca^{2+} conductance causing membrane depolarization, there are emerging data consistent with possible metabotropic functions of this ionotropic receptor, including some that may not be dependent on ion flux (Paulo et al. 2009). For example, using "affinity-immobilized" α-bungarotoxin (α-BGT)-conjugated beads, high-affinity bound "α_7 nAChR-protein binding complexes" were isolated from mouse brain homogenates, which were eluted with carbachol, a low molecular weight cholinergic agonist. Following carbachol elution, these membrane protein complexes were solubilized with Triton X-100 detergent, and the α_7 nAChR-binding proteins were fractionated with SDS-PAGE, digested in-gel with trypsin, and the peptide fragments characterized by nano-electrospray ionization (nano-ESI) mass spectrometry. "Nonspecific" proteins that were eluted and fractionated but not members of the "α_7 nAChR interactome" were identified by performing these exact procedures in parallel on brain homogenates prepared from α_7 nAChR knockout mice of the same genetic background. The α_7 subunit was not detected in the solubilized, carbachol-eluted protein complexes derived from the membranes of the knockout mice using anti-α_7 nAChR antibody. The M3-M4 linker, the intracellular region linking the third and fourth transmembrane helices that display variability with respect to amino acid sequence and length across all nAChR subunits, is the most likely site for interactions of the α_7 nAChR with its protein binding partners. Fifty five potential "candidate" binding protein members of the α_7 nAChR interactome were identified with this proteomic methodology (Paulo et al. 2009). Twenty six of the 55 candidate members of the α_7 nAChR interactome overlapped with 698 proteins identified in a prior study of the "post-synaptic density proteome"; they included proteins whose functions include "cell structure and protein trafficking" (e.g., Homer protein homolog 1 and microtubule associated tau), "chaperones" (e.g., 14-3-3 protein eta and gelsolin; actin-depolymerizing factor brevin), and proteins involved in "signal transduction" (e.g., cAMP-dependent protein kinase A; $G_{i,}$ α2 subunit; G_o α1 subunit; and G_q α subunit) (Paulo et al. 2009). Of course, these interactome candidates must be verified and their possible metabotropic roles investigated. Importantly, metabotropic roles of the NMDA receptor have been reported that are not dependent on ionic flux and involve alignment of binding partners with the C-terminal regions of specific NMDA receptor subunits and dependent on PDZ-binding domains (Chung 2013; Dore et al. 2017; Burket and Deutsch 2019).

3 Allosteric and Other Endogenous Modulators of Nicotinic Receptors

Conventional orthosteric agonists of the α_7 nAChR bind to an extracellular domain at the interface of two adjacent subunits and cause both rapid receptor activation and desensitization (Gill et al. 2013). The effects of conventional orthosteric agonists are blocked by methyllycaconitine (MLA), the α_7-selective nAChR competitive antagonist. In contrast, many positive allosteric modulators (PAMs), of which there are two primary types (i.e., type I and type II), bind to a common or overlapping site in the receptor's intra-subunit transmembrane region and lack intrinsic efficacy in the absence of orthosteric agonist or endogenous ligand. Type II PAMs antagonize agonist-induced desensitization, whereas type I PAMs have minimal effects on receptor desensitization. In addition to PAMs, there are compounds that bind to the allosteric transmembrane site and activate the receptor in the absence of conventional orthosteric agonist or endogenous ligand (e.g., ACh and choline); some of these compounds may activate the receptor without causing desensitization and have been termed allosteric agonists (Gill et al. 2013). A variety of orthosteric and allosteric ligands are becoming available and should stimulate translational exploration of their possible therapeutic indications. TQS (4-(1-napthyl)-3*a*, 4,5,9*b*-tetrahydro-3*H*-cyclopenta[*c*]quinolone-8-sulfonamide) is a type II PAM because it reduces the rate of desensitization when co-administered with an orthosteric agonist, while potentiating agonist-induced responses (Gill et al. 2013). 4BP-TQS (4-(4-bromophenyl)-3*a*,4,5,9*b*-tetrahydro-3*H*-cyclopenta[*c*]quinolone-8-sulfonamide) binds to the allosteric transmembrane site and activates the receptor when administered alone with only minimal effects on receptor desensitization (Gill et al. 2013). An additional feature TQS and 4BP-TQS share, in addition to binding to the same transmembrane allosteric site, is that they facilitate recovery from the desensitization caused by orthosteric agonists (Gill et al. 2013). MLA was shown to antagonize the allosteric agonist 4BP-TQS in primary rat hippocampal neurons but did so noncompetitively because they bind to different sites. The ability of allosteric ligands to both antagonize agonist-induced desensitization and promote recovery from desensitization may offer a therapeutic advantage over orthosteric agonists in cases of haploinsufficient expression of, or impaired signal transduction by, α_7 nAChRs.

Unlike heteropentameric nAChRs that have two specialized ACh binding sites per pentamer and assume a high-affinity conformation for ACh upon agonist-induced desensitization, the homopentameric α_7 nAChR has five agonist binding sites, and its affinity for agonist does not change significantly upon desensitization (Papke et al. 2014). Because the probability of agonist-induced ion channel opening is much reduced in the desensitized state of the α_7 nAChR, the ability of α_7-selective type II PAMs to "destabilize" the desensitized state and promote ion channel opening may afford them therapeutic advantages over orthosteric agonists. The (+)-enantiomer of racemic 4BP-TQS is the active stereoisomer (referred to as GAT107) and was shown to have properties of both an allosteric agonist and an allosteric modulator of orthosteric agonist-evoked responses when administered both concurrently with and after the orthosteric agonist (Papke et al. 2014). Additionally, GAT107 can cause long-lasting "priming" of the α_7 nAChR (Papke et al. 2014). Data suggest that the site mediating direct activation of the receptor by GAT107 may be distinct from both the primary PAM site and the orthosteric agonist binding site. Moreover, because the site responsible for direct activation is sensitive to antagonism by MLA, whereas primed potentiation is not MLA-sensitive, it suggests that these two sites differ in the homopentameric α_7 nAChR. The existence of these various sites and their implications for manipulating functioning of native receptors may lead to development of highly selective and effective medications (Papke et al. 2014).

Kynurenic acid (KYNA), a naturally occurring metabolite in the kynurenine pathway of tryptophan metabolism, may have important roles in the regulation of $GABA_A$-, α_7 nACh-, and NMDAR-mediated neurotransmission (Alkondon et al. 2004). Astrocytes are primarily responsible for the irreversible transamination of L-kynurenine to KYNA; greater than 70% of KYNA production in adult brain is catalyzed by kynurenine aminotransferase II (KAT II), whereas the remainder is produced by KAT I. Depending on its endogenous concentration, KYNA may affect NMDAR-mediated neurotransmission in its role as a competitive antagonist of glycine/D-serine's binding to a site created by the NR1 NMDAR subunit. However, data also suggest that, under normal physiological conditions, the low nanomolar to low micromolar concentrations of KYNA may be insufficient to affect the endogenous tone of NMDAR-mediated neurotransmission (i.e., below its IC_{50} values for inhibiting the NMDAR); thus, its physiological role as a regulator of NMDAR-mediated neurotransmission is uncertain (Alkondon et al. 2004). Developmental changes in the effect of diminished KYNA production on α_7-containing nAChRs and their regulation of GABAergic projections in hippocampus were studied in mice homozygous for deletion of KAT II (*mKat-2*$^{-/-}$ mice) (Alkondon et al. 2004). Mice with targeted *null* mutations of *mKat-2*$^{-/-}$ were made, and, in order to minimize the possibility of spurious or inconclusive results due to genetic drift, every fourth-generation *mKat-2*$^{-/-}$ mouse maintained on a 129SvEv genetic background was bred with genetically inbred 129SvEv wild type mice to create breeders for subsequent generations. As expected, hippocampal KAT II activity was disrupted by the homozygous *mKat-2*$^{-/-}$ deletion. However, whereas KAT II activity accounts for ~70% of total KAT activity in adult mouse brain, hippocampal KAT II activity represented only ~10% of total KAT activity in the adult 60-day-old wild type control mice in this study (Alkondon et al. 2004). Further, hippocampal KAT II activity accounted for ~40% of the total KAT II activity in the wild type mice at age 21 days. A metabolic consequence of KAT II activity accounting for a greater percentage of total hippocampal KAT activity in the 21-day-old wild type versus the 60-day-old wild type mice was reflected in an ~55% reduction of hippocampal KYNA levels in the 21-day-old homozygous *mKat-2*$^{-/-}$ mice, compared to 21-day-old wild type controls. Hippocampal KYNA levels did not differ between the older adult 60-day-old homozygous *mKat-2*$^{-/-}$ mice and their 60-day-old wild type controls. A functional behavioral significance to the reduced KYNA levels in the 21-day-old homozygous *mKat-2*$^{-/-}$ mice was suggested by their increased level of spontaneous locomotor activity, compared to wild type controls, which was not seen in the 60-day-old *mKat-2*$^{-/-}$ mice.

Data consistent with activation of somatodendritic α_7-containing nAChRs in hippocampal CA1 interneurons were found with whole-cell patch-clamp recordings in hippocampal slices (Alkondon et al. 2004). Specifically, choline-elicited currents that decayed during the agonist pulse (i.e., fast inactivation) were blocked irreversibly and reversibly by α-BGT (α-BGT) and MLA, respectively. Although the decay phase of the recorded choline-evoked currents in the hippocampal slices did not differ between the 21-day-old homozygous *mKat-2*$^{-/-}$ and wild type mice, the amplitude and net charge of the currents recorded from the interneurons in the *mKat-2*$^{-/-}$ mice were significantly larger (Alkondon et al. 2004). The greater choline-evoked currents in the 21-day-old *mKat-2*$^{-/-}$ mice were not due to an increased density of α_7-containing nAChRs on the somatodendritic surface of their CA1 interneurons, as measured with ^{125}I-α-BGT binding; thus, the higher α_7-containing nAChR activity was most likely related to decreased levels of hippocampal KYNA in these animals. Importantly, in the 21-day-old homozygous *mKat-2*$^{-/-}$ mice with lowered hippocampal levels of KYNA, the frequency and amplitude of inhibitory postsynaptic currents (IPSCs) recorded from CA1 pyramidal neurons were higher than in the wild type mice. The frequency and amplitude of IPSCs did not differ between the two groups of 60-day-old mice, which is consistent with the lack of significant differences in hippocampal KYNA levels between these two groups of animals. NMDAR activity (e.g., increased activity as a result of lowered hippocampal KYNA levels in the 21-day-old *mKat-2*$^{-/-}$ mice) did not appear to be significantly involved in modulating the frequency and amplitude of IPSCs recorded from CA1 pyramidal neurons. Specifically, incubating hippocampal slices with APV, a competitive NMDA receptor antagonist, did not alter interevent intervals or amplitudes of IPSCs recorded from either group of 21-day-old mice. Incubating hippocampal slices from 21-day-old wild type mice in the presence or absence of α-BGT did not affect the interevent intervals and amplitudes of IPSCs recorded from CA1 pyramidal neurons; however, similar perfusion of hippocampal slices from 21-day-old homozygous *mKat-2*$^{-/-}$ mice with α-BGT decreased the amplitude and frequency of IPSCs recorded from pyramidal neurons, normalizing them to the recordings from age-matched wild type control mice (Alkondon et al. 2004). Moreover, perfusion of hippocampal slices from 21-day-old homozygous *mKat-2*$^{-/-}$ mice with KYNA reduced the frequency of IPSCs recorded from CA1 pyramidal neurons; in contrast, similar in situ exposure of hippocampal slices from wild type mice to KYNA did not affect the cumulative distribution of their interevent intervals of IPSCs. The authors also showed that the increased "activity" of α_7-containing nAChRs on the CA1 interneurons of the homozygous *mKat-2*$^{-/-}$ mice was not an epiphenomenon of changes in their dendritic length; prior work by these authors showed relationships between dendritic lengths of cultured hippocampal neurons and hippocampal CA1 interneurons in slices and the density of α_7-containing nAChRs (Alkondon et al. 2004). Overall, these data are consistent with a developmentally dependent modulatory role of KYNA, a naturally occurring tryptophan metabolite, on α_7 nAChR-mediated inhibitory input onto CA1 pyramidal neurons. The data also highlight the important role the astrocyte plays, via its production and local release of KYNA, in regulating the synchronous oscillatory output of CA1 pyramidal neurons specifically and, perhaps, assemblies of neocortical pyramidal neurons in general.

KYNA has low penetrability across the blood-brain barrier; thus, its levels in brain are derived locally from kynurenine, its "brain-penetrable" precursor (Albuquerque and Schwarcz 2013). As discussed, kynurenine is irreversibly transaminated by the KAT II isoenzyme enriched in astrocytes, which store this precursor substrate and serve as the source of the neuroactive pool of KYNA. The "probenecid-sensitive organic acid transporters" located on brain capillary endothelial cells mediate efflux of KYNA out of the brain and, thereby, regulate its extracellular levels in brain (Albuquerque and Schwarcz 2013). Importantly, the local concentration of KYNA in the "tripartite synapse" is likely to be higher than the low nanomolar range measured in CSF and microdialysis studies. Although not consistently reported across all laboratories, KYNA was reported to noncompetitively inhibit (IC_{50} in the low μM range) α_7 nAChRs on cultured hippocampal neurons; data also suggest that the inhibitory effect of KYNA on α_7 nAChRs could be blocked by galantamine, which possesses allosteric agonist properties at nAChRs (Albuquerque and Schwarcz 2013). Differences in methodological issues, such as the method used for application of the agonist pulse, site on the neuron where agonist pulses are applied (i.e., dendrite versus soma), and nature of the neuron (e.g., GABA inhibitory neuron versus pyramidal neuron), could account for some of the failures to replicate KYNA's inhibitory activity on α_7 nAChRs. Data suggest that α_7 nAChRs located on CA1 pyramidal neurons may be more sensitive to inhibitory effects of KYNA than α_7 nAChRs on GABAergic neurons. Moreover, as CA1 interneurons age (i.e., PD23-PD35 versus PD10-PD18), they become more sensitive to inhibitory effects of KYNA (Albuquerque and Schwarcz 2013). Elucidation of the neuromodulatory role of KYNA and factors influencing susceptibility of α_7 nAChRs to its inhibitory modulation may lead to development of KYNA-based targeted interventions for selective indications.

PAM-2 (3-furan-2-yl-*N*-*p*-tolyl-acrylamide), an α_7 nAChR-selective type II PAM that reactivates desensitized α_7 nAChRs, and DMXB-A, a partial α_7 nAChR-selective agonist, were shown to attenuate ketamine-induced cognitive and social deficits in male rats (Potasiewicz et al. 2017). These selective α_7 nAChR agonist interventions improved a ketamine-induced deficit of cognitive flexibility in an attentional set-shifting task and a ketamine-induced deficit of novel object recognition. Further, PAM-2 and DMXB-A also reversed ketamine-induced social withdrawal when social interaction was assessed in an open-field arena (Potasiewicz et al. 2017). The preclinical data showing therapeutic effects of acute administration of PAM-2 and DMXB-A in a "rat ketamine model of NMDA receptor hypofunction" encourage therapeutic exploration of targeted α_7 nAChR-selective agonist interventions for an ASD indication. Presumably, the type II PAM would have a lesser liability for changing the sensitivity of the α_7 nAChR upon chronic administration (Deutsch et al. 2008a, b, 2011, 2013, 2014, 2015, 2016; Potasiewicz et al. 2017).

4 Nicotinic Receptor-Mediated Regulation of GABAergic Interneurons

Diminished expression of PV and biomarkers consistent with dysfunctional GABAergic interneurons were observed in mice with *null* expression of *Chrna7*, the gene encoding the α_7 nAChR subunit in mice (Lin et al. 2014). For in vitro studies, primary cortical cultures were prepared from embryonic (E17–19) brain of α_7 nAChR *knockout* (α7-KO) and wild type control mice, and the cultures were studied at 21–28 *days* in vitro (DIV), using western blotting, immunocytochemistry, and patch-clamp recording. Immunochemical studies were also conducted in post-natal cortical tissue homogenates and coronal brain slices of α7-KO and wild type mice of both sexes. In cortical homogenates, western blots revealed significantly reduced immunoreactive content of PV at postnatal day (PD) 21 and PD56 in α7-KO mice, which persisted when examined again at age 9 months, compared to wild type homogenates (Lin et al. 2014). Similarly, the immunoreactive content of GAD65 and the α_1 subunit of the $GABA_A$ ($GABA_A\alpha_1$) receptor was significantly reduced on PD21 and PD56 in cortical homogenate of α7-KO mice, compared to cortical homogenate from wild type controls. Confocal microscopic examination of immunofluorescently stained GABAergic markers in mouse prefrontal cortex revealed that immunostaining intensity for PV, GAD65, GAD67, and $GABA_A\alpha_1$ was reduced in the α7-KO mice at PD56 and PD90, compared to wild type littermates.

The reduced immunoreactive content of $GABA_A\alpha_1$ in prefrontal cortex was observed in both PV-expressing GABA interneurons and pyramidal neurons (Lin et al. 2014). The reduced expression on pyramidal neurons is consistent with a disrupted inhibitory input onto assemblies of pyramidal neurons in α7-KO mice that could adversely affect their synchronous oscillatory output. Importantly, although the immunoreactive content of PV was significantly reduced, the number of PV-expressing GABAergic neurons in prefrontal cortex in the α7-KO mice did not differ significantly from wild type controls (Lin et al. 2014). This latter finding of reduced PV expression but an unaltered number of PV-expressing GABAergic neurons has been described in a number of validated mouse models of ASD (Peñagarikano et al. 2011; Cea-Del Rio and Huntsman 2014; Cellot and Cherubini 2014; Tomassy et al. 2014; Filice et al. 2016; Burket et al. 2017). Moreover, these data are consistent with a regulatory role of the α_7 nAChR in the function of PV-expressing GABAergic interneurons (i.e., decreased expression of PV, an important intracellular Ca^{2+}-binding protein). Also, the GABAergic deficit appeared to have selectivity for PV-expressing neurons as the immunoreactive content of somatostatin in prefrontal cortex did not differ between the α7-KO mice and their wild type littermates. Deletion of *Chrna7* (and *null* expression of the α_7 nAChR subunit) was associated with disruption of the presynaptic terminal in the prefrontal cortex of α7-KO as shown by a significantly reduced immunoreactive content of VGAT, the vesicular GABA transporter, a marker of presynaptic terminals in GABAergic neurons (Lin et al. 2014). In vitro immunocytochemical and electrophysiological studies in cultured cortical cells from α7-KO mice and wild type littermates confirmed the creation of dysfunctional synapses and impaired GABA-mediated inhibitory tone in the α7-KO mice. Thus, immunostaining of these cultured cortical cells showed reduced content of VGAT and $GABA_A\alpha_1$ in α7-KO cortical cultures. Further, whole-cell voltage clamp recordings showed reduced frequency and decay time of "spontaneous inhibitory postsynaptic currents (sIPSC)" in α7-KO cortical cultures (Lin et al. 2014). Immunocytochemical examination of the cultured cortical cells double-labeled with antibodies to PV and GAD65 showed impairment of both development of PV-expressing basket cells and formation of perisomatic GABAergic synapses in the α7-KO cultured cortical cells. Finally, immunostaining of the cultured cortical cells showed that the NR1 subunit of the NMDA receptor was expressed by GABAergic neurons, and its immunoreactive content is reduced in α7-KO cultured cortical cells (Lin et al. 2014). The data are consistent with "cooperativity" and "cross-talk" between the α_7 nAChR and NMDAR in their contributions to the regulation of firing of the PV-expressing GABAergic inhibitory basket cell and, ultimately, the synchronous oscillatory output of cortical pyramidal neurons (Mastropaolo et al. 2004; Lewis and González-Burgos 2008; Deutsch et al. 2008a, 2010; Gonzalez-Burgos and Lewis 2012; Gonzalez-Burgos et al. 2015; Enwright et al. 2016; Pafundo et al. 2018). Diminished expression of the α_7 nAChR may lead to decreased expression of functional NMDARs, which is predicted to be associated with increased frequency of nonsynchronous pyramidal cell firing and adverse functional consequences in human (e.g., increased likelihood of seizure activity and impairments of working memory and, perhaps, social cognition).

Preclinical exploration of α_7 nAChR agonists for potential pro-cognitive therapeutic indications led to the observation and description of "priming." Essentially, at doses often significantly below the EC_{50} for a biological effect in the preclinical screen (e.g., induction of long-term potentiation [LTP] in rat brain septo-hippocampal slice preparations), the test compound can increase (i.e., "prime") the response to acetylcholine, the natural endogenous ligand (Townsend et al. 2016). A recent study showed that at low nanomolar concentrations, FRM-17848 [(*R*)-7-cyano-*N*-quinuclidin-3-yl]benzo[*b*]thiphene-2-carboxamide], a selective α_7 nAChR agonist, enhanced LTP production in a hippocampal brain circuit mediated by $GABA_A$ α_5-receptors ($GABA_A\alpha_5R$) at a low nanomolar concentration (3.16 nM), consistent with the proposed mechanism of priming. The EC_{50} for FRM-17848 (= 455 nM) and the K_i (= 11 nM) were determined with ^{125}I- α-BGT using an in vitro filtration binding assay with rat brain homogenate (Townsend et al. 2016). Priming was demonstrated in whole-cell recordings made from *Xenopus* oocytes expressing the human α_7 nAChR, whereby the addition of the 3.16 nM concentration of FRM-17848 to the perfusate enhanced the currents evoked in response to 40 μM of acetylcholine. Recordings from CA1 pyramidal neurons in septo-hippocampal slices showed that the 3.16 nM concentration of FRM-17848 caused small but significant hyperpolarization of most cells and increased the frequency and amplitude of IPSCs; the latter was reversed 20 min after the washout of FRM-17848 and inhibited by MLA (50 μM). Although counterintuitive, this α_7 nAChR-mediated inhibitory influence on CA1 pyramidal neurons was shown to be involved in its enhancement of LTP via the probable mechanism of priming. Specifically, a $GABA_A$ α_5R antagonist that had no effect by itself on the induction of LTP and essentially showed no interaction with the α_7 nAChR at the dose studied in this experiment inhibited the enhancement of LTP by FRM-17848. Consistent with the involvement of the $GABA_A$ α_5R in the enhancement of LTP by FRM-17848, a positive allosteric modulator (PAM) of the $GABA_A$ α_5R was similarly shown to enhance LTP. The data suggest that selective activation of a specific subpopulation of α_5-containing $GABA_A$ receptors contributes to the enhancement of LTP and is predicted to have pro-cognitive effects. The unexpected and somewhat paradoxical effects of potentiating $GABA_A$ α_5R-mediated neurotransmission by FRM-17848, as well as the FRM-17848-induced hyperpolarization and increased IPSCs in CA1 pyramidal neurons, are difficult to reconcile with its enhancement of LTP. However, there are data that the authors reference reporting a $GABA_A$ α_5R PAM increasing the bursting activity of neocortical neurons. In any event, the data clearly implicate the α_7 nAChR's involvement in the output of CA1 pyramidal neurons, as reflected in enhancement of LTP and increased frequency of IPSCs by FRM-17848 (Townsend et al. 2016).

The intensity of immunostained L-glutamic acid decarboxylase-65 (GAD65), a GABA synthetic enzyme, was significantly reduced in the hippocampal CA3 region in adult male and female (age 5–6 months) mice with heterozygous deletions of *Chrna7* (Het-*Chrna7*$^{+/-}$), the gene encoding the α_7 nAChR; comparisons were made to same-age, same-sex wild type controls (Adams et al. 2012). Autoradiographic intensity of the binding of ^{35}S-TBPS, a cage convulsant that labels the $GABA_A$ receptor's chloride ion channel, was significantly reduced in the dentate gyrus and CA3 and CA1 hippocampal regions of adult male Het-*Chrna7*$^{+/-}$ mice, compared to same-age male wild type controls. The intensity of immunohistochemical staining of GABA and GAT-1, the vesicular GABA transporter, revealed no significant differences between genotypes and gender in hippocampus of adult mice (Adams et al. 2012). The authors suggested that the reduced intensity of GAD65 immunostaining in the mice with heterozygous deletions of *Chrna7* could result in decreased activity-dependent release of GABA in hippocampus and, thereby, explain their earlier report of deficits in hippocampal auditory sensory processing in these mutant mice (Adams et al. 2008). The mechanisms and implications for the gender-specific decrease in hippocampal $GABA_A$ receptor binding in adult male Het-*Chrna7*$^{+/-}$ mice are not known but are consistent with deficits in central inhibitory tone in the male mutant mice.

Data describing a relationship between lowered density of hippocampal α_7 nAChRs and deficient hippocampal auditory gating (Adams et al. 2008), as well as the location of this receptor on GABAergic neurons, prompted a study of the effects of PNU-282987 (*N*-[3(*R*)-1-azabicyclo[2.2.2]oct-3-yl]-4-chlorobenzamide hydrochloride), a selective α_7 nAChR agonist, on evoked currents in cultured rat hippocampal neurons, modulation of GABAergic synaptic activity recorded from CA1 pyramidal neurons in hippocampal slices, and auditory gating disrupted by amphetamine in the hippocampal CA3 region and reticular thalamic nucleus of anesthetized rats (Hajós et al. 2005). PNU-282987 has high affinity for rat α_7 nAChRs ($K_i = 26$ nM) and negligible interaction with other pentameric nAChRs. Whole-cell currents were evoked by 1-second applications of PNU-282987 in hippocampal neurons cultured from PD3 Sprague-Dawley rats that were antagonized by MLA. Moreover, PNU-282987 increased the frequency of spontaneous GABAergic synaptic events recorded from CA1 pyramidal neurons. PNU-282987 also normalized the ratio disrupted by amphetamine of the amplitude of the evoked response by the second of a pair of tones to the response evoked by the first tone; the evoked responses were recorded in the hippocampal CA3 region of anesthetized rats and measured by the potential difference between the positive P20 and negative N40 deflections. The disruptive effects of amphetamine on the evoked amplitudes to the first (decreased) and second (increased) of the pair of tones were also attenuated by PNU-282987. Disrupted auditory gating of reticular thalamic neurons by amphetamine, as measured by an increased ratio of recorded spikes evoked by the second tone to the number of spikes evoked by the first tone, was also normalized by PNU-282987 (Hajós et al. 2005). Consistent with an emerging literature, the data suggest that acute administration of a selective α_7 nAChR agonist can stimulate GABAergic inhibitory influences onto pyramidal neurons and restore disrupted auditory sensory gating in hippocampus. Of course, complex changes in receptor sensitivity and expression related to the duration and dose of the agonist "pulse" and chronicity of "treatment" must be understood as translational therapeutic implications are explored.

C3H mice with heterozygous deletions of the gene coding the α_7 nAChR subunit were used to explore the role the α_7 nAChR plays in "hippocampal circuit function"; specifically, the excitability of hippocampal CA3 pyramidal neurons and auditory gating were studied in "C3H α7 receptor null heterozygous" and C3H wild type mice (Adams et al. 2008). Previous work from these investigators showed that DBA/2 mice, a genetically inbred mouse strain with a 35% reduction in the density of hippocampal α_7 nAChRs, compared to the C3H strain, as measured with radiolabeled α-BGT, displayed impaired hippocampal auditory gating, which improved with administration of α_7 nAChR agonists (Stevens et al. 1998). The α_7 nAChR is present on GABAergic neurons, and its activation caused release of GABA in cultured hippocampal neurons and hippocampal slices. Hippocampal auditory gating was measured in CA3 pyramidal neurons of anesthetized C3H α7 heterozygote and C3H wild type mice; auditory-evoked potentials (i.e., P20 and N40) elicited by a pair of tones presented 500 milliseconds apart were measured. Normal auditory gating was defined as the ratio of the amplitude of the N40 wave measured relative to the peak of the P20 wave evoked by the second tone to the similarly measured amplitude evoked by the first tone ≤ 0.5, whereas deficient auditory gating was defined as the ratio >0.5. Relative to the C3H wild type mice, the density of α-BGT binding in the C3H α7 heterozygous mice was significantly reduced in the dentate gyrus and CA1 and CA3 regions of the hippocampus (Adams et al. 2008). The ratio of the auditory-evoked potentials to the second versus the first tone of the stimulus pair was 1.17 ± 0.15 for the group of 10 C3H α7 heterozygous mice, consistent with deficient auditory gating, whereas the ratio in the group of 10 C3H wild type mice was 0.46 ± 0.06, consistent with normal auditory gating (Adams et al. 2008). The amplitudes of the evoked auditory potentials were significantly greater in the C3H α7 heterozygous mice, which is consistent with an increased responsivity of the CA3 pyramidal neurons in the heterozygous mice. The data show that the α_7 nAChR contributes to the regulation of CA3 pyramidal neuron responsivity and hippocampal auditory gating (Adams et al. 2008).

The C3H *Chrna7* heterozygous (C3H *Chrna7* Het) and homozygous KO (C3H *Chrna7* KO) mice were also used in a follow-up study to examine relationships between α_7 nAChR expression and PV expression by a subtype of hippocampal GABAergic neuron (Bates et al. 2014). Interestingly, and somewhat unexpectedly, PV immunoreactive protein content in hippocampal lysate was "inversely" related to the "dose" of *Chrna7* in male mice; that is, lowest expression of PV was in the wild type controls, and expression was increased by 39% and 52% in C3H *Chrna7* Het and C3H *Chrna7* KO mice, respectively (Bates et al. 2014). Further, there was a significant interaction between genotype and gender: male C3H *Chrna7* KO mice had (29%) higher PV immunoreactive protein content in hippocampal lysate than female C3H *Chrna7* KO mice. Similarly, reduced dosage of *Chrna7* was associated with higher immunoreactive protein content of hippocampal GAD67, and the effect of reduced gene dosage on increased hippocampal GAD67 expression appeared to be greater in female C3H *Chrna7* Het mice than male C3H *Chrna7* Het mice. There were complex interactive effects of *Chrna7* gene dosing and gender on the expression of specific $GABA_A$ receptor subunits. Although somewhat unexpected, the data are consistent with significant α_7 nAChR-mediated effects on hippocampal $GABA_A$ receptor-mediated neurotransmission and GABA interneuron function (Bates et al. 2014; Burket and Deutsch 2019). (Perhaps, some of the unexpected findings reflect compensatory changes within critical hippocampal circuits.) Again, GABA release from GABA interneurons synchronizes the oscillatory output of pyramidal outflow neurons in neocortex.

The laminar location of interneurons within rat prefrontal neocortex can determine their sensitivity to subtype-selective nAChR activation; specifically, the activation and synchronization of a network of layer I interneurons, which contains the highest laminar densities of cholinergic axons and varicosities, were dependent on the heteropentameric $\alpha_4\beta_2$ nAChR (Bandyopadhyay et al. 2006). This $\alpha_4\beta_2$ nAChR-dependent activation of GABA waves enjoyed spatial selectivity and was most intense in the superficial layers of cortex, and GABA wave activity spread horizontally in both directions of a coronally sectioned brain slice. These horizontally spreading GABA waves in the upper layers of the neocortex were evoked and enhanced by neostigmine, an acetylcholinesterase inhibitor, and 1,1-dimethyl-4-phenyl-piperazinium iodide (DMPP), a nAChR agonist, and were not antagonized by atropine, consistent with their induction dependent on nAChR, as opposed to mAChR, stimulation. However, their evocation and enhancement by neostigmine were not blocked by MLA but were antagonized by dihydro-β-erythroidine (DHβE), a selective antagonist of heteropentameric $\alpha_4\beta_2$ nAChRs (Bandyopadhyay et al. 2006). Not surprisingly, the spatial distribution or laminar enrichment of homopentameric α_7 nAChRs and heteropentameric $\alpha_4\beta_2$ nAChRs is not random. Given the complexity of the phenotype and dependence of horizontal spreading of GABA waves in the upper layers of neocortex on $\alpha_4\beta_2$ nAChRs, it may be naïve to assume that autism-like psychopathology could result from an exclusive pathologic involvement of only one nAChR subtype or even only one neurotransmitter system. The lesson of clozapine's effectiveness in medication-refractory schizophrenia is sobering and relevant to medication development strategies for ASD; clozapine enjoys little in the way of target specificity. Complex neuropsychiatric disorders will almost certainly involve an imbalance of neurotransmission mediated by multiple neurotransmitters; moreover, targeting only one neurotransmitter receptor in ASD and other neuropsychiatric disorders should alert clinicians to be vigilant for the possible emergence of side effects and other unanticipated findings due to compensatory changes in the sensitivities of a variety of neurotransmitter receptors. Thus, the potential role of $\alpha_4\beta_2$ nAChRs in higher cortical-dependent cognitive and social functions must always be considered as targeted α_7 nAChR agonist therapeutic strategies are explored.

5 15q13.3 Copy Number Variants, α_7 Nicotinic Acetylcholine Receptors, and the Pathogenesis of Neurodevelopmental Disorders

Copy number variants (CNVs), including microdeletions and less commonly microduplications, at chromosome 15q13.3 (the genetic locus containing *CHRNA7*, the gene encoding the α_7 nAChR subunit), and sequence variants at the promoter region of *CHRNA7* are rarely implicated in the etiology of ASD, intellectual disability (ID), seizures, and schizophrenia (Leonard et al. 2002; Deutsch et al. 2010; Bacchelli et al. 2015). The extent of the recurrent 15q13.3 microdeletion is usually about 1.5 Mb and encompasses six genes: *MTMR15*, *MTMR10*, *TRPM1*, *KLF13*, *OTUD7A*, and *CHRNA7*, in addition to hsa-mir-211, a miRNA gene. Dose-dependent changes in *CHRNA7* expression are thought to be most responsible for the neurodevelopmental phenotypes associated with 15q13.3 microdeletions because smaller deletions encompassing essentially only *CHRNA7* have been associated with neurodevelopmental phenotypes (Deutsch et al. 2016). *CHRNA7* sequence variants that may be etiologically associated with ASD and other neurodevelopmental phenotypes are hard to detect because of the existence of *CHRFAM7A*, a hybrid gene containing exons A-E of *FAM7A* fused to a 300-kb-long duplication of exons 5–10 and the 3′ end of *CHRNA7* (Bacchelli et al. 2015; Deutsch et al. 2016). Thus, it is difficult to resolve sequence changes specific to *CHRNA7* that may be associated with neurodevelopmental abnormalities because of the nearly identical sequences of *CHRFAM7A*, a hybrid gene of uncertain significance. Rare CNVs at chromosome 15q13.3 and sequence variants at the *CHRNA7* locus were investigated in 135 Italian subjects with ASD from 133 families, whose diagnoses were confirmed with the Autism Diagnostic Interview-Revised (ADI-R) and the Autism Diagnostic Observation Schedule (ADOS); the rates of genetic variations in the affected sample were compared to 174 unaffected controls (Bacchelli et al. 2015). A paternally inherited microduplication was detected in a single affected male subject (age 5 years and 5 months) diagnosed with a regressive presentation of ASD; after a period of apparently normal development, regression of language, refusal of physical contact, feeding disturbance, and hyperactivity were appreciated at 18 months in this subject. A follow-up telephone inquiry disclosed the onset of complex partial seizures with generalization at age 13 years. The microduplication was detected in the unaffected father, and the paternal grandfather's sister and brother were reported to have ID and speech delay, respectively. The microduplication was approximately 500 kb and spanned the entire *CHRNA7* gene and exon 1 of the longer isoform of *OTUD7A* (Bacchelli et al. 2015). Interestingly, the rare, but more commonly noted 15q13.3 microdeletion was not detected in this sample of 135 Italian subjects with ASD. Because sequence identity between exons 5–10 in *CHRNA7* and the 300 kb duplication in *CHRFAM7A* is >99%, a "long-range PCR sequencing strategy" was employed that took advantage of one of three small insertions (i.e., a 3' UTR 36 bp insertion) discovered in *CHRNA7* and absent in *CHRFAM7A* to design a primer and amplify exons 5–10 specific for *CHRNA7*, enabling resolution of *CHRNA7* and *CHRFAM7A* from each other (Bacchelli et al. 2015). As noted, no deletions and only one paternally inherited microduplication, and four rare sequence variants (<1% minor allele frequency) were found in the sample of 135 Italian subjects with ASD. One of these four rare variants in the affected sample was a maternally inherited non-synonymous variant in exon 10 resulting in a glutamate to lysine substitution that was felt to be of no or little pathogenic significance because, in addition to the proband's mother being unaffected, three individuals out of 125 unrelated and unaffected Italian controls that underwent this "long-range PCR sequencing strategy" had this same rare sequence variant. The other three rare variants found in the affected sample were in the promoter region, and one of these rare promoter variants (located at −241 bp from ATG) was previously reported to interact epistatically with a "more frequent variant in the 5' UTR region" to decrease transcriptional expression of *CHRNA7* (Leonard et al. 2002; Bacchelli et al. 2015). Interestingly, one subject in the affected sample carried this double variant with the rare −241 bp sequence variant on the paternal chromosome and the more common variant in the 5' UTR region (−86 bp from ATG) on the maternal chromosome (Bacchelli et al. 2015). The coexistence of both promoter variants that suppress transcription of *CHRNA7* (i.e., −241 bp and −86 bp from ATG) appears to be a very rare event in the general population (Bacchelli et al. 2015).

Low copy repeats (LCRs) occurring at six breakpoint regions (BP1–BP6) and LCRs of a distal area of *CHRNA7* that is located proximally to the BP5 breakpoint within the region of the first exon of *OTUD7A* (i.e., distal *CHRNA7*-LCR) account for the instability of the genomic architecture in the region spanning q11 through q13 of chromosome 15; recurrent CNVs in this region result from non-allelic homologous recombination (NAHR) between LCRs (Gillentine and Schaaf 2015; Deutsch et al. 2016; Gillentine et al. 2017). Descriptions of the pathogenic phenotype associated with deletion of a critical region encompassing *CHRNA7* and, possibly, the first exon of *OTUD7A*, its proximal neighboring gene, strongly support relationships between reduced genetic expression of the α_7 nAChR and a range of neuropsychiatric disorders, including ID, ASD, seizures, and schizophrenia (i.e., the 15q13.3 microdeletion syndrome) (Deutsch et al. 2010, 2011, 2015, 2016). However, the pathogenicity and phenotypes reliably associated with microduplications ascribed to *CHRNA7* have been harder to identify because these duplications are often found with similar frequencies in reported samples of unaffected "controls" and referred patient samples (Gillentine et al. 2017). 15q13.3 microduplications are usually detected by chromosome microarray analysis (CMA) in samples of patients clinically referred for genetic evaluation of developmental delays/disabilities (DD), ID, congenital anomalies, and ASD (Gillentine et al. 2017). Determining associations of 15q13.3 microduplications with cognitive and behavioral phenotypes and ascertainment of their true prevalence, especially in control populations, are difficult and confounded by incomplete penetrance and variable expressivity. The majority of the 15q13.3 microduplications in referred patient populations, resulting in an increased gene dosage of *CHRNA7*, seem to result from NAHR between BP5 and the distal *CHRNA7*-LCR. The frequency of this microduplication in referred patient populations is about 0.8% (1 in 125); however, some estimates in control samples range from 0.55% to 0.62% (Gillentine et al. 2017). Eighteen patients (ages 5–14 years) referred for CMA because of a variety of conditions, including CNS disorder, encephalopathy, ASD, DD, short stature, speech delay, dysmorphic features, ADHD, ID, dyslexia, and macrocephaly, among other conditions, and found to have 15q13.3 microduplications underwent 2-day comprehensive assessments in order to better characterize phenotypes associated with duplications of 15q13.3 (Gillentine et al. 2017). The ~145 kb duplications due to NAHR between LCRs at the BP5 and distal *CHRNA7*-LCR loci were detected in 16 patients. The microduplication was inherited from either the father ($N = 9$) or mother ($N = 8$) in the 17 subjects with parental CMA data. Eleven of the 17 parental carriers of the duplications had neuropsychiatric histories, including histories of bipolar disorder, depression, ID, speech delay, dyslexia, ADHD, schizophrenia, learning disability, and apraxia. IQ scores of the 17 probands tested with the *Differential Ability Scale, Second Edition* (DAS-II) were shifted to the left when compared to curves of normally distributed scores with a mean score of 100 [SD = 15], and the full-scale ratio IQ (FSRIQ) scores of five probands were below 70. Probands also met criteria for language impairment (11 out of 18), ADHD (8 out of 18), and ASD ($N = 6$ out of 17 meeting both ADI-R and ADOS criteria). The study was consistent with incomplete penetrance of the 15q13.3 duplication as it was inherited in 6 of 17 probands from a seemingly unaffected parent and one affected proband had an unaffected sibling with the duplication. Variable expressivity of the phenotypes was observed in the 16 probands with the ~145 kb CHRNA7-LCR/BP5 duplications. The variable expressivity could be due to single nucleotide sequence variants or differences in genetic backgrounds between probands. Schizophrenia was not diagnosed in any of the probands, which may reflect that all of them were aged 14 years or less and, thus, had not reached age of greatest risk. Overall, the clinical phenotypes of the affected probands with the 15q13.3 microduplications were less severe than the clinical phenotypes observed in patients diagnosed with the 15q13.3 microdeletion syndrome (Deutsch et al. 2010, 2011, 2014, 2015, 2016; Gillentine et al. 2017). The data do suggest that a heterogeneous range of neuropsychiatric phenotypes can be associated with increased, as well as decreased, dosage of *CHRNA7* and/or altered transcriptional efficiency related to the presence of these CNVs.

The phenotypes of 42 probands with "small" heterozygous deletions spanning no more than all of *CHRNA7*$^{+/-}$ and the first exon of *OTUD7A* were summarized in order to characterize genotype-phenotype relationships (Gillentine and Schaaf 2015). LCR elements of distal *CHRNA7* containing NAHR "hotspots" occur proximal to the BP5 breakpoint and lie within exons 1 and 2 of *OTUD7A*; the latter gene is proximal to *CHRNA7* (Gillentine and Schaaf 2015). As noted, NAHR events between D-*CHRNA7*-LCR and BP5 can lead to small microdeletions encompassing the entire *CHRNA7* gene; based on enzyme kinetic considerations, haploinsufficient expression of the deubiquitinase encoded by *OTUD7A* was not thought to have pathogenic consequences. (However, very recent data that will be reviewed below challenge the assumption that haploinsufficient expression or DNA sequence variants of *OTUD7A* lack pathogenic consequences (Uddin et al. 2018; Yin et al. 2018).). The summary of this series of 42 probands with 15q13.3 heterozygous microdeletions showed that haploinsufficient expression was associated with pathogenic consequences and variable expressivity, including unaffected neurotypical presentations in 6 of 18 parental carriers of the deletion. Thirty seven of the 42 probands were reported to be "neuroaffected," but there was marked phenotypic variability among probands, many of whom had more than one clinical condition (Gillentine and Schaaf 2015). Thus 19 out of the 37 (51%) neuroaffected probands had cognitive deficits (including ID, developmental delay, and language difficulties); 13 (35%) had seizures, epilepsy, and/or EEG abnormalities; 6 (16%) were reported to have ASD; 5 (13.5%) were reported to have ADHD or attention difficulties; and 5 (13.5%) were reported to have language or speech impairments (Gillentine and Schaaf 2015). Theoretically, early identification of a 15q13.3 deletion and early intervention with an α_7 nAChR-selective PAM may improve the phenotype and functional outcome of neuroaffected probands. Also, although it may be an infrequent cause of ASD, the 15q13.3 microdeletion syndrome suggests that diminished or impaired transduction of the acetylcholine signal by the α_7 nAChR can be a pathogenic mechanism (Deutsch et al. 2010, 2011, 2015, 2016).

As noted, the extent of the 15q13.3 microdeletion syndrome between breakpoints BP4-BP5 can be as large as 1.53 Mb (i.e., the "typical" deletion) encompassing as many as seven protein-coding genes, one microRNA, and two "pseudogenes"; clinically, the deletion is typically heterozygous (Uddin et al. 2018; Yin et al. 2018). The heterozygous microdeletion syndrome is associated with ID, ASD, epilepsy, and schizophrenia and has focused attention on decreased expression of the α_7 nAChR,and *CHRNA7* as the "driver" gene most responsible for the pathogenesis of the variably expressed neurodevelopmental phenotypes. However, the identification of unaffected control subjects carrying *CHRNA7* deletions and clinically affected subjects, whose *CHRNA7* deletions overlap the first exon of *OTUD7A*, the gene encoding OTU deubiquitinase 7A, has stimulated interest in possible pathogenic contributions of this latter gene and its haploinsufficient expression to the 15q13.3 microdeletion syndrome (Uddin et al. 2018; Yin et al. 2018). In fact, behavioral studies, neuronal cultures, and transcriptome and proteome data analyses conducted in mouse models, including a mouse with a syntenic heterozygous deletion (i.e., *Df(h15q13)*/+) and an *Otud7a*-null mouse, as well as whole genome and exome sequencing data of clinically affected subjects implicate a pathogenic role for *OTUD7A*. The deubiquitinating enzyme localizes to dendritic spine compartments, and its protein-protein co-expression network is relevant to development and function of synapses (Uddin et al. 2018).

Importantly, there can be marked variability in the expression of ASD-like behavioral phenotypes in mouse models of ASD and behavioral assays may be too insensitive to detect significant differences between mouse "models" and appropriate comparator or wild type strains. This was most dramatically shown in a recent comparison of *Chrna7*-deficient mice bred on a C57BL/6J background and wild type controls on a variety of behavioral measures (Yin et al. 2017). Specifically, with the sole exception of a significant genotype x sex interaction on the marble burying test, an example of a repetitive behavior (male *Chrna7* KO mice performing significantly worse than male wild type controls), the three genotypes (i.e., groups of Chrna7 heterozygous [HET] and homozygous [KO] mutant and wild type mice) did not differ on a variety of other behaviors "mimicking" or related to ASD (Yin et al. 2017). Behavioral measures that failed to distinguish the phenotypic groups included self-grooming, social preference in the three-chamber apparatus (i.e., measuring preference for an enclosed socially salient mouse versus an inanimate inverted cup), and reciprocal social interactions. These data draw attention to the possible etiopathogenic significance of other genes within the microdeletion, epigenetic modifications, and/or neuroplastic changes within circuits that may have "compensated" for deficits of signal transduction by α_7 nAChRs, in addition to other possibilities.

RNA sequencing was used to measure cortical gene expression at three developmental stages in *Df(h15q13)*/+ and wild type mice: embryonic day 16 (E16), postnatal day 21 (P21), and adult (P63) (Uddin et al. 2018). The transcriptional data suggested disruption of cortical development in *Df(h15q13)*/+ mice. Histological analysis of layers 2/3 of prefrontal cortex (PFC) revealed reduction in both spine density and mature mushroom-shaped spines; reduction in spine length and spine neck length; and decreased dendritic arborization in the *Df(h15q13)*/+ mice, compared to wild type controls. Similar histological findings were observed in cultured primary cortical neurons derived from *Df(h15q13)*/+ mouse embryonic brains (E16), compared to cultured cortical neurons from wild type embryonic mice. This group explored the pathogenicity of the 15q13.3 microdeletion in clinical microarray data of 38,325 affected subjects with neurodevelopmental disorders and a control population of 22,241 subjects: 156 affected subjects were identified with typical BP4-BP5 15q13.3 microdeletions, whereas only 1 control subject had the typical BP4-BP5 deletion (a highly significant difference $p < 1.30 \times 10^{-29}$) (Uddin et al. 2018). However, when smaller deletions were determined in order to delineate the "minimal region" and narrow the potential gene candidates contributing to the pathogenesis, 52 controls were found to possess microdeletions confined to *CHRNA7*, and a 5-year-old girl with global developmental delay had a BP4-BP5 genetic deletion that included *OTUD7A* but not *CHRNA7* (Uddin et al. 2018). Again, the minimal region deletions in 43 affected subjects were "overlapping," including *CHRNA7* and also impacting *OTUD7A*. DNA sequence-level mutations of the genes contained within the 15q13.3 microdeletion were explored in affected subjects with neurodevelopmental disorders, who did not have a 15q13.3 microdeletion; three of eight de novo sequence-level mutations involved *OTUD7A* (Uddin et al. 2018). *OTUD7A* was shown to possess a brain-specific mRNA expression pattern. Further, *OTUD7A* was the only gene within the 15q13.3 microdeletion region shown to be part of the "brain-specific protein module" in a "weighted gene co-expression network analysis (WGCNA)" (Uddin et al. 2018). Expression data showed that *Otud7a* and a transfected human transcript are expressed in soma and dendrites and a fraction colocalized with PSD-95, consistent with expression in the postsynaptic excitatory synapse (Uddin et al. 2018; Yin et al. 2018). In utero electroporation to transfect developing layers 2/3 PFC neurons of *Df(h15q13)*/+ mice with human *OTUD7A*, as well as transfection of cultured *Df(h15q13)*/+ cortical neurons with human *OTUD7A*, rescued the morphological deficits associated with haploinsufficient expression of *Otud7a*, including rescuing of reduced spine density, reduced spine length, reduced proportion of mature mushroom spines, and reduced dendritic branching (Uddin et al. 2018). The data clearly support critical roles of *OTUD7A* in normal synapse/spine development and the pathogenesis of neurodevelopmental disorders associated with 15q13.3 microdeletions (Uddin et al. 2018; Yin et al. 2018).

Homozygous knockout (KO) and heterozygous (HET) mice with deletions of *Otud7a* on a C57BL/6J genetic background and C57BL/6J wild type mice were studied to determine the pathogenic contribution of *Otud7a* to the 15q13.3 microdeletion syndrome (Yin et al. 2018). The *Otud7a* homozygous null mice showed marked preweaning (30%) weight reduction, compared to the HET and wild type mice, consistent with growth delay. The *Otud7a* homozygous KO mice were delayed in the following milestones, compared to HET and wild type mice: negative geotaxis (i.e., homozygous KO pups were delayed in turning around when placed facedown on a 30^0 incline), cliff aversion (i.e., homozygous KO pups were delayed in avoiding falling off the edge of a paper box), and incisor eruption (Yin et al. 2018). Compared to wild type mice, the *Otud7a*-null mice showed abnormal repetitive spike events most prominently in the frontal cortex and parietal cortex and seizure-like behavioral arrest. A gene dosage-dependent decrease in ultrasonic vocalizations (i.e., KO > HET) was observed in individually separated pups on P6, P8, and P10. The *Otud7a*-null mice had reduced grip strength and impaired rotorod performance; the latter is consistent with deficits in motor coordination and motor learning, compared to HET and wild type mice. Although *Otud7a*-null and HET mice had no significant hearing loss, there was a gene dosage effect with respect to reduction of their acoustic startle response (i.e., KO > HET), and female homozygous KO mice showed significant reduction of the prepulse inhibition (PPI) of their acoustic startle response, consistent with possible impairment of sensorimotor gating (however, interpretation could be confounded by a floor effect resulting from the reduced acoustic startle response in the mice with homozygous deletion of *Otud7a*). Importantly, the number of dendritic spines in primary cortical neurons derived from *Otud7a*-null mice was significantly reduced, compared to primary cortical neurons cultured from wild type littermates; further, the reduced spine density in the mice with the homozygous deletion could be rescued by transfection with the human *OTUD7A* transcript (Yin et al. 2018). The *Otud7a* KO mice did not display deficits of novel object recognition or conditioned fear and did not display deficits of social behavior in the three-chamber apparatus; thus, selective deletion of *OTUD7A* or a loss of function-mutated *OTUD7A* sequence may contribute to a permissive background enabling expression of the 15q13.3 microdeletion syndrome; alternatively, they may contribute to a limited component of the possible variably expressed phenotypes. In any event, transcriptomic data show that *OTUD7A* has higher expression in neuron and oligodendrocyte progenitor cells, consistent with a role in synapse/spine development (Uddin et al. 2018; Yin et al. 2018).

The feasibility of very early intervention to improve signal transduction by the α_7 nAChR was shown in approximately 1-month-old infants using suppression of the P50-evoked potential to the second of a pair of identical auditory stimuli as the outcome measure (Ross et al. 2013, 2016). Expression of the α_7 nAChR is about tenfold higher in fetal, as compared to adult, hippocampus, and its fetal expression facilitates the developmental switch GABA undergoes from an excitatory neurotransmitter in fetal life to the major inhibitory neurotransmitter in the adult brain (Ross et al. 2013, 2016; Deutsch et al. 2017). High millimolar concentrations of choline, a full α_7 nAChR agonist, in amniotic fluid is thought to be associated with facilitation of expression of *KCC2*, the membrane chloride transporter that mediates GABA's "developmental" switch in the fetus. Metabolic demands for choline are very high in the fetus because it is a major constituent of membrane lipids and participant in one-carbon metabolism. Thus, satisfying the in utero neurotransmitter requirement of choline is dependent on adequate maternal dietary intake during pregnancy. Because dietary administration of free choline is catabolized to trimethylurea in the gut by intestinal flora, which has an intolerable foul fish odor, pregnant mothers were orally administered 6.3 g of phosphatidylcholine/day, a dietary source of serum choline, in two divided doses, or placebo from 17.2 ± 2.1 weeks after the last menstrual period through delivery (~900 mg of choline supplementation/day). After birth, the infants themselves were orally supplemented with either 100 mg of phosphatidylcholine/day or matching placebo to about 3 months of age. Mothers and their infants were randomized to receive either phosphatidylcholine or matching placebo (Ross et al. 2013, 2016). Importantly, a separate cohort of 24 infants with impaired P50 suppression (i.e., P50 inhibition ratio ≥ 0.5) was shown to have problems on the attention subscale of the Child Behavior Checklist at age 3.5 years, compared to 26 children with intact P50 suppression (ratio < 0.5). Auditory-evoked potentials were recorded from 86 total children at ~1 month of age, and parental ratings of behaviors on the Child Behavior Checklist were obtained on a subset of 49 of these children ($N = 23$ receiving phosphatidylcholine and $N = 26$ receiving placebo) at ~age 40 months (Ross et al. 2016). A significantly greater proportion of infants (mean age 33 days) treated with phosphatidylcholine supplementation (76%) had normal P50 inhibition ratios (<0.5), compared to infants treated with placebo (43%) (Ross et al. 2013, 2016). Moreover, infants in the placebo group that were homozygous for a sequence variant of *CHRNA7* associated with schizophrenia showed higher P50 inhibition ratios (i.e., more impaired sensory inhibition) than infants homozygous for this same sequence variant treated with phosphatidylcholine. The attention and social withdrawal subscale scores on the Child Behavior Checklist of the infants treated with phosphatidylcholine were lower (i.e., less severe) than for the infants treated with placebo. Normal p50 suppression ratios (i.e., <0.5) are mediated, at least in part, by central GABAergic inhibitory neurons, and there are data supporting an imbalance of central inhibitory and central excitatory tone in persons with ASD. In summary, these data suggest that early interventions can improve signal transduction by the α_7 nAChR; moreover, genotypic data may be useful in selecting both the nature of the intervention (e.g., phosphatidylcholine, α_7 nAChR PAM, or combination of the two) and the timing of its administration, including, possibly, during fetal brain development.

Maternal immune activation (MIA) is elicited by injecting poly(I:C), a viral mimic compound, intraperitoneally into pregnant female mice at embryonic day 12.5 (E12.5). MIA is an etiological environmental risk factor for induction of autism-like behaviors in the offspring, which is mediated by pro-inflammatory cytokines, especially interleukin-6 (IL-6) (Smith et al. 2007; Hsiao and Patterson 2011; Wu et al. 2015). Consistent with a pathogenic role of IL-6 in MIA, anti-IL-6 antibody was protective against MIA induction of autism-like behaviors in offspring (Smith et al. 2007). The α_7 nAChR was implicated in the MIA induction of autism-like behaviors in offspring in a study that explored the effect of administering choline, a selective α_7 nAChR agonist, to pregnant female mice during the course of fetal brain development and early postnatal period (beginning at E0.5 and continuing through the period of gestation and lactation) and its interaction with haploinsufficient expression of *Chrna7* (i.e., *Chrna7*$^{+/-}$) in offspring on expression of autism-like behaviors and their induction by MIA in these offspring (for evaluating effects on the behavior of offspring, the timed mating pair was male *Chrna7*$^{+/-}$ and female wild type mice). Moreover, in studies exploring the effect of maternal choline supplementation on cytokine expression in fetal brain and its interaction with offspring genotype, three genotypes were studied (i.e., *Chrna7*$^{+/-}$, *Chrna7*$^{-/-}$, and wild type) that resulted from a timed mating pair of male *Chrna7*$^{+/-}$ and female *Chrna7*$^{+/-}$ mice (Wu et al. 2015). MIA is known to decrease prepulse inhibition (PPI), consistent with impairment of sensorimotor gating; decrease entry into the center zone of an open field, consistent with increased anxiety; and increase marble burying, consistent with increased repetitive behavior in 6-, 7-, and 8-week-old wild type offspring, respectively (Wu et al. 2015). Maternal choline supplementation during gestation and through the period of lactation in the wild type offspring prevented the MIA induction of reduced entries into the center zone of the open field with no change in the center zone entries of the wild type offspring that did not receive exposure to MIA. Compared to the marble burying behavior of the wild type offspring exposed to MIA alone, the marble burying behavior of the wild type offspring exposed to MIA and whose mothers received choline supplementation was significantly reduced. However, maternal choline supplementation did not antagonize the reduction in PPI in the wild type offspring exposed to MIA (Wu et al. 2015). Expression of functional α_7 nAChRs in hippocampus, as determined by autoradiographic analysis with one concentration of ^{125}I-α-BGT (5 nM), was not changed in wild type offspring exposed to MIA or maternal choline supplementation. However, 3 h after pregnant dams were exposed to poly(I:C) to cause MIA, levels of *IL-6* and *Chrna7* mRNA were increased in wild type fetal brain, which were reduced by maternal choline supplementation (Wu et al. 2015). The induction of mRNA levels of *IL-6* in fetal brain by MIA, assessed 3 h after maternal injection of poly(I:C), was dependent on the *Chrna7* genotype of the offspring [i.e., wild type (+/+), heterozygous (+/−) and homozygous (−/−)]. Specifically, only the wild type fetal brain (*Chrna7*$^{+/+}$) showed an induction of *IL-6* mRNA expression; moreover, not surprisingly, only the wild type fetal brain was able to mount a significant increase in *Chrna7* mRNA expression. However, reduced or absent expression of *Chrna7* in heterozygous and homozygous mice, respectively, was associated with increased basal expression of *IL-6* mRNA in the absence of MIA, compared to the wild type control fetal brain. Importantly, MIA could not induce further enhancement of IL-6 mRNA expression in the fetal brains of *Chrna7*$^{+/-}$ and *Chrna7*$^{-/-}$ mice (Wu et al. 2015). Thus, *Chrna7* expression is a major regulator of *IL-6* expression; again, IL-6 is a major mediator of MIA in response to mid-gestational injection of poly(I:C). With respect to behavioral outcomes, offspring with haploinsufficient expression of *Chrna7* exposed to MIA by mid-gestational injection of poly(I:C) spent less time in the center zone of the open field. Also, the heterozygous offspring (*Chrna7*$^{+/-}$) not exposed to MIA showed higher levels of repetitive marble burying than wild type offspring not exposed to MIA. Finally, MIA-stressed heterozygous offspring showed a significantly greater deficit in PPI than the heterozygous offspring not exposed to MIA (Wu et al. 2015). Thus, the *Chrna7* genotype status regulates or moderates some of the responses to MIA in the offspring, including behaviors that reflect deficits in neurodevelopment. The data suggest that efficient signal transduction by the α_7 nAChR during fetal and early postnatal brain development is protective against at least some of the effects of MIA. Moreover, the accumulating data contribute to interest in investigating possible early "fetal" interventions, delivered via a dietary source of choline to pregnant and lactating mothers, for offspring at risk for neurodevelopmental disorders, especially ASD (Ross et al. 2010).

Spatial working memory in primates is dependent on the persistent firing of cortical pyramidal neurons, referred to as "delay cells," in deep layer III of the dorsolateral prefrontal cortex (dlPFC). These delay cells fire during the delay period in which there is an absence of sensory stimulation, preserving a temporary representation of visual space that can guide proactive motor action (Yang et al. 2013). Electron microscopic investigations of α_7 nAChRs in layer III of dlPFC, using monoclonal antibodies and α-BGT to label the receptor, showed that staining was most intense in dendritic spines in association with the postsynaptic density (Yang et al. 2013). Single neuron recordings from dlPFC in the monkey show that activation of postsynaptic α_7 nAChRs located in glutamatergic (excitatory) synapses is necessary for the persistent firing of delay cells during a 2.5 s delay, which is the electrophysiological substrate underlying spatial working memory, before eyes are moved in the preferred direction toward a brief visual cue in order to receive a reward (Yang et al. 2013); the task is referred to as a "spatial oculomotor delayed response task." Iontophoretic application of the selective α_7 nAChR antagonist reduced delay-related firing during the delay for the neurons' preferred direction, which is the period when the spatial location of the cue is maintained in working memory. Further, "low doses" of three highly selective α_7 nAChR agonists increased delay-related firing of delay cells in their preferred direction, which was antagonized by MLA. The anatomic co-localization of α_7 nAChRs and NR2B subtype-containing NMDARs in excitatory synapses on dendritic spines was supported by observations that iontophoresis of a selective α_7 nAChR agonist overcame the reduction in delay-related firing caused by iontophoretic application of Ro25-6981, a selective NR2B-NMDAR antagonist (Yang et al. 2013). Similarly, application of MLA blocked the ability of NMDA to increase delay-related firing. The data clearly implicate a role for the α_7 nAChR in a higher executive function in primates (i.e., spatial working memory), which is related to its localization within pyramidal cell circuits in layer III of the dlPFC. Importantly, cognitive dysfunction is commonly observed in neurodevelopmental disorders, and, in particular, deficits of spatial working memory, as reflected in reduced spontaneous alternations in the Y-maze, were characterized in the Balb/c mouse model of ASD (Burket et al. 2015a).

6 Nicotinic Receptors and Autism

6.1 Histochemical, Receptor Binding, Gene Expression, and Electrophysiological Studies

Abnormalities of cholinergic nuclei, cholinergic projections, and cholinergic receptors, as well as abnormalities of growth factors involved in the maturation and maintenance of cholinergic neurons, have been described in postmortem brains of persons with ASD (Perry et al. 2001; Martin-Ruiz et al. 2004; Ray et al. 2005; Deutsch et al. 2010). The interpretation of these findings and their specificity for ASD are confounded by age, medication histories, and comorbidities such as seizure disorders and ID. Also, tissue sample sizes in many of these studies were small, and binding sites were often interrogated with only single concentrations of radiolabeled ligands. Importantly, independently of any consistent findings of cholinergic dysfunction in ASD, the involvement of the α_7 nAChR in regulating the balance between central inhibitory and central excitatory tone, sensory inhibition, and the synchronous oscillatory output of neocortical pyramidal neurons, which underlie processes of attention and cognition, would justify therapeutic exploration of α_7 nAChR agonist interventions (Perry et al. 2001; Martin-Ruiz et al. 2004; Ray et al. 2005; Deutsch et al. 2010; Záborszky et al. 2018). Nonetheless, there is an emerging catalog of provocative findings consistent with cholinergic abnormalities occurring in at least some clinical presentations of ASD, many of which have been reviewed (Deutsch et al. 2010, 2011, 2015, 2016), and some of which will be selectively presented below.

An early study explored cholinergic markers and brain-derived neurotrophic factor (BDNF) in the basal forebrain, a site of cholinergic nuclei and projections, and frontal and parietal cerebral cortex of seven deceased adults with autism (many of whom had ID and seizures), ten normal controls without ID, six subjects with other congenital cerebral disorders and ID, and three subjects with Down syndrome (Perry et al. 2001). In this study, subject groups did not significantly differ in their ChAT activity, a presynaptic marker of cholinergic neurons, in frontal cortex, parietal cortex, or basal forebrain, nor did AChE activity in basal forebrain significantly differ between subjects with autism and normal controls. However, BDNF levels were significantly higher (~threefold) in the basal forebrain of subjects with autism compared to the normal controls. Specific binding of ^{3}H-pirenzepine, a muscarinic M_1 cholinergic receptor ligand, was significantly reduced in the parietal cortex of the subjects with autism, compared to the normal controls and subjects with other congenital cerebral disorders and ID. IC_{50} values for displacement of ^{3}H-pirenzepine by cold ligand were measured in two of the autistic and three normal controls and did not appear to differ, suggesting that the change in specific binding was not due to a change in the affinity of the muscarinic M_1 receptor for ligand. Specifically, with respect to nAChRs, ^{125}I-α-BGT binding, a marker of the α_7 nAChR, did not differ between groups; however, the specific binding of ^{3}H-epibatidine, which labels heteropentameric nAChRs containing α_3 or α_4 and β_2 subunits, was significantly lower in frontal and parietal cortex of both the subjects with autism and subjects with other congenital cerebral disorders and ID, compared to the normal control group (Perry et al. 2001). Western blotting of homogenate samples of parietal cortex from subjects with autism and normal controls showed reduced immunoreactive protein content of the α_4 and β_2 nAChR subunits in the subjects with autism. ^{3}H-Nicotine binding did not reveal differences between the subjects with autism and normal controls in the basal forebrain. Again, group sizes were small, and there was often marked variability in the measures. Nonetheless, the data are suggestive: increased levels of BDNF in the basal forebrain of the autistic subjects could reflect its compensatory role in the maintenance of cholinergic nuclei and their projections; and the changes in specific binding of muscarinic M_1 receptors and ^{3}H-epibatidine and expression levels of the α_4 and β_2 nAChR subunits in the autistic subjects could be consistent with disruption of cholinergic transmission, but these changes may also be affected by comorbid ID and seizures (Perry et al. 2001).

Histopathologic investigation of postmortem hippocampus obtained from persons with autism showed evidence of reduced size and increased packing density of neurons as well as decreased "dendritic complexities" (Blatt et al. 2001). High-affinity choline uptake sites (HACUs) labeled with a single concentration of ^{3}H-hemicholinium (10 nM) and the M_1 muscarinic site labeled with a single concentration of ^{3}H-pirenzepine (10 nM) were studied in hippocampi in an autoradiographic study of four male subjects with autism (age range 19–22 years); all had ID and three had histories of seizures, and three male controls (age range 16–24 years), two died from gunshot wounds, and one from a motor vehicle accident (Blatt et al. 2001). Although this was not an exhaustive investigation of cholinergic neurotransmission and sample sizes were small, no differences were detected between groups in the labeling of HACUs and M_1 muscarinic binding sites in any of the hippocampal subfields or laminae. However, differences were suggested with respect to possible decreased density and altered distribution of hippocampal $GABA_A$ receptors in the autistic subjects (Blatt et al. 2001). Conceivably, changes in $GABA_A$ receptors could reflect, at least in part, altered nicotinic cholinergic input to selected subpopulations of GABAergic interneurons.

Reduced number of cerebellar Purkinje cells and altered size of the cerebellar vermal lobules are common findings in histological and anatomic investigations of ASD. Importantly, the Purkinje cell loss and structural abnormalities of the cerebellar vermal lobules may be influenced by seizures, IQ, and gender (Lee et al. 2002). In addition to cholinergic input originating in the vestibular nuclei that supports the cerebellar role in the control of motor behavior, the cerebellum has a role in cognition mediated by "indirect" projections it receives from the cerebral cortex via the "corticopontocerebellar system" and output of deep cerebellar nuclei that reach the cerebral cortex after processing in the red nucleus and thalamus (Lee et al. 2002). Cognitive functions involving cerebellar processing are thought to include executing rapid shifts of attention and efficient orienting of attention (Lee et al. 2002). Cerebellar investigations of cholinergic markers were conducted in postmortem tissue from eight adults with autism (mean age = 24.63 years ±5.32[SD]), ten controls (mean age = 27.90 years ±6.19[SD]), eight persons with other congenital cerebral disorders and ID (mean age = 31.13 years ±6.85[SD]), and three persons with Down syndrome (mean age = 40.00 years ±7.00[SD]). There was no history of ID in the ten control subjects. In the eight subjects with autism, verbal IQ ranged from <20 to 40, and five had histories of seizures. Cerebellar ChAT activity, the presynaptic marker of cholinergic neurons, did not differ between groups; however, quantitative receptor autoradiography showed that the specific binding of ^{3}H-epibatidine (1 nM) was significantly reduced in the granule cell layer, Purkinje layer, and molecular layer, and the specific binding of ^{125}I-α-BGT (1.2 nM) was significantly increased in the granule cell layer, compared to the control group (Lee et al. 2002). The decrease and increase of cerebellar ^{3}H-epibatidine and ^{125}I-α-BGT binding in autism, respectively, are consistent with dysregulated nicotinic neurotransmission. Again, the presence of ID could have independently contributed to the significance of these findings. Western blotting revealed that the immunoreactive content of the α_4 nAChR subunit was significantly reduced in the subjects with autism compared to the control subjects. The subjects with autism did not differ from the controls and other groups with respect to M_1 and M_2 muscarinic receptor binding; again, this failure to detect group differences may have been influenced by the small number of tissue samples and variability within measures, as well as other confounding variables, such as IQ, seizure history, subject age, and gender. In their conclusion, given the possibility of nicotinic cholinergic abnormalities, the authors suggested consideration of a targeted nicotinic agonist intervention in persons with autism to improve attentional performance (Lee et al. 2002).

Diminished surface expression of functional nAChR pentameric receptors in persons with ASD may result from dysregulation of transcription, translation of mRNA in the periphery of the neuron, post-translational processing of the polypeptide subunits, membrane trafficking, or membrane insertion of functional receptors. Thus, there can be lack of correlations between levels of mRNA for individual receptor subunits, immunoreactive content of individual receptor subunits, and binding that depends on radiolabeling the pentameric receptor in homogenates of plasma membranes. In order to further characterize the basis of the nAChR abnormalities in autism, "parallel measurements" were obtained in the parietal cortex and cerebellum of levels of mRNA and immunoreactive protein content of the α_4, α_7, and β_2 nAChR subunits and autoradiographic analysis of receptor binding using single concentrations of ^{3}H-epibatidine (1 nM) and ^{125}I-α-BGT (1.2 nM) (Martin-Ruiz et al. 2004). Frozen cortical and cerebellar autopsy tissue was obtained from six adults with autism (mean age = 26.5 years ±4.8[SD]; five males and one female) and eight age-matched controls with no histories of ID (mean age = 30.5 years ±7.8 [SD]; five males and three females). mRNA levels of the α_4 nAChR subunit, normalized to levels of the GAPDH housekeeping gene mRNA, were significantly reduced (~threefold) in the parietal cortex (Brodmann area 39) of the subjects with autism, compared to the controls (Martin-Ruiz et al. 2004). However, mRNA expression levels of the β_2 subunit, normalized to GAPDH, and the α_7 subunit, normalized to actin, in the parietal cortex, did not differ between groups. Similarly, in the cerebellum, mRNA expression levels of the α_4 subunit, normalized to GAPDH mRNA levels, were reduced (~twofold) in the autistic subjects, whereas there was no significant change in hippocampal mRNA expression levels of the α_7 or β_2 subunits. The immunoreactive protein content of the α_4 nAChR subunit was significantly reduced in the parietal cortex and cerebellum of the autistic subjects; there were no significant between-group changes in the immunoreactive protein content of the β_2 and α_3 subunits in parietal cortex and no significant between-group changes in the α_7 and β_2 immunoreactive protein content in the cerebellum (Martin-Ruiz et al. 2004). Autoradiographic analysis revealed significant reductions of ^{3}H-epibatidine binding in the autistic subjects, compared to controls, in the following regions: frontal, parietal and occipital cortex, and cerebellum (Martin-Ruiz et al. 2004). Moreover, ^{125}I-α-BGT binding was significantly increased (~threefold) in the cerebellum of autistic subjects, whereas there were no significant between-group changes in the frontal and parietal cortex (Martin-Ruiz et al. 2004). The data are consistent with possible autism-related changes in nicotinic cholinergic neurotransmission, reflected in changes in transcription and translation of nAChR subunits and surface expression of functional receptors. However, the data remain suggestive because of small samples of tissues examined, ages of subjects (i.e., adults), and possible confounding effects of ID and seizures (the authors did not mention presence or absence of ID and seizures in their brief case descriptions) (Martin-Ruiz et al. 2004).

The immunohistochemical labeling of neurons, including their distribution and intensity of staining within the microscopic field, with antibodies directed against the α_7 nAChR subunit was reduced in the paraventricular nucleus and nucleus reuniens of the thalamus in postmortem sections of three adults with autism, compared to three adult controls (Ray et al. 2005). The reduction of immunohistochemical staining of the α_7 nAChR subunit was not seen in cell processes in either the paraventricular nucleus or nucleus reuniens. Two of the three adult males with autism (mean age = 29.3 years ±9.3[SD]; age range 29–32 years) had ID, and one of these two subjects had a history of seizures. The control subjects consisted of two men and one woman (age range 19–37 years), none of whom had ID; however, two of the controls had chronic illnesses that may have affected CNS function (i.e., metastatic Hodgkin's lymphoma and alcoholic liver disease). Consistent with disruption of nicotinic cholinergic neurotransmission in general in the three subjects with autism, there was also a reduction in immunohistochemical staining of the β_2 nAChR subunit, including both the percentage of fields covered by β_2 immunostained neurons and intensity of staining, in both the thalamic paraventricular nucleus and nucleus reuniens. The authors also observed an increased immunoreactivity of the α_7 nAChR subunit in astrocytes in both the paraventricular nucleus and nucleus reuniens in the subjects with autism, which was thought to be "reactive" because it was most apparent in the subject with a history of epilepsy (Ray et al. 2005). Interestingly, there was also an increase in β_2 nAChR subunit immunoreactivity in the paraventricular nucleus and nucleus reuniens in this one autistic subject with a history of epilepsy. The paraventricular nucleus and nucleus reuniens are referred to as "nonspecific" midline thalamic nuclei, which are known to have "diffuse and diverse" afferent and efferent connections (Ray et al. 2005). These thalamic nuclei have reciprocal connections with corticolimbic areas implicated in the pathogenesis of ASD, including prefrontal cortex, amygdala, and hippocampal formation (Ray et al. 2005). The data are certainly suggestive of both general dysregulation of nicotinic cholinergic transmission and specific involvement of the α_7 nAChR in the thalamus. However, the interpretation is confounded by small sample size, age, ID, and seizures in the patient group and chronic disease in the control subjects.

In addition to the potential for "error" in histopathological studies of human postmortem brain due to small sizes of tissue samples and possible confounding effects of age, gender, intellectual disability (ID), history of seizures, medications, and chronic illnesses, among other variables, an additional unexpected source of concern has been raised in immunohistochemical studies using nAChR subtype-selective antibodies to detect anatomic distribution, cellular and subcellular localization, and quantity of expressed nAChR subunits. Specifically, commercially available antibodies raised against specific nAChR subunits (including α_3, α_4, α_7, β_2, and β_4) and shown to have specificity in oocyte expression systems and little or no cross-reactivity with other nAChR subunits showed "immunopositive" signals when tested against brain tissue from knockout mice with selective deletion of the subunit against which the antibody was directed (Moser et al. 2007). Thus, complementary and, oftentimes, collaborative approaches between laboratories must be employed to ensure the validity of expression profiles and construction of reliable anatomic maps that localize nAChR subtypes, including PCR, in situ hybridization, immunohistology, immunoprecipitation, northern and western blotting, ligand binding, and electrophysiology (Moser et al. 2007).

Binding of MeCP2 to methyl CpG-binding domains in the 15q13.2 and 15q13.3 region appears necessary for an interaction between this region and the Prader-Willi imprinting center (PWS-IC) that is located more than 7 Mb away; this interaction serves as a long-range epigenetic regulator of chromatin remodeling that influences expression of *CHRNA7* (Yasui et al. 2011). The PWS-IC is the smallest region of overlap of paternal deletions resulting in Prader-Willi syndrome (PWS) and participates in regulating *cis*-expression of genes in the ~13 Mb region between 15q11.2 and 15q13.3, including the developmentally regulated distal genes *GABRB3*, *UBE3A*, and *CHRNA7*. MeCP2 loss of function mutations are causally associated with the X-linked dominant Rett syndrome, and disruption of this MeCP2-mediated molecular bridging interaction in Rett syndrome between the PWS-IC and distal sites in the 15q11.2 and 15q13.3 region may be responsible for some phenotypic similarities between PWS, Rett syndrome, and autism (Yasui et al. 2011). The developmental regulation of *CHRNA7* expression and the regulatory role of the MeCP2-mediated interaction between the PWS-IC and 15q11.2-15q13.3 locus were studied and confirmed in vitro by culturing SH-SY5Y neuroblasts in the presence of phorbol 12-myristate 13-acetate (PMA), which stimulates their neuronal differentiation (Yasui et al. 2011). Interestingly, of all the 15q subregions, the expression of *CHRNA7* in the 15q13.2-15q13.3 region was the most likely to be affected by MeCP2 binding. Examination of human postmortem cerebral cortex (Brodmann area 9) revealed that *CHRNA7* transcript levels were about tenfold higher in typically developing infants less than 1 year of age and level off at around age 20 years ($N = 20$ typically developing controls), whereas levels of *CHRNA7* mRNA in frontal cortex obtained from patients with Rett syndrome ($N = 7$) and autism ($N = 11$) were significantly reduced, compared to controls (Yasui et al. 2011). Preclinical studies suggest that MeCP2 is a positive modulator of *Chrna7* expression. Overall, the data suggest that diminished α_7 nAChR-mediated neurotransmission may contribute to the pathogenesis of ASD and support interventional therapeutic strategies to selectively stimulate α_7 nAChRs in at least some patients with ASD (Deutsch et al. 2010, 2015).

6.2 Behavioral and Pharmacological Studies

Chronic 4-week exposure to nicotine, a nonselective nAChR agonist, administered in drinking water was shown to dose-dependently improve social interactions and decrease spontaneous self-grooming in the BTBR mouse model of ASD (Wang et al. 2015). Chronic nicotine administration has been reported to upregulate both the functional homopentameric α_7 nAChR and nAChRs containing α_4 and β_2 subunits. In the three-chamber apparatus, nicotine (100 μg/mL or 15 mg/kg/day corrected for water consumption) significantly increased the amount of time BTBR mice spent in the chamber containing an enclosed same-sex, same-age novel stimulus mouse, versus an identical inverted cup, and also increased the amount of time BTBR mice spent sniffing the novel mouse versus sniffing the inanimate inverted wire cup. The C57BL/6J comparator strain showed both differing sensitivity to prosocial effects of nicotine with a dose-dependent "dissociation" between sensitivity measured with time spent in the social chamber and time spent sniffing the novel mouse. Specifically, the highest dose of nicotine (400 μg or 55 mg/kg/day corrected for water consumption) increased the time C57BL/6J mice spent in the social chamber, whereas the 50 μg/mL and 100 μg/mL doses (10 and 15 mg/kg/day corrected for water consumption, respectively) increased time spent sniffing the novel stimulus mouse (Wang et al. 2015). The two highest doses of nicotine (200 μg/mL and 400 μg/mL or 35 and 55 mg/kg/day, respectively) significantly decreased the repetitive self-grooming behavior of BTBR mice (Wang et al. 2015). The 200 μg/mL dose of nicotine did not make BTBR mice more anxious (i.e., did not change center time in the open field) but did decrease horizontal locomotor activity, which could confound interpretation of prosocial effects at this dose because sociability assessed in the three-chamber apparatus is dependent on horizontal locomotor activity. Additionally, in the BTBR strain, potentially confounding metabolic effects of the higher 200 μg/mL and 400 μg/mL doses may be present as mice treated with these doses either gained less weight or actually lost weight, respectively, compared to the vehicle condition. The data are consistent with nAChR abnormalities in the BTBR mouse model of ASD and support exploration of therapeutic targeting of nAChRs in ASD (Wang et al. 2015).

AVL-3288, an α_7 nAChR-selective "positive allosteric modulator (PAM)," improved sociability and repetitive self-grooming behavior, therapeutic effects that were not confounded by, or an epiphenomenon of, increased locomotor activity, in the BTBR mouse model of ASD; male BTBR mice were tested at 10–16 weeks of age (Yoshimura et al. 2017). AVL-3288 increased the current flow induced by application of an α_7 nAChR agonist at its EC_{10} concentration (i.e., the agonist concentration that evoked 10% of the maximum response) to the human receptor expressed in *Xenopus* oocytes by more than 600%; 2-electrode voltage clamp oocyte electrophysiology was used to measure currents, and the EC_{50} for AVL-3288's positive modulatory effect was 0.7 μM. In the standard three-chamber apparatus, BTBR mice treated with AVL-3288 (3 mg/kg, intraperitoneally) spent significantly more time sniffing/exploring the chamber containing an enclosed novel mouse than the chamber containing a novel object, whereas no social preference was observed in the vehicle-only condition (Yoshimura et al. 2017). This positive allosteric effect of AVL-3288 on social preference was antagonized by pretreatment with 3 mg/kg of MLA, an α_7 nAChR-selective antagonist. Further, AVL-3288 (3 and 10 mg/kg) significantly reduced the amount of time BTBR mice engaged in repetitive self-grooming (Yoshimura et al. 2017). The data clearly support therapeutic exploration of α_7 nAChR PAMs and the α_7 nAChR as a therapeutic target (Deutsch et al. 2010, 2011, 2015, 2016).

Mice with conditional knockout of *Mecp2* in ChAT (choline acetyltransferase)-expressing cholinergic neurons (referred to as *Chat-Mecp2*$^{-/y}$ mice), especially in basal forebrain (BF), mimicked many phenotypic features of Rett syndrome; moreover, PV-expressing GABAergic neurons in the CA1 region of hippocampus in these *Chat-Mecp2*$^{-/y}$ mice, which receive dense projections from BF cholinergic neurons, showed diminished expression of the α_7 nAChR (Zhang et al. 2016). The data suggest that the reduced hippocampal expression of α_7 nAChRs was responsible for the deficit of social memory, increased excitability of hippocampal pyramidal neurons, and mood dysregulation displayed by the *Chat-Mecp2*$^{-/y}$ mice, which were "rescued" by injection of a viral vector containing the *Mecp2* transcript into BF or injection of PNU-282987, a selective α_7 nAChR agonist, or nicotine itself into the CA1 region of hippocampus (Zhang et al. 2016). The *Chat-Mecp2*$^{-/y}$ mice showed decreased anxiety- and depressive-like behavior on a variety of behavioral measures that are often used in the screening of compounds for antianxiety and antidepressant efficacy, such as spending more time in the open arms of the elevated plus maze and lit compartment of the light-dark box, and decreased immobility in the forced swim test, compared to appropriate controls. The *Chat-Mecp2*$^{-/y}$ mice showed normal social preference in the standard three-chamber paradigm, preferring the chamber containing an enclosed stimulus mouse to the chamber containing the empty cage. However, unlike control mice, they showed a deficit in social memory showing no preference for a novel, unfamiliar "stranger" mouse over a familiar mouse. The *Chat-Mecp2*$^{-/y}$ mice also showed less social interaction with stranger mice in an open-field arena. Interestingly, the *Chat-Mecp2*$^{-/y}$ mice showed no deficits of spatial learning and memory when tested in the Morris water maze. The *Chat-Mecp2*$^{-/y}$ knockout mice with essentially absent expression of Mecp2 in BF had markedly downregulated expression of ChAT in BF cholinergic neurons but normal expression of GAD65 (Zhang et al. 2016). As noted, western blot analysis revealed that *Chat-Mecp2*$^{-/y}$ mice showed a significant reduction of α_7 nAChR expression in hippocampus, which is ordinarily co-expressed with PV in a subset of GABAergic neurons. The excitability of this subpopulation of PV-expressing GABAergic neurons was decreased in the *Chat-Mecp2*$^{-/y}$ mice, compared to controls, as reflected in prolongation of their interspike interval (ISI) and decreased firing frequency; presumably, the reduced ISI and decreased firing frequency of the PV-expressing GABAergic neurons accounted for the increased excitability (i.e., disinhibition) of CA1 pyramidal neurons in the *Chat-Mecp2*$^{-/y}$ mice (Zhang et al. 2016). This latter observation could also explain the high prevalence (~80%) of recurrent seizures in patients with Rett syndrome. Behaviorally, the increased pyramidal cell excitability in the *Chat-Mecp2*$^{-/y}$ mice was reflected in "frequent hyperexcitable discharges" recorded electroencephalographically in freely moving mice and increased susceptibility to pilocarpine-induced seizures (Zhang et al. 2016). The data clearly support a role for MeCP2 as an epigenetic regulator of acetylcholine (ACh) synthesis and also show that a "downstream" consequence of diminished ChAT expression is downregulated α_7 nAChR expression. Importantly, the data strongly support therapeutic targeting of the α_7 nAChR in order to address key primary and secondary symptom domains in Rett syndrome and ASD, including mood dysregulation, deficits of sociability, and seizure proneness.

Pentylenetetrazol (PTZ)-kindled male ICR mice with a lowered threshold and greater sensitivity to PTZ-elicited clonic seizure activity showed both reduced social preference and reduced social recognition memory in the standard three-chamber apparatus, when compared to vehicle-treated controls (Takechi et al. 2016). Specifically, the PTZ-kindled mice showed no preference for an enclosed stimulus mouse over an empty inverted cup and no preference for an enclosed novel stranger mouse over an enclosed familiar mouse. Prior treatment of the PTZ-kindled mice with ABT-418, a compound with 10,000-fold greater selectivity for the heteropentameric $\alpha_4\beta_2$ nAChR subtype, normalized both the social preference and social recognition memory of the PTZ-kindled mice (Takechi et al. 2016). The piriform cortex is anatomically situated between limbic and cortical networks and is pathologically implicated in propagation of seizure activity (Takechi et al. 2016). Immunolabeling of the piriform cortex showed that the α_4 nAChR subunit and neuroligin3 (NLG3), a postsynaptic transmembrane adhesion molecule associated with the excitatory synapse and, when deleted or mutated, with ASD, were colocalized in this structure in both vehicle-treated control and PTZ-kindled mice. However, α_4 subunit expression was significantly reduced and NLG3 expression significantly increased in the PTZ-kindled mice, compared to vehicle-treated controls. The PTZ-kindled mouse model of ASD clearly shows that the heteropentameric $\alpha_4\beta_2$ nAChR subtype can be involved in increased central excitability, as reflected in a lowering of the threshold for clonic seizure activity and impaired sociability. Moreover, disturbed expression of the α_4 nAChR subunit is associated with probable disturbance of the trans-synaptic architecture of the excitatory synapse, as reflected in the increased expression of NLG3. Thus, evaluation of possible adverse effects of a targeted α_7 nAChR agonist therapeutic strategy for ASD on expression and function of $\alpha_4\beta_2$ nAChR-mediated processes must occur.

The aggression displayed by male mice with heterozygous deletions of *Chrna7* (i.e., α7 HET mice) and, thereby, haploinsufficient expression of the α_7 nAChR was increased, compared to their wild type littermates, in the "resident-intruder" behavioral paradigm (Lewis et al. 2018). In this paradigm, placing an intruder mouse into the home cage of a resident mouse elicits bouts of repeated aggression in the resident mouse. Compared to wild type controls, the α7 HET mice "initiated attack in more bouts" and "attacked significantly faster," which was not an epiphenomenon of increased locomotor activity (Lewis et al. 2018). High density of α_7 nAChRs was expressed in the hippocampus of the wild type mice, and, in particular, the granule cell layer of the dentate gyrus was "activated" in the aggressive resident mice; activation was assessed by expression of Arc, an immediate early gene, in granule cells (Lewis et al. 2018). A non-GABA-positive granule cell population was activated in the dentate gyrus by intruder mice; moreover, CaMKII expression colocalized with immunofluorescent staining of Arc in the granule cells activated by intruder mice. Both nicotine and GTS-21, an α_7 nAChR-selective agonist, suppressed aggression of resident α7 HET mice (Lewis et al. 2018). Moreover, bilateral stereotactic hippocampal infusion of adeno-associated virus 2 (AAV2) vector containing/expressing small hairpin RNA (shRNA) targeting α_7 nAChRs blocked the anti-aggressive, "serenic" effects of nicotine and GTS-21 in wild type male Balb/c mice, which are more aggressive in this paradigm than male C57BL/6 mice; the latter C57BL/6 mouse strain served as the genetic background for the original *Chrna7* deletion (Lewis et al. 2018). The data suggest that the dentate gyrus is a node activated in an aggression circuit, and the α_7 nAChR modulates aggression in the "resident-intruder" paradigm, which can be therapeutically targeted to reduce aggression (Lewis et al. 2018). Further, the data support exploration of α_7 nAChR-selective agonist interventions for the treatment of the secondary symptom of aggression/tantruming/irritability in children with ASD.

6.3 Clinical Studies

The effects of acute, single-dose targeting of the α_7 nAChR with 3-(2,4-dimethoxybenzylidene) anabaseine (DMXB-A), a selective, partial agonist for the α_7 nAChR, were studied in two normally intelligent (verbal comprehension scores of 89 and 93 on the Wechsler Abbreviated Scale of Intelligence, Second Edition and Wechsler Adult Intelligence Scale, Fourth Edition, respectively) adult (aged 50 and 24 years, respectively) males with ASD (Olincy et al. 2016). The diagnoses were supported with completion of Module 4 of the Autism Diagnostic Observation Schedule, Second Edition. Placebo and two doses (75 mg and 150 mg followed by 37.5 mg and 75 mg, respectively, 2 h later, the half-life of the first dose) were randomly administered orally on successive days, and the entire protocol was completed within 3 h during each of the 3 days. Because α_7 nAChRs located on GABAergic inhibitory neurons contribute to the "P50 measure of sensory inhibition, " this measure was used as a biomarker of target engagement by DMXB-A. Specifically, sensory inhibition was measured by the amplitude of the P50-evoked response (a positive potential appearing ~50 ms post-stimulus) to each of a pair of identical auditory stimuli delivered 0.5 s apart; it is recorded from an electrode affixed to the vertex of the head according to a standard protocol (Olincy et al. 2016). Normally, the ratio of the evoked amplitude of the P50 response to the second of the pair of identical auditory stimuli is ≤ 0.50 of the amplitude evoked by the first auditory stimulus of the pair. The P50 ratios in the placebo conditions of both patients were elevated (0.99 and 0.61, respectively). The higher dose of DMXB-A reduced these ratios in patients 1 and 2 to 0.29 and 0.03, respectively, consistent with a centrally effective stimulatory effect on α_7 nAChR. Similarly, the higher dose was more prominently associated with improvement on objective and patient-rated measures of attention. Perhaps, not surprisingly, there was no obvious beneficial effect in this single, acute dose study of DMXB-A on self-rated measures of social responsiveness in two adult males with ASD (Olincy et al. 2016).

Thirteen children and adolescents with autism (10 males and 3 females; mean age = 8.8 years ±3.5; 7 of whom had mild to moderate ID), whose DSM-IV diagnoses were supported by the ADI-R, were treated with galantamine in an open-label prospective design for up to 12 weeks (Nicolson et al. 2006). Patients were medication-free for at least 4 weeks prior to initiation of galantamine, which was titrated to a maximally tolerated total daily dose of 24 mg (total daily doses ranged between 12 and 24 mg; mean = 18.4 mg ±4.3). Ten subjects completed the full 12-week trial; two withdrew after 8 weeks because of clinical worsening, and one withdrew after 11 weeks because of headaches. Clinician-rated anger on a subscale of the Children's Psychiatric Rating Scale and parent-ratings on the irritability and social withdrawal subscales of the Aberrant Behavior Checklist (ABC) showed significant improvement at the end of treatment (Nicolson et al. 2006). Thus, improvement was detected with two commonly used rating instruments; galantamine was safely administered and, with one exception, tolerated. However, the study was of short duration (i.e., 12 weeks), included a small number of subjects ($N = 13$), and was open-label; nonetheless, the study encourages exploration of cholinergic interventions for persons with ASD, especially for secondary symptoms of aggression and temper outbursts.

Forty children (aged 4–12 years) fulfilling DSM-IV-TR criteria for autistic disorder completed a parallel-group, placebo-controlled, double-blind 10-week clinical trial that explored adjuvant therapeutic effects of galantamine, an AChE inhibitor that possesses positive allosteric modulatory effects on nAChRs, when added to risperidone (Ghaleiha et al. 2014). All children were free of psychotropic medications for at least 6 weeks and met the severity criterion of ≥12 on the irritability subscale of the Aberrant Behavior Checklist-Community scale (ABC-C) prior to study entry. Doses of risperidone, galantamine, and galantamine's matching placebo were titrated according to subjects' weight, and the 58-item ABC-C completed by parents at baseline and weeks 5 and 10 served as the outcome measure of efficacy; the primary outcome measure was the irritability subscale, and the four other subscales served as secondary measures (i.e., lethargy/social withdrawal, stereotypic behavior, hyperactivity/noncompliance, and inappropriate speech subscales). A general linear model with repeated measures was used to analyze the data. By week 10, the daily doses of the children receiving galantamine were distributed as follows: three children received 24 mg; two children received 20 mg; 12 children received 16 mg; and three children received 12 mg. The mean daily doses of risperidone did not differ between groups: 1.32 mg for the galantamine and 1.15 mg for the placebo groups. The children receiving galantamine showed significant improvement on the irritability subscale at weeks 5 and 10 and significant improvement on the lethargy/social withdrawal subscale at week 10; there was no significant improvement on other subscales (Ghaleiha et al. 2014). Importantly, no significant adverse events were reported by parents. These data are promising, but, as noted by the investigators, the study was limited by its small sample sizes and short 10-week duration.

7 Conclusion

Given the rich cholinergic innervation of the forebrain, it is not surprising that a data-driven rationale for exploring possible therapeutic effects of targeted α_7 nAChR-agonist interventions in persons with ASD is evolving. Thus, two "proof of concept/

principle" exploratory clinical trials of galantamine, a compound that increases the lifetime of ACh within the synaptic cleft and likelihood that its binding results in opening of the ACh/choline-gated nAChR, were included in the review. The availability of selective α_7 nAChR agonists, partial agonists, allosteric agonists, and PAMs will undoubtedly stimulate additional trials. Cholinergic interventions have the ability to restore disrupted delicate balance between central excitatory and

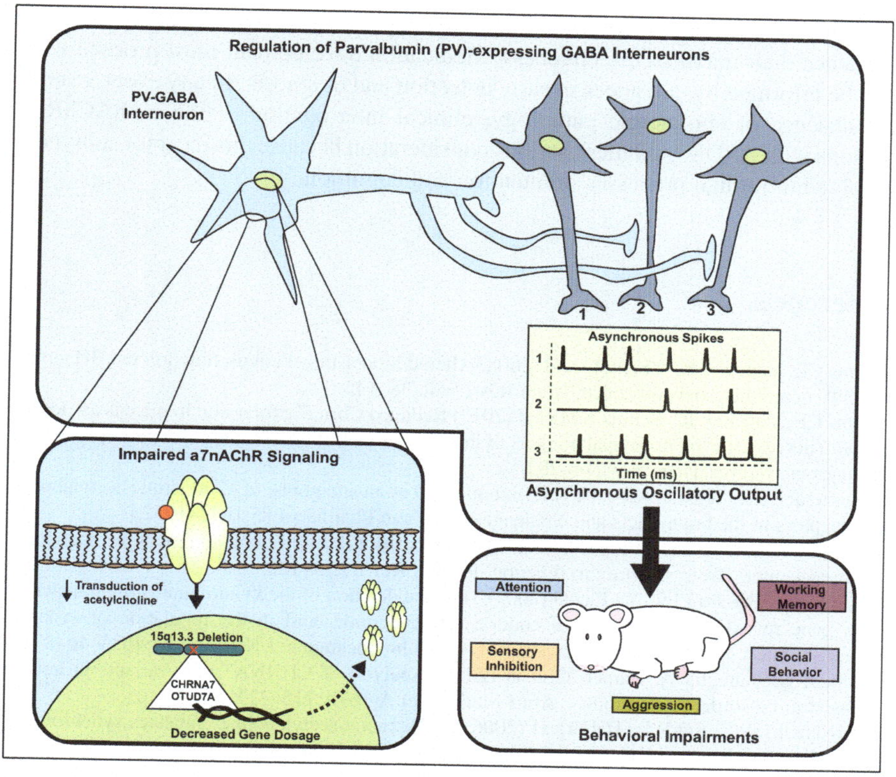

Fig. 2 Impaired signal transduction by the α_7 nicotinic acetylcholine receptor contributes to pathogenesis of autism spectrum disorder. The α_7 nicotinic acetylcholine receptor (α_7 nAChR) regulates inhibitory output of GABAergic interneurons that project onto assemblies of pyramidal outflow neurons, synchronizing their oscillatory output. Synchronous oscillatory output is important for higher executive functions, such as working memory, and maintenance of functional connectivity between discrete and anatomically noncontiguous areas of the brain. Haploinsufficient or diminished expression of the α_7 nAChR, due to 15q13.3 microdeletions or promoter variants, respectively, or inefficient transduction of the acetylcholine (ACh)/choline signal (e.g., due to DNA sequence variants of *CHRNA7*, the gene encoding the α_7 nAChR), may be associated with a variety of disrupted behavioral readouts in various mouse models of autism spectrum disorder (ASD). Preclinical data and a few "proof-of-principle/concept" clinical trials support an evolving rationale for exploring therapeutic targeting of the α_7 nAChR in persons with ASD, using selective α_7 nAChR-agonists, partial agonists, allosteric agonists, and partial allosteric modulators (PAMs) (see text for details)

inhibitory transmission in persons with ASD, which is necessary for optimal information transfer within critical circuits that support attention, cognition, language, and affiliative social behaviors. Figure 2 summarizes the hypothesized role that the α_7 nAChR plays in regulating firing of GABAergic inhibitory projections and synchronizing oscillatory output of assemblies of neocortical pyramidal neurons. Theoretically, selective-α_7 nAChR agonist interventions would "synchronize" disrupted oscillatory output of pyramidal outflow neurons and normalize behavioral impairments of the ASD phenotype. Clearly, the effectiveness of medications may depend on the timing of their initiation; thus, developmental age may determine and influence their maximal effectiveness. Medication development must proceed with, and be informed by, advances in early detection and diagnosis. In any event, even in the absence of cholinergic pathology, clinical interrogation of the α_7 nAChR in persons with ASD would merit serious consideration because this receptor subtype is involved in normal processes of attention and cognition.

References

Adams CE, Yonchek JC, Zheng L et al (2008) Altered hippocampal circuit function in C3H alpha7 null mutant heterozygous mice. Brain Res 1194:138–145

Adams CE, Yonchek JC, Schulz KM et al (2012) Reduced Chrna7 expression in mice is associated with decreases in hippocampal markers of inhibitory function: implications for neuropsychiatric diseases. Neuroscience 207:274–282

Albuquerque EX, Schwarcz R (2013) Kynurenic acid as an antagonist of α7 nicotinic acetylcholine receptors in the brain: facts and challenges. Biochem Pharmacol 85:1027–1032

Albuquerque EX, Pereira EFR, Alkondon M, Rogers SW (2009) Mammalian nicotinic acetylcholine receptors: from structure to function. Physiol Rev 89:73–120

Alkondon M, Pereira EFR, Yu P et al (2004) Targeted deletion of the kynurenine aminotransferase ii gene reveals a critical role of endogenous kynurenic acid in the regulation of synaptic transmission via alpha7 nicotinic receptors in the hippocampus. J Neurosci 24:4635–4648

Bacchelli E, Battaglia A, Cameli C et al (2015) Analysis of CHRNA7 rare variants in autism spectrum disorder susceptibility. Am J Med Genet A 167A:715–723

Bandyopadhyay S, Sutor B, Hablitz JJ (2006) Endogenous acetylcholine enhances synchronized interneuron activity in rat neocortex. J Neurophysiol 95:1908–1916

Bates RC, Stith BJ, Stevens KE, Adams CE (2014) Reduced CHRNA7 expression in C3H mice is associated with increases in hippocampal parvalbumin and glutamate decarboxylase-67 (GAD67) as well as altered levels of GABA(A) receptor subunits. Neuroscience 273:52–64

Bear MF, Connors BW, Paradiso MA (2016) Neuroscience: exploring the brain, 4th edn. Wolters Kluwer, Philadelphia

Blatt GJ, Fitzgerald CM, Guptill JT et al (2001) Density and distribution of hippocampal neurotransmitter receptors in autism: an autoradiographic study. J Autism Dev Disord 31:537–543

Brodkin ES (2008) Social behavior phenotypes in fragile X syndrome, autism, and the Fmr1 knockout mouse: theoretical comment on McNaughton et al. (2008). Behav Neurosci 122:483–489

Burket JA, Deutsch SI (2019) Metabotropic functions of the NMDA receptor and an evolving rationale for exploring NR2A-selective positive allosteric modulators for the treatment of autism spectrum disorder. Prog Neuro-Psychopharmacol Biol Psychiatry 90:142–160

Burket JA, Benson AD, Tang AH, Deutsch SI (2014) Rapamycin improves sociability in the BTBR T(+)Itpr3(tf)/J mouse model of autism spectrum disorders. Brain Res Bull 100:70–75

Burket JA, Benson AD, Green TL et al (2015a) Effects of VU0410120, a novel GlyT1 inhibitor, on measures of sociability, cognition and stereotypic behaviors in a mouse model of autism. Prog Neuro-Psychopharmacol Biol Psychiatry 61:10–17
Burket JA, Benson AD, Tang AH, Deutsch SI (2015b) NMDA receptor activation regulates sociability by its effect on mTOR signaling activity. Prog Neuro-Psychopharmacol Biol Psychiatry 60C:60–65
Burket JA, Urbano MR, Deutsch SI (2017) Sugarcoated Perineuronal nets regulate "GABAergic" transmission: bittersweet hypothesis in autism Spectrum disorder. Clin Neuropharmacol 40:120–130
Cea-Del Rio CA, Huntsman MM (2014) The contribution of inhibitory interneurons to circuit dysfunction in Fragile X Syndrome. Front Cell Neurosci 8:245
Cellot G, Cherubini E (2014) Reduced inhibitory gate in the barrel cortex of Neuroligin3R451C knock-in mice, an animal model of autism spectrum disorders. Physiol Rep 2:e12077
Chung C (2013) NMDA receptor as a newly identified member of the metabotropic glutamate receptor family: clinical implications for neurodegenerative diseases. Mol Cells 36:99–104
Deutsch SI, Rosse RB, Schwartz BL et al (2008a) Effects of CDP-choline and the combination of CDP-choline and galantamine differ in an animal model of schizophrenia: development of a selective alpha7 nicotinic acetylcholine receptor agonist strategy. Eur Neuropsychopharmacol J Eur Coll Neuropsychopharmacol 18:147–151
Deutsch SI, Schwartz BL, Schooler NR et al (2008b) First administration of cytidine diphosphocholine and galantamine in schizophrenia: a sustained alpha7 nicotinic agonist strategy. Clin Neuropharmacol 31:34–39
Deutsch SI, Urbano MR, Neumann SA et al (2010) Cholinergic abnormalities in autism: is there a rationale for selective nicotinic agonist interventions? Clin Neuropharmacol 33:114–120
Deutsch SI, Urbano MR, Burket JA et al (2011) Pharmacotherapeutic implications of the association between genomic instability at chromosome 15q13.3 and autism spectrum disorders. Clin Neuropharmacol 34:203–205
Deutsch SI, Schwartz BL, Schooler NR et al (2013) Targeting alpha-7 nicotinic neurotransmission in schizophrenia: a novel agonist strategy. Schizophr Res 148:138–144
Deutsch SI, Burket JA, Benson AD (2014) Targeting the α7 nicotinic acetylcholine receptor to prevent progressive dementia and improve cognition in adults with Down's syndrome. Prog Neuro-Psychopharmacol Biol Psychiatry 54:131–139
Deutsch SI, Burket JA, Urbano MR, Benson AD (2015) The α7 nicotinic acetylcholine receptor: a mediator of pathogenesis and therapeutic target in autism spectrum disorders and down syndrome. Biochem Pharmacol 97:363–377
Deutsch SI, Burket JA, Benson AD, Urbano MR (2016) The 15q13.3 deletion syndrome: deficient α(7)-containing nicotinic acetylcholine receptor-mediated neurotransmission in the pathogenesis of neurodevelopmental disorders. Prog Neuro-Psychopharmacol Biol Psychiatry 64:109–117
Deutsch SI, Kreiser NL, Urbano MR et al (2017) Autism presenting in the context of a genetic variant of CFTR and early HSV exposure confounded by chronic pain, altered gut microbiota and paternal abandonment; limitations of current pharmacotherapy and barriers to personalized treatment recommendations. Pers Med Psychiatry 3:24–29
Dore K, Stein IS, Brock JA et al (2017) Unconventional NMDA receptor signaling. J Neurosci 37:10800–10807
Ehninger D, Silva AJ (2011) Rapamycin for treating tuberous sclerosis and autism spectrum disorders. Trends Mol Med 17:78–87
Enwright JF, Sanapala S, Foglio A et al (2016) Reduced labeling of Parvalbumin neurons and Perineuronal nets in the dorsolateral prefrontal cortex of subjects with schizophrenia. Neuropsychopharmacology 41:2206–2214
Filice F, Vörckel KJ, Sungur AÖ et al (2016) Reduction in parvalbumin expression not loss of the parvalbumin-expressing GABA interneuron subpopulation in genetic parvalbumin and shank mouse models of autism. Mol Brain 9:10

Garg S, Plasschaert E, Descheemaeker M-J et al (2015) Autism spectrum disorder profile in neurofibromatosis type I. J Autism Dev Disord 45:1649–1657
Ghaleiha A, Ghyasvand M, Mohammadi M-R et al (2014) Galantamine efficacy and tolerability as an augmentative therapy in autistic children: a randomized, double-blind, placebo-controlled trial. J Psychopharmacol (Oxf) 28:677–685
Gill JK, Chatzidaki A, Ursu D et al (2013) Contrasting properties of α7-selective orthosteric and allosteric agonists examined on native nicotinic acetylcholine receptors. PLoS One 8:e55047
Gillentine MA, Schaaf CP (2015) The human clinical phenotypes of altered CHRNA7 copy number. Biochem Pharmacol 97:352–362
Gillentine MA, Berry LN, Goin-Kochel RP et al (2017) The cognitive and behavioral phenotypes of individuals with CHRNA7 duplications. J Autism Dev Disord 47:549–562
Gonzalez-Burgos G, Lewis DA (2012) NMDA receptor hypofunction, Parvalbumin-positive neurons, and cortical gamma oscillations in schizophrenia. Schizophr Bull 38:950–957
Gonzalez-Burgos G, Cho RY, Lewis DA (2015) Alterations in cortical network oscillations and Parvalbumin neurons in schizophrenia. Biol Psychiatry 77:1031–1040
Hajós M, Hurst RS, Hoffmann WE et al (2005) The selective alpha7 nicotinic acetylcholine receptor agonist PNU-282987 [*N*-[(3R)-1-Azabicyclo[2.2.2]oct-3-yl]-4-chlorobenzamide hydrochloride] enhances GABAergic synaptic activity in brain slices and restores auditory gating deficits in anesthetized rats. J Pharmacol Exp Ther 312:1213–1222
Hsiao EY, Patterson PH (2011) Activation of the maternal immune system induces endocrine changes in the placenta via IL-6. Brain Behav Immun 25:604–615
Jones CK, Byun N, Bubser M (2012) Muscarinic and nicotinic acetylcholine receptor agonists and allosteric modulators for the treatment of schizophrenia. Neuropsychopharmacology 37:16–42
Lee M, Martin-Ruiz C, Graham A et al (2002) Nicotinic receptor abnormalities in the cerebellar cortex in autism. Brain J Neurol 125:1483–1495
Leonard S, Gault J, Hopkins J et al (2002) Association of promoter variants in the alpha7 nicotinic acetylcholine receptor subunit gene with an inhibitory deficit found in schizophrenia. Arch Gen Psychiatry 59:1085–1096
Lewis DA, González-Burgos G (2008) Neuroplasticity of neocortical circuits in schizophrenia. Neuropsychopharmacology 33:141–165
Lewis AS, Pittenger ST, Mineur YS et al (2018) Bidirectional regulation of aggression in mice by hippocampal Alpha-7 nicotinic acetylcholine receptors. Neuropsychopharmacology 43:1267–1275
Lin H, Hsu F-C, Baumann BH et al (2014) Cortical parvalbumin GABAergic deficits with α7 nicotinic acetylcholine receptor deletion: implications for schizophrenia. Mol Cell Neurosci 61:163–175
Magdalon J, Sánchez-Sánchez SM, Griesi-Oliveira K, Sertié AL (2017) Dysfunctional mTORC1 signaling: a convergent mechanism between syndromic and nonsyndromic forms of autism Spectrum disorder? Int J Mol Sci 18:E659
Martin-Ruiz CM, Lee M, Perry RH et al (2004) Molecular analysis of nicotinic receptor expression in autism. Brain Res Mol Brain Res 123:81–90
Mastropaolo J, Rosse RB, Deutsch SI (2004) Anabasine, a selective nicotinic acetylcholine receptor agonist, antagonizes MK-801-elicited mouse popping behavior, an animal model of schizophrenia. Behav Brain Res 153:419–422
Moser N, Mechawar N, Jones I et al (2007) Evaluating the suitability of nicotinic acetylcholine receptor antibodies for standard immunodetection procedures. J Neurochem 102:479–492
Nicolson R, Craven-Thuss B, Smith J (2006) A prospective, open-label trial of galantamine in autistic disorder. J Child Adolesc Psychopharmacol 16:621–629
Oginsky MF, Cui N, Zhong W et al (2014) Alterations in the cholinergic system of brain stem neurons in a mouse model of Rett syndrome. Am J Physiol Cell Physiol 307:C508–C520
Olincy A, Blakeley-Smith A, Johnson L et al (2016) Brief report: initial trial of Alpha7-nicotinic receptor stimulation in two adult patients with autism Spectrum disorder. J Autism Dev Disord 46:3812–3817

Pafundo DE, Miyamae T, Lewis DA, Gonzalez-Burgos G (2018) Presynaptic effects of N-methyl-D-aspartate receptors enhance Parvalbumin cell-mediated inhibition of pyramidal cells in mouse prefrontal cortex. Biol Psychiatry 84:460–470

Papke RL, Horenstein NA, Kulkarni AR et al (2014) The activity of GAT107, an allosteric activator and positive modulator of α7 nicotinic acetylcholine receptors (nAChR), is regulated by aromatic amino acids that span the subunit interface. J Biol Chem 289:4515–4531

Paulo JA, Brucker WJ, Hawrot E (2009) Proteomic analysis of an alpha7 nicotinic acetylcholine receptor interactome. J Proteome Res 8:1849–1858

Peñagarikano O, Abrahams BS, Herman EI et al (2011) Absence of CNTNAP2 leads to epilepsy, neuronal migration abnormalities, and core autism-related deficits. Cell 147:235–246

Perry EK, Lee ML, Martin-Ruiz CM et al (2001) Cholinergic activity in autism: abnormalities in the cerebral cortex and basal forebrain. Am J Psychiatry 158:1058–1066

Potasiewicz A, Hołuj M, Kos T et al (2017) 3-Furan-2-yl-N-p-tolyl-acrylamide, a positive allosteric modulator of the α7 nicotinic receptor, reverses schizophrenia-like cognitive and social deficits in rats. Neuropharmacology 113:188–197

Ray MA, Graham AJ, Lee M et al (2005) Neuronal nicotinic acetylcholine receptor subunits in autism: an immunohistochemical investigation in the thalamus. Neurobiol Dis 19:366–377

Ross RG, Stevens KE, Proctor WR et al (2010) Research review: cholinergic mechanisms, early brain development, and risk for schizophrenia. J Child Psychol Psychiatry 51:535–549

Ross RG, Hunter SK, McCarthy L et al (2013) Perinatal choline effects on neonatal pathophysiology related to later schizophrenia risk. Am J Psychiatry 170:290–298

Ross RG, Hunter SK, Hoffman MC et al (2016) Perinatal phosphatidylcholine supplementation and early childhood behavior problems: evidence for CHRNA7 moderation. Am J Psychiatry 173:509–516

Sahin M (2012) Targeted treatment trials for tuberous sclerosis and autism: no longer a dream. Curr Opin Neurobiol 22:895–901

Sharma A, Hoeffer CA, Takayasu Y et al (2010) Dysregulation of mTOR signaling in fragile X syndrome. J Neurosci 30:694–702

Smith SEP, Li J, Garbett K et al (2007) Maternal immune activation alters fetal brain development through interleukin-6. J Neurosci 27:10695–10702

Stevens KE, Kem WR, Mahnir VM, Freedman R (1998) Selective alpha7-nicotinic agonists normalize inhibition of auditory response in DBA mice. Psychopharmacology 136:320–327

Takechi K, Suemaru K, Kiyoi T et al (2016) The α4β2 nicotinic acetylcholine receptor modulates autism-like behavioral and motor abnormalities in pentylenetetrazol-kindled mice. Eur J Pharmacol 775:57–66

Tomassy GS, Morello N, Calcagno E, Giustetto M (2014) Developmental abnormalities of cortical interneurons precede symptoms onset in a mouse model of Rett syndrome. J Neurochem 131:115–127

Townsend M, Whyment A, Walczak J-S et al (2016) α7-nAChR agonist enhances neural plasticity in the hippocampus via a GABAergic circuit. J Neurophysiol 116:2663–2675

Tsai PT, Hull C, Chu Y et al (2012) Autistic-like behaviour and cerebellar dysfunction in Purkinje cell Tsc1 mutant mice. Nature 488:647–651

Uddin M, Unda BK, Kwan V et al (2018) OTUD7A regulates neurodevelopmental phenotypes in the 15q13.3 microdeletion syndrome. Am J Hum Genet 102:278–295

Wang L, Almeida LEF, Spornick NA et al (2015) Modulation of social deficits and repetitive behaviors in a mouse model of autism: the role of the nicotinic cholinergic system. Psychopharmacology 232:4303–4316

Winden KD, Ebrahimi-Fakhari D, Sahin M (2018) Abnormal mTOR activation in autism. Annu Rev Neurosci 41:1–23

Wu W-L, Adams CE, Stevens KE et al (2015) The interaction between maternal immune activation and alpha 7 nicotinic acetylcholine receptor in regulating behaviors in the offspring. Brain Behav Immun 46:192–202

Yang Y, Paspalas CD, Jin LE et al (2013) Nicotinic α7 receptors enhance NMDA cognitive circuits in dorsolateral prefrontal cortex. Proc Natl Acad Sci U S A 110:12078–12083
Yasui DH, Scoles HA, Horike S-I et al (2011) 15q11.2-13.3 chromatin analysis reveals epigenetic regulation of CHRNA7 with deficiencies in Rett and autism brain. Hum Mol Genet 20:4311–4323
Yin J, Chen W, Yang H et al (2017) Chrna7 deficient mice manifest no consistent neuropsychiatric and behavioral phenotypes. Sci Rep 7:39941
Yin J, Chen W, Chao ES et al (2018) Otud7a knockout mice recapitulate many neurological features of 15q13.3 microdeletion syndrome. Am J Hum Genet 102:296–308
Yoshimura RF, Tran MB, Hogenkamp DJ et al (2017) Allosteric modulation of nicotinic and GABAA receptor subtypes differentially modify autism-like behaviors in the BTBR mouse model. Neuropharmacology 126:38–47
Záborszky L, Gombkoto P, Varsanyi P et al (2018) Specific basal forebrain-cortical cholinergic circuits coordinate cognitive operations. J Neurosci 38:9446–9458
Zhang Y, Cao S-X, Sun P et al (2016) Loss of MeCP2 in cholinergic neurons causes part of RTT-like phenotypes via α7 receptor in hippocampus. Cell Res 26:728–742

Activators of α7 nAChR as Potential Therapeutics for Cognitive Impairment

Xiaohai Wang, Ian M. Bell, and Jason M. Uslaner

Contents

Abstract The α7 nicotinic acetylcholine receptor (nAChR) is a promising target for the treatment of cognitive deficits associated with psychiatric and neurological disorders, including schizophrenia and Alzheimer's disease (AD). Several α7 nAChR agonists and positive allosteric modulators (PAMs) have demonstrated procognitive effects in preclinical models and early clinical trials. However, despite intense research efforts in the pharmaceutical industry and academia, none of the α7 nAChR ligands has been approved for clinical use. This chapter will focus on the α7 nAChR ligands that have advanced to clinical studies and explore the reasons why these agents have not met with unequivocal clinical success.

Keywords Alpha 7 · Alzheimer's · Cognition · Nicotinic · Schizophrenia

X. Wang and J. M. Uslaner (✉)
Department of Neuroscience Research, Merck & Co. Inc., West Point, PA, USA
e-mail: jason_uslaner@merck.com

I. M. Bell
Department of Discovery Chemistry, Merck & Co. Inc., West Point, PA, USA

Curr Topics Behav Neurosci (2020) 45: 209–246
https://doi.org/10.1007/7854_2020_140
Published Online: 26 May 2020

1 Introduction

1.1 Structure and Function of α7 nAChR

Nicotinic acetylcholine receptors (nAChRs) belong to the superfamily of Cys-loop ligand-gated ion channels that also includes the ligand-gated excitatory serotonin receptors (5-HT_3Rs) and the inhibitory $GABA_A$ receptors ($GABA_A$Rs) and glycine receptors (GlyRs) (Connolly and Wafford 2004). The nAChRs can be further divided into muscle and neuronal types. In mammals, the neuronal nAChRs are encoded by 11 genes expressing subunits (α2–α7, α9, α10, β2–β4), which can form heteromers or homomers (Hurst et al. 2013). Among the neuronal nAChRs, heteromeric α4β2 and homomeric α7 receptors predominate in the brain, with α4β2 nAChR having high-affinity and α7 nAChR having low-affinity binding to ACh (Dineley et al. 2015).

The α7 nAChR subunits are arranged as a pentameric structure surrounding a central cation-permeable (Ca^{2+}, Na^+) water-filled channel pore (McGehee and Role 1995; Connolly and Wafford 2004). Each subunit has a long extracellular N-terminal domain, four transmembrane regions (TM1–TM4), a large intracellular loop between TM3 and TM4, and a short C-terminal end. Both the N- and C- termini face the extracellular space, with the N-terminal domain contributing to the extracellular ligand-binding site (Gotti and Clementi 2004). There are five identical ACh binding sites per α7 receptor, each located at the interface between two adjacent subunits. In addition to forming homomeric nAChRs, the α7 subunit has also been shown to co-assemble with other nAChR subunits, such as β2 subunit, forming heteromeric receptors. However, the functional significance of the α7 containing heteromeric receptors is largely unknown (Araud et al. 2011; de Lucas-Cerrillo et al. 2011; Murray et al. 2012).

The α7 nAChR can exist in multiple conformational states, including resting, activated, and desensitized states (Papke 2014). Activation occurs when the agonist ligand binds to the α7 nAChR receptor and causes the ion channel to open. This is rapidly followed by entering into a desensitized state (in milliseconds) in which the ion channel closes with the agonist still bound. Receptor desensitization is a fundamental property of most ligand-gated ion channels and may have important functions. For example, the rapid desensitization of the α7 nAChR has been proposed to shape synaptic signaling and prevent excessive Ca^{2+} influx into the cell (Giniatullin et al. 2005).

Receptor desensitization is an important property to consider regarding drug design. Signaling by endogenous ACh is rapidly terminated via metabolism by acetylcholinesterase and butyrylcholinesterase. However, sustained exposure to exogenous α7 agonists may produce desensitization and a resulting loss of their pharmacology. Importantly, the concentrations of α7 agonists that can induce desensitization upon prolonged application were found to be several orders of magnitude lower than those that cause activation with brief application, and there is little or no overlap between the desensitization and the activation concentration-response curves (Bertrand et al. 2015). Therefore, targeting an unbound steady-state exposure for α7 agonists that is associated with direct α7 nAChR activation is unlikely to produce a sustained response in clinical settings.

High permeability to Ca^{2+} is one of the unique functional properties of the α7 nAChR (Bertrand et al. 1993; Seguela et al. 1993; Delbono et al. 1997). The physiological role of Ca^{2+} entering through the α7 nAChR is important for a number of critical Ca^{2+}-dependent processes including neurotransmitter release, synaptic plasticity, and neurite outgrowth (Role and Berg 1996; Albuquerque et al. 1997). However, excessive Ca^{2+} influx through the α7 nAChR may induce cytotoxicity, and this has been demonstrated in a transgenic mouse model expressing mutant α7 nAChR with reduced receptor desensitization (Orr-Urtreger et al. 2000). In the homozygous mice, the mutations are associated with extensive cell death throughout cortex and are lethal at birth. Cytotoxicity has also been observed in α7 nAChR expressing cell lines upon treatment with α7 modulators that reduced or abolished receptor desensitization (Ng et al. 2007; Dinklo et al. 2011). These observations underscore the important role of desensitization in the normal function of the α7 nAChR.

1.2 Expression and Localization of α7 nAChR

The expression pattern of the α7 nAChR in the central nervous system has been extensively characterized in multiple species using either radiolabeled α7-specific ligands or in situ hybridization (Breese et al. 1997; Whiteaker et al. 1999; Han et al. 2000, 2003). In general, α7 nAChR is highly expressed in brain regions involved in cognitive functions, such as the hippocampus and cerebral cortex. The pattern of expression appears similar across species, though some species differences have been reported (Breese et al. 1997). For example, the expression level of α7 nAChR was found to be extremely high in human reticular thalamic nucleus (RTN), with $[^{125}I]$-α-bungarotoxin ($[^{125}I]$-αBTX) binding detected throughout the nucleus. In contrast, very low levels of α7 nAChR mRNA and $[^{125}I]$-αBTX binding were detected in the rat RTN. The RTN contains a large population of GABAergic neurons and is known to play an important role in attention and sensory gating (Mitrofanis and Guillery 1993). Therefore, activating α7 nAChR in this brain region may have profound effects on attention and sensory gating deficits in schizophrenic patients.

Neuronal α7 nAChRs are highly expressed at presynaptic and postsynaptic structures (Fabian-Fine et al. 2001; Jones and Wonnacott 2004). Presynaptically, α7 nAChR activation has been reported to modulate the release of a variety of neurotransmitters, including glutamate, GABA, dopamine, and ACh (Rousseau et al. 2005; Zhu et al. 2005; Livingstone et al. 2009; Huang et al. 2014a, b). Postsynaptically, high levels of α7 nAChR have been detected on GABAergic interneurons in the hippocampus, cortex, and thalamus across species, indicating a critical role of α7 nAChR in modulating GABAergic tone in the brain (Alkondon et al. 1999, 2000; Reid et al. 2001).

1.3 Therapeutic Potential of α7 nAChR Ligands

1.3.1 Alzheimer's Disease

AD is a progressive neurodegenerative disorder in which cognitive deficits gradually worsen over a number of years. Characteristics of the disease include degeneration of cholinergic neurons in the basal forebrain and reduction of cholinergic innervation of the cerebral cortex, hippocampus, and other brain regions (Whitehouse et al. 1982; Auld et al. 2002). It has been hypothesized that cholinergic hypofunction contributes to the cognitive deficits of patients suffering from AD. This hypothesis is supported by the fact that acetylcholinesterase inhibitors (AChEIs), which inhibit hydrolysis of acetylcholine, provide symptomatic benefit and have been approved for the treatment of the cognitive impairments in AD. The α7 nAChR was shown to be highly expressed in the brain regions associated with cognitive function, and the expression level of α7 nAChRs can be affected by AD pathology (Burghaus et al. 2000). In addition, numerous α7 nAChR ligands have revealed procognitive effects in preclinical models, indicating potential utility in the treatment of cognitive deficits associated with AD (Bertrand et al. 2015; Wallace and Porter 2011).

1.3.2 Schizophrenia

Schizophrenia is a complex disorder that manifests with positive symptoms, negative symptoms, and cognitive deficits. Current antipsychotics are effective at ameliorating the positive symptoms but provide limited clinical benefit in terms of the negative symptoms and cognitive deficits. Accumulating evidence suggests that α7 nAChR is a potential target for the treatment of cognitive impairment in schizophrenia (Wallace and Bertrand 2013; Freedman 2014). Human genetic evidence has revealed that polymorphisms in CHRNA7, the gene encoding the α7 nAChR subunit, and a partial duplication of CHRNA7, CHRFAM7A, are associated with schizophrenia (Freedman et al. 2003; Sinkus et al. 2009). These polymorphisms lead to diminished gene expression, which is confirmed by reduced α7 nAChR mRNA levels and decreased [^{125}I]-αBTX binding in postmortem brain tissues from schizophrenic patients (Court et al. 1999). Additionally, early clinical studies have shown that activation of α7 nAChRs with selective agonists is associated with attenuation of P50 sensory gating deficits as well as cognitive improvement in patients with schizophrenia (Keefe et al. 2015; Olincy et al. 2006; Lieberman et al. 2013).

2 α7 nAChR Ligands: Chemical and In Vitro Characterization

2.1 α7 nAChR Agonists

From a pharmacological viewpoint, the method of receptor activation most similar to endogenous activation is by use of an orthosteric agonist such as nicotine (Chart 1). This agonist can activate all members of the nAChR family. While the natural product nicotine is not selective, a number of synthetic small molecule agonists that show enhanced selectivity for α7 nAChR versus the other nAChRs have been described, and several of these agonists have progressed to clinical trials. The following compound profiles summarize key data for select α7 nAChR agonists, which are defined as either full agonists or partial agonists relative to the effects produced by acetylcholine and/or nicotine.

In terms of chemical structure, most of these naturally occurring orthosteric agonists may be described by a three-component pharmacophoric model (King et al. 2017). Specifically, the agonists usually feature (1) a rigid, bicyclic, basic amine, such as a quinuclidine, which presumably mimics the quaternary ammonium moiety in the endogenous agonist acetylcholine (Chart 1); (2) a heterocyclic or carbonyl-containing hydrogen bond acceptor, which is thought to mimic the ester carbonyl of acetylcholine; and (3) a lipophilic aromatic or heteroaromatic group. These conserved features are, to some extent, a reflection of the nature of the orthosteric binding site to which acetylcholine and nicotine bind.

There is significant sequence homology at the orthosteric binding site among many of these Cys-loop ligand-gated ion channels, and this can present challenges for the design of selective α7 nAChR agonists. In particular, many of the known agonists of the α7 nAChR have high affinity for the 5-HT_3 receptor, at which they are often functional antagonists (Mazurov et al. 2012). This poor selectivity represents a potential concern in terms of tolerability since antagonism of the 5-HT_3R can be associated with gastrointestinal adverse events, such as constipation.

2.1.1 GTS-21

First synthesized by a chemistry team at the University of Florida, Gainesville (Zoltewicz et al. 1993), GTS-21 (Chart 2), also known as DMXB or DMXB-A, is

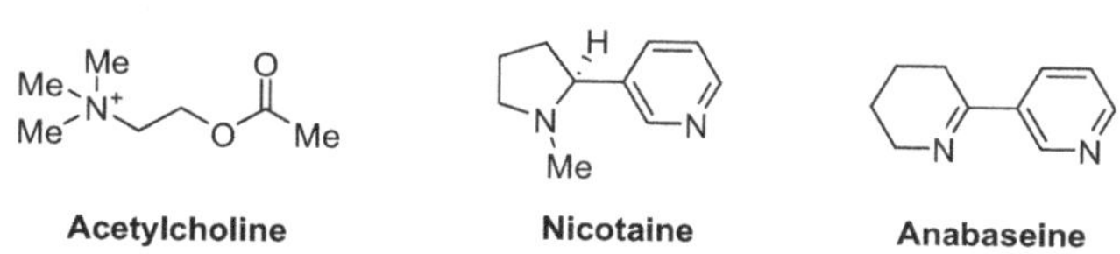

Chart 1 Structures of naturally occurring nicotinic agonists

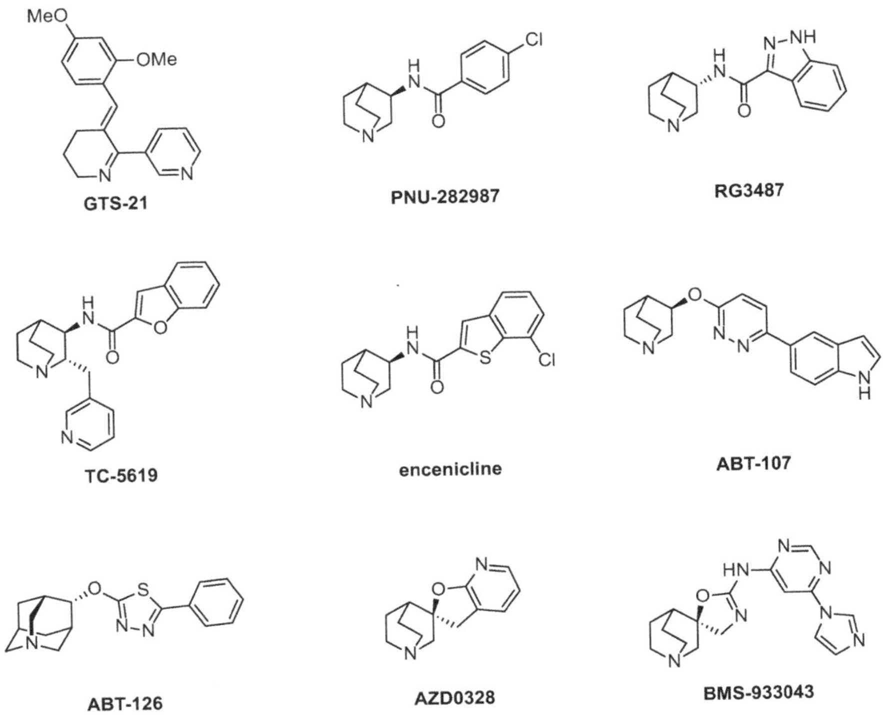

Chart 2 Structures of selected α7 nAChR agonists

structurally related to the natural product and non-selective nicotinic agonist anabaseine (Chart 1) (Wheeler et al. 1981).

GTS-21 was found to displace high-affinity nicotinic [^{125}I]α-bungarotoxin and [^{3}H]acetylcholine binding in rat cortical membrane (Hunter et al. 1994). In electrophysiology studies on human α7 nAChR expressed in *Xenopus* oocytes, GTS-21 was determined to be an agonist at a concentration of 500 μM, but had little effect on oocytes expressing α4β2 nAChR. Taken together, these results suggest that GTS-21 is a selective, albeit non-potent, α7 nAChR agonist at the concentrations that could be tested (Hunter et al. 1994). Subsequent studies showed that GTS-21 is a very weak agonist at human α7 nAChR and at a concentration of 1 mM it only produced 12% of the maximal response seen with nicotine (Briggs et al. 1995). In fact, GTS-21 was found to be more potent as an antagonist of the α7 nAChR, and it blocked the response of nicotine with an IC_{50} value below 10 μM (Briggs et al. 1995).

2.1.2 PNU-282987

PNU-282987 was identified from screening small focused libraries of designed amides against an α7–5-HT_3 chimeric protein. These libraries demonstrated that good agonist activity could be achieved with amides derived from the coupling of 3-aminoquinuclidine with a variety of benzoic acids. The best compound identified was the 4-chlorobenzamide PNU-282987 (Chart 2), which activated the α7–5-HT_3 chimera with an EC_{50} value of 128 nM (Bodnar et al. 2005).

PNU-282987 was found to have high affinity for α7 nAChR in rat brain homogenate ($K_i = 27$ nM) using [^{3}H]methyllycaconitine ([^{3}H]MLA) as the radioligand. It exhibited no agonist activity at both α3β4 and α1β4γδ nAChRs at concentrations up to 100 μM and no antagonism of either receptor ($IC_{50} \geq 60$ μM) (Bodnar et al. 2005). The compound was also assessed in a panel of 32 receptor, ion channel, or enzyme assays and produced <30% inhibition at all targets, except for the 5-HT_3 receptor, at a test concentration of 1 μM. PNU-282987 was found to be a functional antagonist of the 5-HT_3 receptor with moderate affinity in a radioligand binding assay ($K_i = 930$ nM) (Bodnar et al. 2005; Hajos et al. 2005).

When applied to cultured rat hippocampal neurons at concentrations ≥300 nM, PNU-282987 produced a rapidly desensitizing inward current that was blocked by the α7 nAChR antagonist MLA (Hajos et al. 2005). The α7 nAChRs in rat hippocampus are predominantly expressed in GABAergic interneurons, and activation of these receptors can modulate GABAergic synaptic activity. Consistent with this, PNU-282987 was shown to enhance spontaneous GABAergic synaptic activity in rat hippocampal slices (Hajos et al. 2005).

2.1.3 RG3487

RG3487 (MEM3454), a quinuclidine-based compound with a similar structure to PNU-282987 (Chart 2), was discovered at Memory Pharmaceuticals (Xie et al. 2008) and subsequently developed by Hoffmann-La Roche. The compound had high binding affinity for α7 nAChR, displacing the radioligand [^{3}H]MLA from rat brain membranes with a K_i value of 6 nM (Wallace et al. 2011). RG3487 was generally clean when tested in a diverse panel of 90 receptor, ion channel, and enzyme assays, producing <50% inhibition at a concentration of 10 μM. The only potent activity identified in this screening panel was at the 5-HT_3 receptor, for which it had high binding affinity ($K_i = 1.2$ nM). This high binding affinity translated into potent functional antagonism at 5-HT_3R with an IC_{50} value of 2.8 nM (Wallace et al. 2011).

RG3487 was found to be a partial agonist of human α7 nAChR expressed in *Xenopus* oocytes with $EC_{50} = 800$ nM and a maximal effect of 63%, relative to acetylcholine (Wallace et al. 2011). The functional selectivity of RG3487 for α7 nAChR over other nicotinic receptor subtypes was demonstrated in *Xenopus* oocytes, in which the compound did not activate α2β2, α3β2, α4β2, α4β2α5, α4β2α6, or α4β2α6β3 nAChRs at concentrations up to 100 μM (Wallace et al. 2011). Sustained application of higher concentrations (>10 nM) RG3487 to *Xenopus* oocytes expressing α7 nAChR led to receptor desensitization in a dose-dependent fashion, with an IC_{50} value of 40 nM. This observation highlights the loss of activity often seen upon chronic exposure of the receptor to an agonist (Wallace et al. 2011). Interestingly, at low nanomolar concentrations (3–10 nM), sustained application of RG3487 produced a potentiation of the effects of acetylcholine.

2.1.4 TC-5619

Many of the known 3-substituted quinuclidine-based α7 nAChR agonists also have high affinity for the $5\text{-}HT_3$ receptor (Mazurov et al. 2012). The Targacept team identified that a (pyridin-3-yl)methyl substituent at the 2-position of the quinuclidine led to significantly reduced affinity for both $5\text{-}HT_3R$ and the hERG potassium channel (Mazurov et al. 2012). Optimization of the amide substituent led to the identification of TC-5619 (Chart 2) (Mazurov et al. 2012).

In rat brain membranes, TC-5619 potently blocked binding of [^{3}H]MLA to α7 nAChR ($K_i = 1$ nM) and exhibited lower affinity for α4β2 nAChR ($K_i = 2{,}100$ nM) (Hauser et al. 2009). In electrophysiology studies on human α7 nAChR expressed in *Xenopus* oocytes, TC-5619 was determined to be a potent full agonist with $EC_{50} = 33$ nM and a maximal effect of 100%, relative to acetylcholine (Hauser et al. 2009).

TC-5619 was reported to have no significant agonist or antagonist activity at α4β2 nAChR and produced little activation (<25%) of muscle or ganglion nAChRs up to 100 μM, suggesting good functional selectivity for α7 nAChR (Hauser et al. 2009). The compound also had little affinity for human $5\text{-}HT_3R$ ($IC_{50} > 10$ μM), validating the team's design strategy (Hauser et al. 2009). TC-5619 was also assessed in a panel of more than 65 receptor and enzyme assays and was found to have weak activity in a non-selective opioid receptor assay ($K_i = 13$ μM) and at the sodium channel site 2 ($K_i = 13$ μM). Therefore, TC-5619 exhibits >1,000-fold selectivity for α7 nAChR, based on the binding assay K_i of 1 nM. Like other orthosteric agonists, TC-5619 demonstrates desensitization upon prolonged administration (Bristow et al. 2016).

2.1.5 Encenicline

The quinuclidinyl amide encenicline (EVP-6124; Chart 2) was invented at Bayer (Hendrix et al. 2010) and subsequently developed by FORUM Pharmaceuticals. It was found to have high affinity for α7 nAChR in rat brain membranes ($K_i = 9.98$ nM) using [^{3}H]MLA as the radioligand (Prickaerts et al. 2012). In electrophysiology studies on human α7 nAChR expressed in *Xenopus* oocytes, encenicline was found to be a potent partial agonist with $EC_{50} = 160$ nM and a maximal effect of 42%, relative to acetylcholine (Prickaerts et al. 2012).

The compound exhibited good functional selectivity against rat α3β4, α4β2, and muscle α1β4γδ nAChRs, with no detectable agonist activity at concentrations up to 100 μM, and weak antagonism at rat α3β4 ($IC_{50} = 16$ μM) (Prickaerts et al. 2012). Encenicline was also assessed in a panel of more than 60 radioligand binding assays and was found to have potent inhibitory activity at human 5-HT_3 receptor ($IC_{50} \approx 10$ nM) and the human 5-HT_{2B} receptor ($K_i = 14$ nM).

Interestingly, in *Xenopus* oocytes expressing α7 nAChR, very low concentrations of encenicline were found to potentiate the effects of acetylcholine. Specifically, at a concentration of 0.3 nM, encenicline produced a significant and sustained increase in the acetylcholine-evoked response. In contrast, 3 nM encenicline only potentiated the first application of acetylcholine, and subsequent applications led to a reduced acetylcholine-evoked response, consistent with receptor desensitization.

2.1.6 ABT-107

ABT-107 (Chart 2) is a potent and selective agonist that was discovered at Abbott (Stoner et al. 2010). It showed high-affinity binding to α7 nAChR in rat cortex ($K_i = 7.2$ nM) using [^{3}H]MLA as the radioligand (Malysz et al. 2010). ABT-107 activated both human and rat α7 nAChRs expressed in *Xenopus* oocytes with $EC_{50} = 47$ nM (human) and $EC_{50} = 91$ nM (rat) and a maximal effect of 80%, relative to acetylcholine (Malysz et al. 2010).

In terms of selectivity, ABT-107 was profiled in a panel of 81 radioligand binding assays and was found to have inhibitory activity at a number of them, including rat sigma receptor (89% @ 10 μM), human 5-HT_{2A} receptor (80% @ 10 μM), human H_3 receptor (79% @ 10 μM), and human M_3 receptor (67% @ 10 μM). The functional selectivity was assessed for ABT-107 in *Xenopus* oocytes expressing other human nAChRs, and ABT-107 was shown to exhibit weak agonism of α4β4 ($EC_{50} = 2{,}700$ nM; maximal effect = 12% relative to nicotine) and weak antagonism at α3β4 ($IC_{50} = 1{,}200$ nM) and at α4β2 ($IC_{50} = 6{,}700$ nM) (Malysz et al. 2010). Like other orthosteric agonists, ABT-107 produces desensitization upon chronic activation (Malysz et al. 2010).

2.1.7 ABT-126

ABT-126 (Chart 2) was discovered at AbbVie (Schrimpf et al. 2012). It was reported to have high binding affinity at human, rat, and mouse α7 nAChRs ($K_i = 11$–14 nM) with good selectivity vs. other nAChR subtypes (Bitner et al. 2013). The compound was also evaluated in a panel of radioligand binding assays and was found to be generally selective for α7 nAChR, but was an antagonist of the 5-HT_3 receptor, for which it had tenfold reduced binding affinity compared with α7 nAChR (Hudzik et al. 2017). ABT-126 was found to be a partial agonist of human α7 nAChR expressed in *Xenopus* oocytes, with a maximal effect of 74%, relative to acetylcholine (Hudzik et al. 2017).

2.1.8 AZD0328

AZD0328 (Chart 2) was invented at Astra AB (Phillips et al. 2000) and is an example of a spiroquinuclidine-based α7 nAChR agonist. It had high affinity for native rat α7 nAChR in rat hippocampal membranes (K_i = 4.7 nM) and human α7 nAChR expressed in HEK-293 cells (K_i = 3.0 nM) using [^{125}I]α-bungarotoxin as the radioligand (Sydserff et al. 2009). In other radioligand binding experiments, AZD0328 exhibited high affinity for human 5-HT_{3A}R (K_i = 12 nM), moderate affinity for rat α4β2 nAChR (K_i = 140 nM), and low affinity for rat α3 and mouse α1β1γδ nAChRs ($K_i \geq$ 2,500 nM).

In electrophysiology studies on human α7 nAChR expressed in *Xenopus* oocytes, AZD0328 was found to be a partial agonist with EC_{50} = 338 nM and a maximal effect of 65%, relative to acetylcholine. In other functional studies on receptors expressed in *Xenopus* oocytes, AZD0328 activated human 5-HT_{3A}R, with EC_{50} = 135 nM and a maximal effect of 56%, relative to serotonin, but was not potent at human α3β4 and α4β2 nAChRs (EC_{50} > 10 μM) (Sydserff et al. 2009). Overall, these binding and functional assays revealed that AZD0328 is a partial agonist of both α7 nAChR and 5-HT_{3A}R, in contrast with other α7 nAChR agonists like RG3487 that were found to be antagonists of 5-HT_3R.

2.1.9 BMS-933043

The Bristol-Myers Squibb team targeted α7 nAChR partial agonists, with the expectation that this profile would lead to reduced receptor desensitization compared with a full agonist (King et al. 2017). They also sought to identify compounds with a good selectivity profile, especially with respect to selectivity vs. the 5-HT_3 receptor that was often a key issue for agonists of α7 nAChR. The team identified a novel chemotype, based on a quinuclidine derivatized with a spiroaminooxazoline, that allowed them to identify α7 nAChR partial agonists with suitable selectivity vs. 5-HT_3R. Further optimization focused on improved selectivity with respect to the hERG potassium channel, and these efforts led to the identification of BMS-933043 (King et al. 2017).

BMS-933043 had high affinity for native rat α7 nAChR in rat brain membranes (K_i = 3.3 nM) and human α7 nAChR expressed in HEK-293 cells (K_i = 8.1 nM) using [^{125}I]α-bungarotoxin as the radioligand (King et al. 2017). In electrophysiology studies on human α7 nAChR expressed in HEK293 cells, the compound was determined to be a partial agonist with EC_{50} = 290 nM and a maximal effect of 78%, relative to acetylcholine (Bristow et al. 2016).

Consistent with the team's design goals, BMS-933043 exhibited very good selectivity vs. 5-HT_{3A}R in terms of binding affinity (K_i = 2.5 μM). Evaluation of functional selectivity in FLIPR assays demonstrated that the compound was 345-fold selective for α7 nAChR vs. 5-HT_{3A}R and that it was a functional antagonist at 5-HT_{3A}R (Bristow et al. 2016). In other functional assays, BMS-933043 was found to have no agonist or antagonist activity at α3β4, α4β2, and α1β1δε nAChRs up to 100 μM (King et al. 2017). The compound was also determined to have no activities of concern ($IC_{50} > 10$ μM) in a panel of 37 receptor, transporter, and enzyme assays (Bristow et al. 2016). In analogy with other partial agonists of α7 nAChR, BMS-933043 does appear to produce desensitization, although recovery from desensitization was reportedly faster as compared to full agonists (Bristow et al. 2016).

2.2 *α7 nAChR Positive Allosteric Modulators*

The α7 nAChR may be activated with a variety of orthosteric agonists, as described in the previous section. An alternative approach to activation of a receptor is to target an allosteric site that increases the affinity or efficacy of the endogenous agonist. A compound that has this profile is referred to as a positive allosteric modulator (PAM). It is also possible to identify compounds that are negative allosteric modulators (NAMs), which bind allosterically but inhibit the effects of agonists. As such, NAMs can produce a similar inhibition to receptor antagonists.

Compared with an orthosteric agonist of the receptor, an α7 nAChR PAM may have a number of advantages. Perhaps the most obvious difference is that chronic exposure of the receptor to most full or partial agonists leads to desensitization of α7 nAChR and this may limit efficacy or lead to a very narrow range of efficacious doses and exposures for an agonist. In contrast, most known α7 nAChR PAMs do not cause desensitization of the receptor. Additionally, a PAM will only enhance receptor activation in the presence of an agonist, such as the endogenous agonist acetylcholine (Chart 1). Therefore, an α7 nAChR PAM should maintain the natural temporospatial characteristics of neurotransmission (Dinklo et al. 2011). Finally, the orthosteric site to which agonists bind is highly conserved across multiple Cys-loop ligand-gated ion channels, and it can be challenging to design agonists that achieve high levels of selectivity for α7 nAChR over these related receptors. In general, allosteric ligands bind to less conserved sites on the receptor, making it more straightforward to design and discover highly selective α7 nAChR PAMs.

A distinctive feature of α7 nAChR PAMs is that they can have different effects on the desensitization kinetics of the receptor. Although the mechanistic details are not fully understood, some α7 nAChR PAMs, which have been designated Type II PAMs, can delay the normal desensitization of the receptor that follows activation and channel opening. In contrast, a number of PAMs do not significantly alter the natural kinetics of the channel and are designated Type I PAMs. In reality, this strict dichotomy is likely an oversimplification as the different PAM "Types" exist on a continuum and it is possible to identify α7 nAChR PAMs on either extreme as well as those with intermediate effects on the receptor kinetics.

The first characterized α7 nAChR PAMs were compounds already known for other biological activities and functions, such as ivermectin (Krause et al. 1998) and 5-hydroxyindole (Zwart et al. 2002). These compounds served as valuable tools to

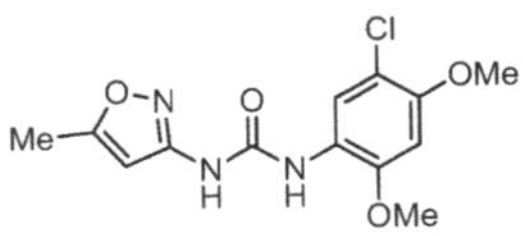

PNU-120596

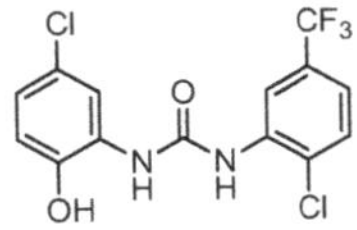

NS1738

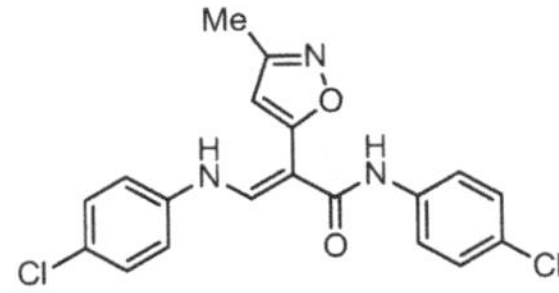

AVL-3288

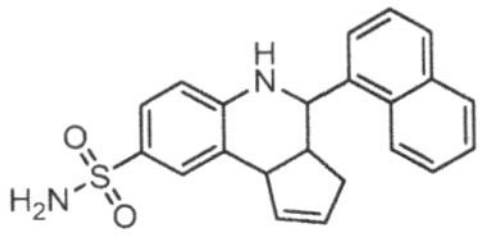

TQS

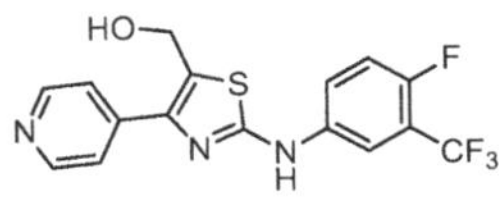

JNJ-1930942

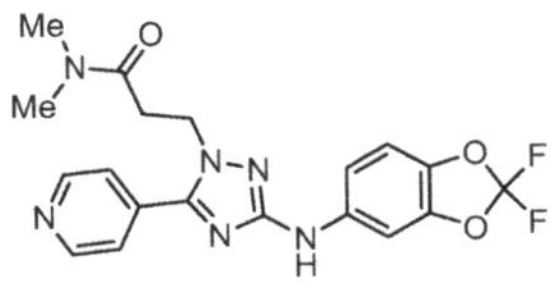

JNJ-39393406

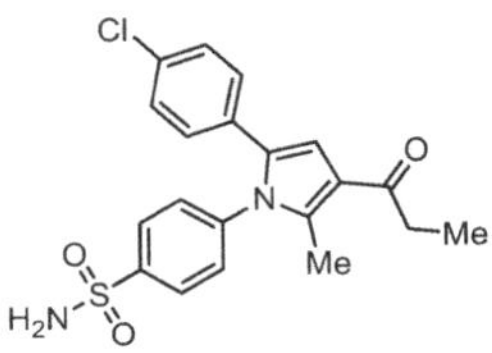

A-867744

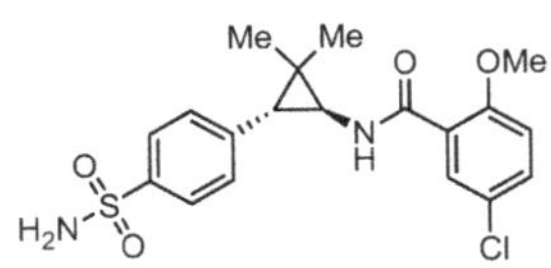

RO5126946

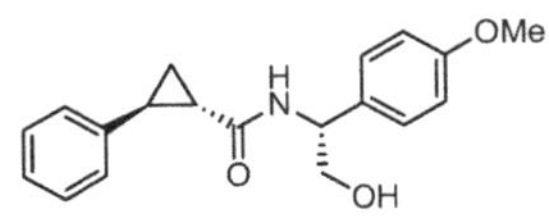

Lu AF58801

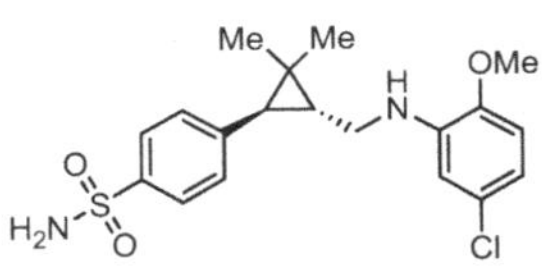

BNC375

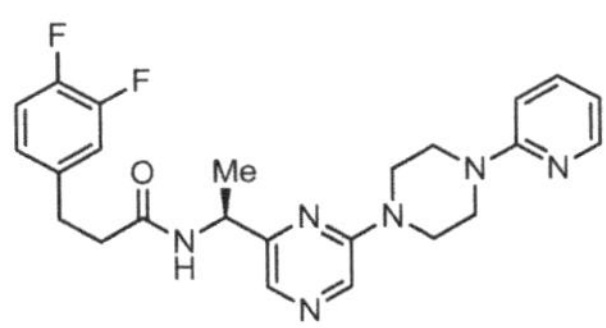

B-973

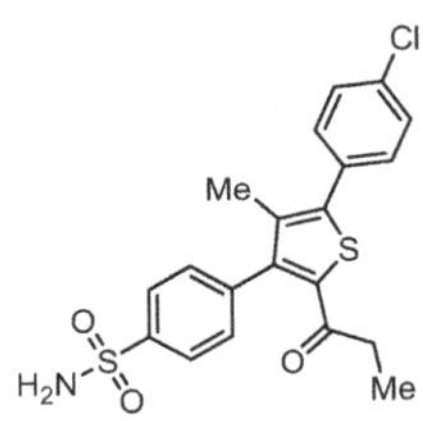

LND101001

Chart 3 Structures of selected α7 nAChR positive allosteric modulators

increase understanding and interest in the field, but poor selectivity and other issues limited their ability to be useful therapeutic agents. Nonetheless, the interest they generated led to the discovery of potent α7 nAChR PAMs with good selectivity and suitable pharmacokinetic properties for in vivo evaluation, and these will be the primary focus of this review.

It is perhaps not surprising that the structures of α7 nAChR PAMs are more diverse than those of the α7 nAChR agonists, which all bind to the same orthosteric site and make largely similar interactions with the receptor. The chemical structures of the various PAMs discussed herein represent a range of chemotypes and appear to rely on varied functionality for their binding interactions with α7 nAChR (see Chart 3). Based on this structural diversity, it seems likely that there is more than one allosteric binding site on the receptor and there is some evidence to support this notion (Malysz et al. 2009). Additional structural and functional studies should shed further light on this topic.

2.2.1 PNU-120596

PNU-120596 (Chart 3) was discovered at Pharmacia & Upjohn as part of FLIPR-based screen for α7 nAChR PAMs (Hurst et al. 2005). In studies on agonist-evoked currents in *Xenopus* oocytes expressing human α7 nAChR, 1 μM PNU-120596 strongly potentiated the effects of 100 μM acetylcholine, increasing both the peak current and the duration of the response (Hurst et al. 2005). The PAM significantly increased the channel mean open time, and this profound change in channel kinetics defines PNU-120596 as a Type II PAM. In fact, PNU-120596 has become the prototypical Type II α7 nAChR PAM and has been extensively used in published studies as an exemplar of this kinetic profile. For example, studies by Papke and co-workers revealed that PNU-120596 exerts its effects, at least in part, by destabilizing a desensitized state of α7 nAChR (Williams et al. 2011).

The effects of PNU-120596 on native α7 nAChR in cultured rat hippocampal neurons were also investigated, and, in general, it produced similar potentiation to that seen with the cloned human receptor. The peak response to acetylcholine observed with native rat α7 nAChR was increased approximately 600% compared with control experiments in the absence of PNU-120596 (Hurst et al. 2005). In contrast to the desensitization often observed with agonists, the robust potentiation produced by PNU-120596 was maintained over multiple pulses of acetylcholine (Hurst et al. 2005). The effects of PNU-120596 were blocked by the selective α7 nAChR antagonist MLA (10 nM), indicating that the PAM exerts its effects through receptors containing the α7 subunit (Hurst et al. 2005).

In other functional studies on neuronal nicotinic receptors expressed in *Xenopus* oocytes, 1 μM, PNU-120596 did not potentiate the effects of acetylcholine on human α4β2, α3β4, or α9α10 nAChRs (Hurst et al. 2005).

2.2.2 NS1738

NS1738 (Chart 3) was one of several diaryl urea compounds identified at NeuroSearch as part of a screening campaign to identify α7 nAChR PAMs (Timmermann et al. 2007). It shares some structural features with PNU-120596, such as the aryl urea substructure, but NS1783 exhibits a distinct kinetic profile from the prototypical Type II PAM. NS1738 did not displace [^{125}I]α-bungarotoxin or [^{3}H] MLA from native rat α7 nAChR in rat brain membranes, at test concentrations up to 100 μM, consistent with the notion that it does not bind to the orthosteric site of the receptor (Timmermann et al. 2007).

In electrophysiology studies on human α7 nAChR expressed in *Xenopus* oocytes, NS1738 potentiated the effects of an EC_{50} concentration of acetylcholine with $EC_{50} = 3.4$ μM and $E_{max} = 322\%$, relative to the acetylcholine EC_{50} response

(Timmermann et al. 2007). Patch clamp electrophysiology was also performed on GH4C1 cells transiently transfected with human α7 nAChR. In these studies, NS1738 alone did not produce any activation of the receptor, but it potentiated the effects of 300 μM acetylcholine with EC_{50} = 1.6 μM and E_{max} = 1,170% (Timmermann et al. 2007). In these electrophysiology experiments, NS1738 appeared to exert its effects primarily by increasing the maximal efficacy of acetylcholine. NS1738 increased the amplitude of the peak currents evoked by acetylcholine but did not significantly change the kinetics of channel opening and desensitization, which is to say that it behaved as a Type I α7 nAChR PAM.

In other functional studies on receptors expressed in *Xenopus* oocytes, at concentrations up to 30 μM, NS1738 did not potentiate the effects of acetylcholine on human α4β2 or α3β4 nAChRs. In fact, NS1738 inhibited these channels weakly with extrapolated IC_{50} values of 89 μM (α4β2) and 27 μM (α3β4). NS1738 did not exhibit any potentiation of human 5-HT_{3A}R at concentrations up to 30 μM (Timmermann et al. 2007).

2.2.3 AVL-3288

A team at the University of California, Irvine, screened their library of $GABA_A$ receptor PAMs and identified an initial hit with PAM activity at both $GABA_A$R and α7 nAChR (Ng et al. 2007). Optimization of this hit was focused on improving selectivity for α7 nAChR vs. $GABA_A$R and on improving pharmacokinetic properties to provide a suitable tool compound for rodent studies. This work led to the identification of the α7 nAChR PAM AVL-3288, also known as CCMI or XY4083 (Chart 3).

Consistent with little or no binding affinity for the orthosteric site, AVL-3288 did not displace [^{125}I]α-bungarotoxin from native rat α7 nAChR in rat brain membranes at test concentrations up to 10 μM (Ng et al. 2007). In electrophysiology studies on human α7 nAChR expressed in *Xenopus* oocytes, AVL-3288 did not activate the receptor itself but potentiated the effects of agonists, including acetylcholine and nicotine. Specifically, AVL-3288 was found to robustly potentiate an EC_5 concentration of acetylcholine with EC_{50} = 0.7 μM and produced a maximal effect of 900%, relative to the acetylcholine EC_5 response (Ng et al. 2007). AVL-3288 exhibited a Type I PAM profile, potentiating agonist-evoked currents while preserving the native channel kinetics.

In other functional studies on receptors expressed in *Xenopus* oocytes, the compound did not exhibit any modulation of human α4β2, rat α3β4, or mouse α1β1γδ nAChRs or of human 5-HT_{3A}R. AVL-3288 was found to exhibit weak potentiation at human $GABA_A$ α1β2γ2L receptors expressed in *Xenopus* oocytes with EC_{50} = 6.3 μM, indicating that it is moderately selective for α7 nAChR (Ng et al. 2007).

One interesting experiment suggested a potential advantage for the Type I over the Type II PAM profile (Ng et al. 2007). In this study, SH-SY5Y-α7 cells were exposed to either AVL-3288 (Type I PAM) or PNU-120596 (Type II PAM) for 24 h at concentrations from 0.3 to 10 μM. While AVL-3288 had no effect on cell viability, PNU-120596 at concentrations ≥3 μM was found to be cytotoxic to these cells. This toxicity was prevented by 10 nM MLA, suggesting an α7-mediated mechanism. The authors proposed that these results indicate that maintaining the natural kinetics of α7 nAChR could prevent toxicity caused by excessive Ca^{2+} influx resulting from Type II PAMs (Ng et al. 2007).

2.2.4 TQS

The first arylsulfonamide-containing α7 nAChR PAM, TQS (Chart 3), was invented by researchers at AstraZeneca (Becker et al. 2004), and it was profiled in detail by a team at Abbott (Gronlien et al. 2007). In electrophysiology studies on human α7 nAChR expressed in *Xenopus* oocytes, TQS potentiated the effects of 100 μM acetylcholine with $EC_{50} = 3.2$ μM and $E_{max} = 420\%$, relative to the acetylcholine control response (Gronlien et al. 2007). TQS was found to induce similar channel kinetics to PNU-120596 and was designated a Type II PAM (Gronlien et al. 2007). In this same study, both TQS and PNU-120596 were shown to be capable of activating desensitized α7 nAChR in the presence of acetylcholine. Other studies have indicated that Type I PAMs like AVL-3288 cannot activate the α7 nAChR once it has been desensitized by exposure to an agonist, suggesting that this ability to reverse desensitization may be specific to Type II PAMs (Ng et al. 2007).

The selectivity of TQS was investigated by examining its effects on human α4β2 or α3β4 nAChRs expressed in HEK-293 cells. In these experiments, TQS did not potentiate the effects of nicotine on either α4β2 or α3β4 nAChRs but did produce some inhibition at α4β2 with $IC_{50} \approx 5$ μM (Gronlien et al. 2007). More recent work has probed the mechanism of action of TQS and provided evidence that it binds to α7 nAChR at a transmembrane allosteric site (Gill et al. 2011).

2.2.5 JNJ-1930942

JNJ-1930942 was first synthesized at Janssen and its PAM activity at α7 nAChR was presumably discovered in a screening campaign. The compound was active as a potentiator of α7 nAChR in a FLIPR assay format with $EC_{50} = 1.9$ μM, using 100 μM choline as agonist (Dinklo et al. 2011). In other FLIPR assays, the compound was shown to produce no activation, potentiation, or inhibition of human α4β2 or α3β4 nAChRs, or of human $5\text{-}HT_{3A}R$, at concentrations up to 30 μM. Based on radioligand binding studies using the orthosteric agonist [^{3}H]A-585539, it was shown that JNJ-1930942 likely binds at an allosteric site on α7 nAChR (Dinklo et al. 2011).

In electrophysiology studies on human α7 nAChR stably expressed in GH4C1 cells, JNJ-1930942 potentiated the effects of 100 μM acetylcholine $E_{max} = 3{,}200\%$, relative to the acetylcholine control response (Dinklo et al. 2011). The compound was found to slow the rapid desensitization of α7 nAChR and is therefore not a Type

I PAM. However, it did not produce the kind of profoundly delayed receptor desensitization observed with Type II PAMs such as PNU-120596, so the Janssen team considered it to have an intermediate profile, between Type I and Type II PAMs. Importantly, the effects of JNJ-1930942 were completely blocked by 30 nM MLA, indicating that the PAM is acting through the α7 nAChR (Dinklo et al. 2011).

2.2.6 JNJ-39393406

Little preclinical work has been published on JNJ-39393406 (Chart 3), but it was reported to potentiate α7 nAChR in a FLIPR assay format with $EC_{50} = 0.66$ μM, using 100 μM choline as agonist (https://ncats.nih.gov/files/JNJ-39393406.pdf). The compound increased the maximum α7 nAChR response to acetylcholine or nicotine by 17- to 20-fold (Winterer et al. 2013). JNJ-39393406 was reported to be selective for α7 nAChR, having no activity at α4β2 or α3β4 nAChRs or at $5\text{-}HT_{3A}R$, and it also had no activities of concern in a panel of 62 receptor and enzyme assays (Winterer et al. 2013).

2.2.7 A-867744

At Abbott, high-throughput screening identified novel pyrrole-based lead compounds that had good potency as α7 nAChR PAMs but exhibited poor pharmacokinetic properties. Optimization focused on improving metabolic stability while maintaining PAM activity, and these efforts led to A-867744 (Chart 3) (Faghih et al. 2009). A-867744 did not displace [^{3}H]MLA from native rat α7 nAChR in rat brain homogenates, at test concentrations up to 10 μM, consistent with an allosteric mode of binding (Malysz et al. 2009).

In electrophysiology studies on human α7 nAChR expressed in *Xenopus* oocytes, A-867744 potentiated the effects of 100 μM acetylcholine with $EC_{50} \approx 1$ μM and $E_{max} = 730\%$, relative to the acetylcholine control response (Malysz et al. 2009). Similar results were observed for rat α7 nAChR, indicating no significant interspecies differences for this PAM. A-867744 was found to delay receptor desensitization and, at $\geq$3 μM, produced acetylcholine-evoked currents that lasted for many seconds. In this regard, A-867744 has a Type II PAM profile similar to PNU-120596. The effects of A-867744 could be blocked by MLA, indicating that the PAM is acting via α7 nAChR (Malysz et al. 2009).

Further evidence for the good selectivity profile of A-867744 was provided by functional studies on related receptors. At concentrations up to 30 μM, the compound did not potentiate the effects of nicotine on human α4β2 or α3β4 nAChRs expressed in HEK-293 cells but did produce some level of inhibition at both with IC_{50} values of 6.4 μM (α4β2) and 20 μM (α3β4). A-867744 had no effect on human $5\text{-}HT_{3A}R$ expressed in *Xenopus* oocytes at concentrations up to 30 μM (Malysz et al. 2009).

2.2.8 RO5126946

A team at Roche and their collaborators discovered RO5126946 (Chart 3) through screening of compound libraries. The compound was shown to potentiate the effects of acetylcholine on human α7 nAChR stably expressed in GH4C1 cells with an $EC_{50} \approx 60$ nM (Sahdeo et al. 2014). In radioligand binding assays, RO5126946 did not displace $[^{125}I]$α-bungarotoxin from human α7 nAChR in GH4C1 cell membranes, at test concentrations up to 100 μM, suggesting that it binds at an allosteric site (Sahdeo et al. 2014).

In electrophysiology studies on human α7 nAChR stably expressed in GH4C1 cells, RO5126946 strongly potentiated the effects of 1 mM acetylcholine with $EC_{50} = 0.89$ μM and $E_{max} = 9{,}200\%$, relative to the acetylcholine control response (Sahdeo et al. 2014). RO5126946 increased both the peak current and the duration of channel opening following agonist stimulation of α7 nAChR. It is therefore classified as a Type II PAM although, in contrast to PNU-120596 and TQS, it was unable to reverse agonist-induced desensitization of α7 nAChR. This observation suggests that RO5126946 may have a kinetic profile somewhat intermediate between Type I and Type II, reminiscent of JNJ-1930942, and it highlights the complexity of the effects of these various allosteric modulators on the receptor (Sahdeo et al. 2014).

On native α7 nAChR in cultured rat hippocampal neurons, RO5126946 produced similar effects to those observed with the cloned human receptor. In these experiments, the peak response to 15 μM nicotine was increased approximately 2,800% in the presence of 1 μM RO5126946. This potentiation could be blocked by 10 nM MLA, indicating that the PAM is acting via α7 nAChR (Sahdeo et al. 2014).

The selectivity of the compound was further confirmed by functional studies on related receptors, in which 1 μM RO5126946 produced no significant effects on either human α4β2 nAChR or human $5\text{-}HT_{3A}R$ expressed in *Xenopus* oocytes (Sahdeo et al. 2014). RO5126946 was also determined to have no activities of concern ($IC_{50} > 10$ μM) in a panel of >70 receptor, transporter, and enzyme assays (Sahdeo et al. 2014).

2.2.9 Lu AF58801

At Lundbeck, high-throughput screening identified a number of α7 nAChR PAMs based on a phenylcyclopropyl substructure that had modest potency and were quite lipophilic. Optimization was focused on improving potency and selectivity, as well as physicochemical and pharmacokinetic properties, and this work led to the discovery of Lu AF58801 (Chart 3) (Eskildsen et al. 2014).

In electrophysiology studies on human α7 nAChR expressed in *Xenopus* oocytes, Lu AF58801 potentiated the effects of 30 μM acetylcholine with $EC_{50} = 1.6$ μM and $E_{max} = 4{,}400\%$, relative to the acetylcholine control response (Eskildsen et al. 2014). Interestingly, while the compound had a Type I PAM profile at the human α7 nAChR, with the same magnitude of increase in peak current and the area under the curve (AUC) of that peak, it appeared to be a Type II PAM of rat α7 nAChR producing a ninefold higher increase in AUC than the increase in peak current (Eskildsen et al. 2014).

The selectivity of Lu AF58801 was examined in functional studies on related receptors expressed in *Xenopus* oocytes, and it produced no significant effects on α3β4, α4β2, or muscle α1β1γδ nAChRs at concentrations up to 30 μM. Additional confirmation of the attractive selectivity profile of Lu AF58801 was provided by screening in a panel of 108 molecular targets, in which it was found to be clean ($IC_{50} > 10$ μM) with the exception that it had modest activity at 5-HT_{2A} receptor ($IC_{50} \approx 10$ μM) (Eskildsen et al. 2014). Lu AF58801 was also reported to be non-cytotoxic up to 100 μM and had no significant effects ($IC_{50} > 10$ μM) in patch clamp studies on key cardiac ion channels, including hERG, $Ca_v1.2$, and $Na_v1.5$ (Eskildsen et al. 2014).

2.2.10 BNC375

Researchers at Bionomics identified lead compounds from screening of small focused libraries, and optimization of these initial hits led to BNC375 (Chart 3) (Harvey et al. 2019). BNC375 was evaluated in electrophysiology studies on rat α7 nAChR stably expressed in GH4C1 cells, and it potentiated the effects of 100 μM acetylcholine with $EC_{50} = 1.9$ μM and $E_{max} = 2{,}700\%$, relative to the acetylcholine control response (Harvey et al. 2019).

The compound was determined to be a Type I α7 nAChR PAM since it maintained the natural kinetics of the channel, increasing current AUC only slightly more than peak current in response to acetylcholine. Notably, the enantiomer of BNC375 was found to be a potent Type II PAM, increasing the current AUC about 25-fold more than the peak, and with $EC_{50} = 0.063$ μM on rat α7 nAChR (Harvey et al. 2019). This observation highlights the subtlety of α7 nAChR PAM structure-activity relationships, since both enantiomers of this structure are potent PAMs but they have very different effects on channel kinetics.

2.2.11 B-973

B-973 (Chart 3) was discovered as part of a high-throughput screening campaign at Bristol-Myers Squibb (Post-Munson et al. 2017). The stereochemistry of B-973 was reported to be (*S*), based on the structure shown in the original publication, and it was later confirmed that the (*S*)-enantiomer is indeed the bioactive form, while the (*R*)-enantiomer was demonstrated to be inactive (Garai et al. 2018). B-973 was reported to be a potent Type II PAM on human α7 nAChR stably expressed in HEK-293 cells. At 1 μM concentration, B-973 potentiated the effects of 100 μM acetylcholine, increasing the peak sixfold (relative to 3 mM acetylcholine) and increasing charge transfer 6,900-fold (Post-Munson et al. 2017).

Interestingly, in addition to its ability to potentiate the response of ACh, at concentrations above 1 μM, B-973 was able to activate α7 nAChR in the absence of orthosteric agonist, producing a prolonged opening of the channel, and these effects could be almost completely blocked by MLA (10 nM). This pharmacology suggests that B-973 is an AgoPAM at the α7 nAChR, having both agonist and positive allosteric modulator effects (Post-Munson et al. 2017). The unusual AgoPAM profile of B-973 should make it an interesting tool compound for a variety of studies.

2.2.12 LND101001

A team at Lupin attempted to identify novel α7 nAChR PAMs with good potency and high brain/plasma ratios (Sinha et al. 2019). They focused their initial efforts on making analogues of A-867744, replacing the pyrrole core with a thiophene to enhance lipophilicity. After exploration of different substitution patterns around the thiophene ring, they identified LND101001, which has the same substituents as A-867744 arranged in a different pattern around the core ring.

LND101001 was active as a potentiator of α7 nAChR in IMR-32 cells (Faghih et al. 2009) in a FLIPR assay format with $EC_{50} = 324$ nM, using 10 μM PNU-282987 as agonist (Sinha et al. 2019). At a concentration of 1 μM, the compound produced a 22-fold increase in the calcium flux signal induced by 10 μM PNU-282987. Based on its profile in this FLIPR assay, it was reported that LND101001 was a Type II α7 nAChR PAM, although no detailed electrophysiological characterization was conducted (Sinha et al. 2019). The effects of LND101001 on PNU-282987-induced calcium flux in IMR-32 cells were blocked by MLA and α-bungarotoxin, suggesting that this PAM acts on the α7 nAChR. Additional data on the selectivity profile of LND101001 have not been described.

3 α7 nAChR Ligands: In Vivo and Clinical Characterization

3.1 α7 nAChR Agonists

The effects of α7 nAChR agonists in preclinical models of cognitive impairment are well characterized, and some of these compounds have advanced to clinical studies for treatment of cognitive deficits associated with schizophrenia and Alzheimer's disease. Despite strong preclinical evidence and the promising early clinical findings, none of the α7 agonists have succeeded in Phase 3 trials. Here we focus on the compounds that have been characterized in both preclinical and clinical studies to illustrate the pharmacological effects of α7 agonists on cognitive function and potential reasons these effects might not have translated to clinical efficacy.

3.1.1 GTS-21

GTS-21 (DMXB-A) was one of the first reported ligands to show binding specificity for the α7 nAChR, and it has demonstrated procognitive effects across multiple cognitive domains in rodents, non-human primates (NHPs), and humans. In both aged and nucleus basalis-lesioned rats, GTS-21 significantly improves learning and memory in a variety of cognitive assays, including Lashley III maze, 17-arm radial maze, and Morris water maze (Arendash et al. 1995; Meyer et al. 1997). GTS-21 also ameliorate sensory gating deficits in rodents. As examples, GTS-21 significantly improved sensory inhibition in DBA/2 mice (Simosky et al. 2001), reduced an auditory gating deficit in isolation-reared rats (O'Neill et al. 2003), and abolished a prepulse inhibition (PPI) impairment induced by MK-801 or apomorphine (Callahan et al. 2014). In addition to enhancing cognition in rodents, GST-21 has demonstrated procognitive effects in non-human primates (NHPs). Pre-treatment of GTS-21 attenuates ketamine-induced cognitive impairments in NHPs in both the delayed matching-to-sample (DMTS) task (Briggs et al. 1997; Buccafusco and Terry 2009) and object retrieval-detour (ORD) task (Cannon et al. 2013). Importantly, an inverted U-shaped dose-effect function was observed in the ORD task when GTS-21 was examined at a wider dose range (Cannon et al. 2013). As such, efficacy was detected at 0.03 mg/kg but not 0.01 and 0.1 mg/kg.

GTS-21 has been evaluated in multiple clinical studies. In the initial Phase 1 trials, GTS-21 was well tolerated up to daily doses of 450 mg (150 mg t.i.d.), with no clinically significant safety findings. In addition, GTS-21 showed statistically significant enhancement in attention, working memory, and episodic memory in healthy subjects (Kitagawa et al. 2003). In the subsequent Phase 2 trials, GTS-21 was evaluated in non-smoking schizophrenic patients taking anti-psychotic drugs (Olincy et al. 2006; Freedman et al. 2008). The first Phase 2 trial demonstrated significant improvement in both cognition and P50 inhibitory gating, especially for the lower GTS-21 dose (75 mg followed by 37.5 mg 2 h later, single dose) compared with placebo (Olincy et al. 2006). However, the second Phase 2 study found no significant differences in the MATRICS cognitive measures between GTS-21 and placebo groups. Patients in this second trial did experience a significant improvement in negative symptoms at higher GTS-21 dose (150 mg b.i.d.) after 4 weeks of treatment and a nearly statistically significant improvement in positive symptoms (Freedman et al. 2008).

There are many issues associated with GTS-21 as an α7 agonist. First, as discussed in the previous section, GTS-21 is a very weak partial α7 agonist, and it is possible that the in vivo pharmacology is driven by activity other than activation of α7 nAChR (Briggs et al. 1995). Second, GTS-21 has higher affinity for α4β2 nAChR, where it is a functional antagonist, compared to its interaction with α7 receptor (Briggs et al. 1997). Third, the human half-life of GTS-21 is extremely short, ~1 h, which makes it difficult to achieve sustained efficacious plasma concentrations (Kitagawa et al. 2003). To assess whether extending plasma exposure GTS-21 can enhance its procognitive effects in schizophrenia, a third Phase 2 trial was conducted using an extended-release formulation (Kem et al. 2018). Both smoking and non-smoking patients were enrolled in this double-blind, randomized, placebo-controlled 1-month trial. GTS-21 (150 mg, q.i.d.) was formulated in hypromellose to produce extended release over 4 h and administered four times a day. No significant effect was observed in cognition or in P50 gating endpoints in either smokers or non-smokers. However, the plasma exposures in non-smokers with the extended-release formulation were comparable to those achieved transiently with 75–150 mg standard formulation. The lack of procognitive effect with this extended-release formulation was interpreted as a result of extensive receptor desensitization with prolonged exposure to GTS-21. This finding led to the conclusion that long duration-acting α7 agonists are not likely to be effective in ameliorating cognitive deficits in patients (Kem et al. 2018).

3.1.2 Encenicline

Encenicline (EVP-6124) has also demonstrated numerous procognitive effects in multiple preclinical models. In a rat novel object recognition (NOR) task, encenicline reversed scopolamine-induced short-term memory deficit. The pharmacology exhibited an inverted U-shaped dose-response curve with peak effect observed at 0.3 mg/kg (Prickaerts et al. 2012). Similarly, the compound prevented natural forgetting in the NOR task at 0.3 mg/kg but not at 0.1 or 1.0 mg/kg. The effect of 0.3 mg/kg in the natural forgetting test was blocked by administration of the selective α7 nAChR antagonist MLA, indicating that the procognitive effect of encenicline was mediated by α7 nAChR (Prickaerts et al. 2012). Based on the pharmacokinetic analysis, the maximum unbound brain drug level of encenicline at 0.3 mg/kg was less than 1 nM. This central exposure was insufficient in vitro to either activate or desensitize the α7 nAChR (Prickaerts et al. 2012). This intriguing observation is consistent with the in vitro finding, as mentioned in previous section, that encenicline potentiates ACh current at sub-nanomolar concentrations (e.g., 0.3 nM). This finding suggests a novel mechanism of action of encenicline as a co-agonist at α7 nAChR (Prickaerts et al. 2012). Indeed, the evidence for co-agonist or "priming" activities has been reported for other nAChR subtypes (Cachelin and Rust 1994; Zwart and Vijverberg 2000; Papke et al. 2011) as well as for α7 nAChR with other partial agonists, including RG3487 (Wallace et al. 2011) and tropisetron (Callahan et al. 2017).

Since encenicline potentiates the effects of ACh in a very narrow concentration range, the observation of steep inverted U-shaped dose-response curves with this compound in multiple in vivo studies is expected. For example, encenicline was found to reverse scopolamine-induced impairment of paired associates learning (PAL) task in NHPs only at 0.01 mg/kg but not at 0.003, 0.03, 0.1, 0.3, and 1.0 mg/kg (Weed et al. 2017). Consistent with the findings in the rat NOR assay, the unbound plasma concentration of encenicline at the effective dose (0.01 mg/kg) in NHPs was in the sub-nanomolar range (Weed et al. 2017). Furthermore, encenicline increased release of dopamine, ACh, and glutamate in medial prefrontal cortex (mPFC) and nucleus accumbens (NAC) in rats with an inverted U-shaped dose-effect relationship (Huang et al. 2014a). The release of these neurotransmitters in mPFC and NAC was optimally stimulated at a maximum unbound plasma concentration of ~0.4 nM, whereas higher exposures (> 1 nM unbound) impaired dopamine and glutamate efflux (Huang et al. 2014a), likely due to the receptor desensitization.

Despite the challenges in dose selection as described above, encenicline advanced to clinical trials for treatment of cognitive impairments associated with schizophrenia and Alzheimer's disease (AD). In the initial single ascending-dose study, encenicline was well tolerated up to 180 mg and demonstrated dose-proportional increase in plasma exposure (Barbier et al. 2015). A Digit Symbol Substitution Test was conducted in healthy volunteers to evaluate the procognitive effects of encenicline over a dose range of 1–180 mg. The optimal cognitive improvements were observed at 20 mg (single dose), which produces a mean exposure extrapolated to a steady-state dosage of 2–4 mg/day (Barbier et al. 2015). The difference in single-dose and multiple-dose extrapolated exposure is due to the long plasma half-life of encenicline (54–62 h) and drug accumulation over chronic dosing. Consistent with the effective drug level in preclinical models, steady-state unbound plasma concentration of encenicline at 2–4 mg/day (q.d.) was expected to be in the sub-nanomolar to low nanomolar range (Keefe et al. 2015). Subsequent Phase 2 trials examined the procognitive effects of encenicline at 0.3 or 1.0 mg/day (q.d.) in schizophrenic patients treated with antipsychotic medication (Keefe et al. 2015; Preskorn et al. 2014). Both smokers and non-smokers were enrolled in these studies. Statistically significant improvements were found across multiple measures of cognition, including Overall Cognition Index (OCI), Schizophrenia Cognition Rating Scale (SCoRS), and PANSS Cognition Impairment Domain (Preskorn et al. 2014; Keefe et al. 2015). These positive findings led to the launch of two global Phase 3 trials aimed at assessing the procognitive effects of encenicline (1.0 or 2.0 mg/day q.d. for 26 weeks) in stable patients with schizophrenia (Brannan 2019). In contrast to the earlier trials, these Phase 3 studies failed to demonstrate a statistically significant difference between encenicline and placebo groups for either NeuroCognitive Composite Score (NCC) or SCoRS end points (Brannan 2019).

Encenicline was also evaluated in patients with mild-to-moderate AD. The initial Phase 1b trial afforded statistically significant improvement in multiple cognitive tests with encenicline treatment at 0.3 and 1.0 mg/day (Deardorff et al. 2015). The subsequent larger Phase 2b study assessed the procognitive effect of encenicline at 0.3, 1.0, or 2.0 mg/day (q.d.) for 6 months in mild-to-moderate AD patients. In this Phase 2b study, the 2.0 mg/day (q.d.) group, but not the 0.3 and 1.0 mg/day groups, demonstrated significant improvement on both Alzheimer's Disease Assessment Scale – Cognitive subscale (ADAS-Cog) and Clinical Dementia Rating – Sum of Boxes (CDR-SB) end points (Deardorff et al. 2015). In addition, a significant relationship was observed between mean trough plasma exposure of encenicline and probability of experiencing an ADAS-Cog-13 improvement of greater than 3 points (Deardorff et al. 2015). Following this successful Phase 2b study, encenicline advanced to the COGNITIV AD Phase 3 trials evaluating the procognitive effect of 2.0 or 3.0 mg/day (q.d.) doses for 6 months in mild-to-moderate AD patients. Unfortunately, the Phase 3 trials were terminated due to severe gastrointestinal (GI) symptoms experienced in a small number of AD patients (Mehta et al. 2017). There are several possible reasons for the severe GI symptoms. First, a contributing factor to the GI adverse event was suspected to be encenicline's off-target activity at the human 5-HT_3 receptor. Antagonists of the 5-HT_3 receptor increase the risk of constipation in human and are actually used to treat chemotherapy-induced emesis and diarrhea-predominant irritable bowel syndrome (Andresen et al. 2008). Additionally, AD patients generally have a higher risk of GI adverse events, particularly since donepezil produces adverse GI effects. Lastly, the encenicline dose was increased to 3.0 mg/day (q.d.) in these Phase 3 studies, whereas lower doses had been evaluated previously.

3.1.3 RG3487

RG3487 was found to improve cognition in a variety of preclinical models representing multiple cognitive domains (Wallace et al. 2011, 2009). For example, in the Morris water maze test, RG3487 improved spatial memory in age-impaired rats. Significant reduction in swim latencies were detected at 0.03–0.3 mg/kg doses of RG3487, whereas lower (0.01 mg/kg) or higher (1 and 10 mg/kg) doses were inactive (Wallace et al. 2011). In addition, an inverted U-shaped dose response with RG3487 was also observed in apomorphine-impaired PPI in rats (Wallace et al. 2011), RG3487-indued DA and ACh release from rat mPFC and hippocampus (Huang et al. 2014b), and cognitive improvement in an NHP model of attention and working memory (Wallace et al. 2009). Pharmacokinetic analysis suggested that the efficacious plasma and brain exposures of RG3487 were in the low nanomolar range. At this concentration range, RG3487 can neither activate nor desensitize the α7 receptor in vitro (Wallace et al. 2011). Similar to encenicline, RG3487 at sustained exposure of low nanomolar concentrations (3 and 10 nM) potentiated ACh-evoked current in oocytes expressing human α7 receptors. At higher concentrations ($>$ 10 nM), RG3487 desensitized the receptor and dramatically reduced ACh-evoked α7 current (Wallace et al. 2011). Therefore, the co-agonist or "priming" effect may also contribute to the mechanism of action of RG3487 in vivo.

RG3487 has been evaluated in a Phase 2a trial in mild-to-moderate AD patients. There has been only limited information released for this Phase 2a trial. Three dose levels of RG3487 were tested, and significant cognitive improvement was reported at the two lower doses (5 and 15 mg/day, q.d.) (Sabbagh 2009). However, the subsequent Phase 2b study failed to show cognitive improvement with RG3487 when added to donepezil in mild-to-moderate AD. It is noteworthy that constipation was the only AE significantly more common in the RG3487 group, likely due to its potent 5-HT_3 receptor antagonist activity (Sabbagh 2009).

RG3487 was subsequently taken into clinical development for the treatment of cognitive impairment associated with schizophrenia (CIAS) in an 8-week Phase 2a study. Patients with stable schizophrenia received RG3487 (5, 15, or 50 mg/day, q.

d.) added to ongoing treatment of second-generation antipsychotics (Umbricht et al. 2014). Dose selection was based on the results of preclinical studies targeting an effective plasma exposure in the low nanomolar range. No significant effect of cognitive improvement was observed with RG3487 as assessed by the MCCB composite score. The effect of smoking status on MCCB was not evaluated because of the small number of non-smokers enrolled (Umbricht et al. 2014).

3.1.4 ABT-126

There is limited published information on the preclinical characterization of ABT-126. One presentation described the procognitive effects of ABT-126 in rodent and primate models that capture domains of working memory, memory consolidation and recall, pre-attention processing, and short-term memory (Bitner et al. 2013). In addition, repeated dosing of ABT-126 did not result in attenuation of efficacy in preclinical models (Bitner et al. 2013). However, it is difficult to determine how well desensitization was interrogated without knowing more about rodent PK of this compound.

ABT-126 has also been characterized clinically. In an initial Phase 2a trial in patients with mild-to-moderate AD, ABT-126 (5 or 25 mg/day, q.d.) was investigated as monotherapy for 12 weeks (Gault et al. 2015). ABT-126 at 25 mg/day (q.d.) was associated with a trend toward improvement in cognition according to the ADAS-Cog total score, while 5 mg/day (q.d.) had no beneficial effect compared to placebo. An exposure-response analysis indicated a significant relationship between ABT-126 exposure and cognitive improvement suggesting higher doses may produce better efficacy (Gault et al. 2015). The steady-state unbound plasma concentration of ABT-126 at 25 mg/day (q.d.) was estimated to be in the low nanomolar range which is close to its binding affinity at the α7 nAChR (Haig et al. 2016b; Liu et al. 2018; Bitner et al. 2013). A subsequent Phase 2b trial was designed to evaluate a higher dose range of ABT-126 (25, 50, or 75 mg/day, q.d.) as monotherapy in subjects with mild-to-moderate AD. The primary endpoint was the change from baseline to week 24 in the ADAS-Cog total score. Although donepezil (10 mg/day) significantly improved the ADAS-Cog score in this trial, no significant improvement was detected at any dose of ABT-126 dose (Gault et al. 2016). Constipation was also reported as one of the most frequent AEs in this study (Gault et al. 2016). Another 24-week Phase 2b study was conducted to assess the efficacy of ABT-126 (25 or 75 mg/day, q.d.) in subjects with mild-to-moderate AD who were taking stable doses of AChEIs. Neither dose of ABT-126 demonstrated significant improvement in cognition at week 24 based on the ADAS-Cog measurement, although 25 mg ABT-126 did show significant improvement at week 4 and a trend toward improvement at week 8 (Florian et al. 2016). Constipation was again observed as a frequent AE, especially in the 75 mg group (Florian et al. 2016), and this is consistent with the compound's relatively low selectivity vs. the 5-HT_3 receptor.

Two Phase 2 trials were conducted to investigate the efficacy of ABT-126 in patients with schizophrenia. The initial study assessed ABT-126 (10 or 25 mg/day, q.d.) in schizophrenic patients for 12 weeks using the MATRICS Consensus Cognitive Battery (MCCB) composite score as the primary endpoint (Haig et al. 2016b). When smokers and non-smokers were combined, ABT-126 only showed a trend of improvement in the MCCB composite score. Subgroup analysis suggested that non-smokers demonstrated a significant improvement at 25 mg and nearly significant improvement at 10 mg, whereas no difference was observed in smokers (Haig et al. 2016b). The following Phase 2 study examined ABT-126 (25, 50, or 75 mg/day, q.d.) in non-smoking patients with schizophrenia for 24 weeks (Haig et al. 2016a). Unfortunately, ABT-126 did not demonstrate a consistent procognitive effect in non-smokers at the higher dose range (Haig et al. 2016a).

3.1.5 TC-5619

Preclinically, TC-5619 ameliorated both PPI and social behavior impairments in a transgenic mouse model of schizophrenia (Hauser et al. 2009). In addition, TC-5619 reversed apomorphine-induced PPI deficits and improved episodic-like memory in a rat NOR assay (Hauser et al. 2009). In the apomorphine-induced PPI assay, where the compound was evaluated at 0.1, 0.3, and 1.0 mg/kg, an inverted U-shaped dose-response curve was observed. Only the middle dose (0.3 mg/kg) was active in this study. The relationship between pharmacokinetic and pharmacodynamic endpoints for TC-5619 cannot be determined as the drug exposure data was not reported from these in vivo studies (Hauser et al. 2009).

In Phase 1 trials, TC-5619 was well tolerated up to a 406 mg single dose and 204 mg multiple dose in healthy subjects (Lieberman et al. 2013). In addition, TC-5619 significantly improved attention in healthy subjects at the 6.8 mg (single dose). Based on these Phase 1 data and the preclinical findings, TC-5619 was expected to produce cognitive benefits at a human dose of approximately 3 mg (Lieberman et al. 2013). Therefore, a dose range of 1–25 mg/day was selected for the subsequent 12-week Phase 2 trial for treatment of cognitive and negative symptoms in patients with schizophrenia (Lieberman et al. 2013) (Lieberman 2013). In this initial Phase 2 study, TC-5619 was dosed at 1 mg/day (q.d.) for the first 4 weeks followed by 5 mg/day (q.d.) from weeks 4 to 8 and 25 mg/day (q.d.) from weeks 8 to 12. The primary endpoint was the Groton Maze Learning Task (GMLT), and the secondary endpoints included the Scale for Assessment of Negative Symptoms (SANS). Statistically significant improvement was observed for GMLT endpoint at week 4 when the patients were taking 1 mg/day TC-5619. At this dose level, the plasma exposure was in the low nanomolar range (Lieberman et al. 2013). The subsequent Phase 2 study evaluated TC-5619 at 5 or 50 mg/day (q.d.) in schizophrenic patients for 24 weeks (Walling et al. 2016). No statistically significant benefit was observed with TC-5619 in this study on primary and secondary measures of cognitive and negative symptoms. The lack of efficacy in the second Phase 2 trial might be due to the receptor desensitization associated with the sustained drug exposure at higher dose levels (Walling et al. 2016). A more comprehensive preclinical characterization of TC-5619, especially the potential co-agonist or "priming" effect of TC-5619 in vitro and in vivo, may help clarify the conflicting clinical findings.

3.2 *α7 nAChR PAMs*

3.2.1 AVL-3288

In preclinical studies, AVL-3288 normalized sensory gating deficits in DBA/2 mice in a dose-dependent manner (Ng et al. 2007). Total brain exposures of AVL-3288 at the behaviorally active dose (0.3 mg/kg) ranged from 0.3 to 1.0 μM during the sensory gating test. These brain concentrations are close to the in vitro potency of AVL-3288 measured in oocytes expressing human α7 nAChR, but it is not possible to draw meaningful conclusions because the unbound brain levels are not known. AVL-3288 reversed MK-801-induced hyperlocomotion in NSA mice (Ng et al. 2007), attenuated ketamine-induced cognitive impairments in NOR task (Nikiforuk and Popik 2014; Nikiforuk et al. 2016), and reversed ketamine-induced cognitive inflexibility (Nikiforuk et al. 2016). These effects are consistent with the hypothesis that positive modulation of α7 receptor is associated with procognitive effects (Ng et al. 2007).

AVL-3288 has been evaluated in a Phase 1 trial for safety and preliminary evidence of procognitive effects in healthy subjects (Gee et al. 2017). The compound was well tolerated at 3, 10, and 30 mg (single dose) without safety concerns. The maximum plasma exposure at 30 mg was ~1 μM which is similar to the efficacious exposure achieved in the preclinical models (Ng et al. 2007). Procognitive effects were investigated at 2-h post administration of AVL-3288 or placebo in healthy subjects. A positive but non-significant effect was observed on Repeatable Battery for the Assessment of Neurocognitive Status (RBANS) Total Scale Score at 10 and 30 mg doses. Furthermore, AVL-3288 showed a trend of increase in P50 inhibition in the 10 and 30 mg groups compared to placebo (Gee et al. 2017). Higher doses could not be assessed with AVL-3288 because of insufficient safety margins with respect to the hERG potassium channel (hERG $IC_{50} = 3$ μM for AVL-3288). Consistent with the off-target activity on hERG channel, a positive correlation between QTc interval and plasma exposure of AVL-3288 was observed in the Phase 1 study (Gee et al. 2017). A Phase 1b trial evaluating AVL-3288 (10 or 30 mg/day) in non-smoking patients with schizophrenia was completed in 2018, but study results had not been released at the time of this writing.

3.2.2 JNJ-39393406

Limited information is available regarding the preclinical characterization of JNJ-39393406. This compound has been reported to improve cognition in animal models assessing various cognitive domains, including attention, executive function, learning and memory, and sensory gating (Winterer et al. 2013). In Phase 1 trials, JNJ-39393406 was well tolerated with no significant safety findings. In addition, JNJ-39393406 afforded good oral bioavailability, and exposure in plasma and cerebrospinal fluid (CSF) exceeded concentrations evaluated in preclinical models (Winterer et al. 2013).

In a Phase 1 proof-of-mechanism study, JNJ-39393406 was assessed for its ability to reverse P50 sensory gating deficits in smoking patients with stable psychotic symptoms (Winterer et al. 2013). The patients received JNJ-39393406 (10, 30, 50, 100, and 200 mg, single dose) as an adjunct treatment to antipsychotics. Sensory gating measurements were performed on day 1 pre-dose and then at 2 and 5 h post-dose. However, JNJ-39393406 failed to reverse sensory P50 gating deficits in schizophrenic patients at any dose tested. It was suspected that smoking status, brain penetration of the compound, and small sample size might confound the study (Winterer et al. 2013).

3.2.3 LND101001

In preclinical studies, LND101001 has demonstrated procognitive effects in both novel object and social recognition tasks in rats (Sinha et al. 2019). When LND101001 was administered acutely, the compound reversed time-delay and scopolamine-induced cognitive deficits at 1 and 3 mg/kg. With a 7-day subchronic dosing paradigm, LND101001 was able to demonstrate efficacy at lower dose levels (i.e., 0.15 and 0.5 mg/kg) in the recognition tasks. There was no inverted U-shaped dose response associated with LND101001 in vivo (Sinha et al. 2019). It is worth noting that the unbound brain exposures of LND101001 were likely in the sub-nanomolar range at the efficacious doses, while the in vitro potency (EC_{50}) of LND101001 is 324 nM. The dissociation between in vitro and in vivo potency has been observed with many α7 PAMs, such as NS1738, AVL-3288, JNJ-193942, A-33867744, and RO5126946. Although the in vitro potency of these compounds is typically in the low μM range, they frequently produce efficacy in vivo at exposures that are several orders of magnitude lower (Timmermann et al. 2007; Malysz et al. 2010; Dinklo et al. 2011; Sahdeo et al. 2014). Exactly how α7 PAMs produce efficacy in vivo with such low exposures is not fully understood but this observation indicates very low levels of target engagement are sufficient to generate robust pharmacodynamic effects and this is consistent across models and species.

In Phase 1 trials, LND101001 has demonstrated encouraging safety profile and linear pharmacokinetics. The compound is being evaluated for efficacy and safety as monotherapy in patients with mild-to-moderate AD (Sinha et al. 2019).

4 Conclusions

The pharmacology of several distinct chemical structures of agonists and PAMs of the α7 nAChR has been characterized extensively in preclinical species and less extensively in humans. Although much has been learned, there are still many unresolved questions. It is not yet clear whether the α7 nAChR has been adequately assessed clinically as a target for improving cognition in disorders such as AD and schizophrenia. We suggest that no definitive answer to this question has been established and that further preclinical and clinical work is justified. Thus far, most compounds profiled clinically and preclinically are agonists that have very complex pharmacology. Many of these compounds appear to demonstrate "priming" at low concentrations and at higher concentrations produce both agonism and desensitization. This complicated pharmacology makes interpretation of the clinical data challenging. In addition, many of these compounds suffer from selectivity issues such as antagonism of the 5-HT_3 receptor that can affect tolerability and limit the clinical doses that may be assessed safely. The lack of generally robust clinical effects observed with α7 nAChR agonists could indicate that this target does not play an important role in promoting cognition in diseased states. Alternatively, the confounding clinical results may be a consequence of the difficulty of achieving appropriate activation of α7 nAChR with an agonist that has a very narrow range of therapeutic concentrations. Due to the inverted U-shaped dose-response pharmacology, the agonist dose must be utilized at which it will robustly activate, but not cause desensitization of the receptor. Studies with α7 nAChR PAMs might be more compelling. These compounds should promote the natural spatiotemporal effects of acetylcholine, but they do not enhance desensitization. As a result, the concentration-effect function seen with α7 nAChR PAMs in preclinical models is generally broader than that observed with agonists, and hopefully this will translate to more robust and reproducible effects in the clinic.

References

Albuquerque EX, Alkondon M, Pereira EF, Castro NG, Schrattenholz A, Barbosa CT, Bonfante-Cabarcas R, Aracava Y, Eisenberg HM, Maelicke A (1997) Properties of neuronal nicotinic acetylcholine receptors: pharmacological characterization and modulation of synaptic function. J Pharmacol Exp Ther 280(3):1117–1136

Alkondon M, Pereira EF, Eisenberg HM, Albuquerque EX (1999) Choline and selective antagonists identify two subtypes of nicotinic acetylcholine receptors that modulate GABA release from CA1 interneurons in rat hippocampal slices. J Neurosci 19(7):2693–2705

Alkondon M, Braga MF, Pereira EF, Maelicke A, Albuquerque EX (2000) alpha7 nicotinic acetylcholine receptors and modulation of gabaergic synaptic transmission in the hippocampus. Eur J Pharmacol 393(1–3):59–67. https://doi.org/10.1016/s0014-2999(00)00006-6

Andresen V, Montori VM, Keller J, West CP, Layer P, Camilleri M (2008) Effects of 5-hydroxytryptamine (serotonin) type 3 antagonists on symptom relief and constipation in nonconstipated irritable bowel syndrome: a systematic review and meta-analysis of randomized controlled trials. Clin Gastroenterol Hepatol 6(5):545–555. https://doi.org/10.1016/j.cgh.2007.12.015

Araud T, Graw S, Berger R, Lee M, Neveu E, Bertrand D, Leonard S (2011) The chimeric gene CHRFAM7A, a partial duplication of the CHRNA7 gene, is a dominant negative regulator of alpha7∗nAChR function. Biochem Pharmacol 82(8):904–914. https://doi.org/10.1016/j.bcp.2011.06.018

Arendash GW, Sengstock GJ, Sanberg PR, Kem WR (1995) Improved learning and memory in aged rats with chronic administration of the nicotinic receptor agonist GTS-21. Brain Res 674(2):252–259. https://doi.org/10.1016/0006-8993(94)01449-r

Auld DS, Kornecook TJ, Bastianetto S, Quirion R (2002) Alzheimer's disease and the basal forebrain cholinergic system: relations to beta-amyloid peptides, cognition, and treatment strategies. Prog Neurobiol 68(3):209–245. https://doi.org/10.1016/s0301-0082(02)00079-5

Barbier AJ, Hilhorst M, Van Vliet A, Snyder P, Palfreyman MG, Gawryl M, Dgetluck N, Massaro M, Tiessen R, Timmerman W, Hilt DC (2015) Pharmacodynamics, pharmacokinetics, safety, and tolerability of encenicline, a selective alpha7 nicotinic receptor partial agonist, in single ascending-dose and bioavailability studies. Clin Ther 37(2):311–324. https://doi.org/10.1016/j.clinthera.2014.09.013

Becker C, Comstock J, Michne WF, Murphy M, Philips E, Rosamond JD, Simpson TR (2004) WO 2004/098600

Bertrand D, Galzi JL, Devillers-Thiery A, Bertrand S, Changeux JP (1993) Mutations at two distinct sites within the channel domain M2 alter calcium permeability of neuronal alpha 7 nicotinic receptor. Proc Natl Acad Sci U S A 90(15):6971–6975. https://doi.org/10.1073/pnas.90.15.6971

Bertrand D, Lee CH, Flood D, Marger F, Donnelly-Roberts D (2015) Therapeutic potential of alpha7 nicotinic acetylcholine receptors. Pharmacol Rev 67(4):1025–1073. https://doi.org/10.1124/pr.113.008581

Bitner R, Anderson D, Drescher K, Kohlhaas K, Gronlien H, Hu M, Li J, Markosyan S, Marsh K, Mohler E, Nikkel A, Radek R, Robb H, Schrimpf M, Waring J, Lee C, Gopalakrishnan M (2013) Preclinical characterization of a selective alpha-7 neuronal nicotinic acetylcholine receptor agonist ABT-126: a novel therapeutic agent for the treatment of cognitive impairment in Alzheimer's disease and schizophrenia. Alzheimers Dement 9(4):P817–P818

Bodnar AL, Cortes-Burgos LA, Cook KK, Dinh DM, Groppi VE, Hajos M, Higdon NR, Hoffmann WE, Hurst RS, Myers JK, Rogers BN, Wall TM, Wolfe ML, Wong E (2005) Discovery and structure-activity relationship of quinuclidine benzamides as agonists of alpha7 nicotinic acetylcholine receptors. J Med Chem 48(4):905–908. https://doi.org/10.1021/jm049363q

Brannan S (2019) Two global phase III trials of encenicline for cognitive impairment in chronic schizophrenia patients: red flags and lessons learned. Schizophr Bull 45(Suppl 2):S141–S142

Breese CR, Adams C, Logel J, Drebing C, Rollins Y, Barnhart M, Sullivan B, Demasters BK, Freedman R, Leonard S (1997) Comparison of the regional expression of nicotinic acetylcholine receptor alpha7 mRNA and [125I]-alpha-bungarotoxin binding in human postmortem brain. J Comp Neurol 387(3):385–398

Briggs CA, McKenna DG, Piattoni-Kaplan M (1995) Human alpha 7 nicotinic acetylcholine receptor responses to novel ligands. Neuropharmacology 34(6):583–590. https://doi.org/10.1016/0028-3908(95)00028-5

Briggs CA, Anderson DJ, Brioni JD, Buccafusco JJ, Buckley MJ, Campbell JE, Decker MW, Donnelly-Roberts D, Elliott RL, Gopalakrishnan M, Holladay MW, Hui YH, Jackson WJ, Kim DJ, Marsh KC, O'Neill A, Prendergast MA, Ryther KB, Sullivan JP, Arneric SP (1997) Functional characterization of the novel neuronal nicotinic acetylcholine receptor ligand GTS-21 in vitro and in vivo. Pharmacol Biochem Behav 57(1–2):231–241. https://doi.org/10.1016/s0091-3057(96)00354-1

Bristow LJ, Easton AE, Li YW, Sivarao DV, Lidge R, Jones KM, Post-Munson D, Daly C, Lodge NJ, Gallagher L, Molski T, Pieschl R, Chen P, Hendricson A, Westphal R, Cook J, Iwuagwu C, Morgan D, Benitex Y, King D, Macor JE, Zaczek R, Olson R (2016) The novel, nicotinic

Alpha7 receptor partial agonist, BMS-933043, improves cognition and sensory processing in preclinical models of schizophrenia. PLoS One 11(7):e0159996. https://doi.org/10.1371/journal.pone.0159996
Buccafusco JJ, Terry AV Jr (2009) A reversible model of the cognitive impairment associated with schizophrenia in monkeys: potential therapeutic effects of two nicotinic acetylcholine receptor agonists. Biochem Pharmacol 78(7):852–862. https://doi.org/10.1016/j.bcp.2009.06.102
Burghaus L, Schutz U, Krempel U, de Vos RA, Jansen Steur EN, Wevers A, Lindstrom J, Schroder H (2000) Quantitative assessment of nicotinic acetylcholine receptor proteins in the cerebral cortex of Alzheimer patients. Brain Res Mol Brain Res 76(2):385–388. https://doi.org/10.1016/s0169-328x(00)00031-0
Cachelin AB, Rust G (1994) Unusual pharmacology of (+)-tubocurarine with rat neuronal nicotinic acetylcholine receptors containing beta 4 subunits. Mol Pharmacol 46(6):1168–1174
Callahan PM, Terry AV Jr, Tehim A (2014) Effects of the nicotinic alpha7 receptor partial agonist GTS-21 on NMDA-glutamatergic receptor related deficits in sensorimotor gating and recognition memory in rats. Psychopharmacology 231(18):3695–3706. https://doi.org/10.1007/s00213-014-3509-2
Callahan PM, Bertrand D, Bertrand S, Plagenhoef MR, Terry AV Jr (2017) Tropisetron sensitizes alpha7 containing nicotinic receptors to low levels of acetylcholine in vitro and improves memory-related task performance in young and aged animals. Neuropharmacology 117:422–433. https://doi.org/10.1016/j.neuropharm.2017.02.025
Cannon CE, Puri V, Vivian JA, Egbertson MS, Eddins D, Uslaner JM (2013) The nicotinic alpha7 receptor agonist GTS-21 improves cognitive performance in ketamine impaired rhesus monkeys. Neuropharmacology 64:191–196. https://doi.org/10.1016/j.neuropharm.2012.05.003
Connolly CN, Wafford KA (2004) The Cys-loop superfamily of ligand-gated ion channels: the impact of receptor structure on function. Biochem Soc Trans 32(Pt3):529–534. https://doi.org/10.1042/BST0320529
Court J, Spurden D, Lloyd S, McKeith I, Ballard C, Cairns N, Kerwin R, Perry R, Perry E (1999) Neuronal nicotinic receptors in dementia with Lewy bodies and schizophrenia: alpha-bungarotoxin and nicotine binding in the thalamus. J Neurochem 73(4):1590–1597. https://doi.org/10.1046/j.1471-4159.1999.0731590.x
de Lucas-Cerrillo AM, Maldifassi MC, Arnalich F, Renart J, Atienza G, Serantes R, Cruces J, Sanchez-Pacheco A, Andres-Mateos E, Montiel C (2011) Function of partially duplicated human alpha77 nicotinic receptor subunit CHRFAM7A gene: potential implications for the cholinergic anti-inflammatory response. J Biol Chem 286(1):594–606. https://doi.org/10.1074/jbc.M110.180067
Deardorff WJ, Shobassy A, Grossberg GT (2015) Safety and clinical effects of EVP-6124 in subjects with Alzheimer's disease currently or previously receiving an acetylcholinesterase inhibitor medication. Expert Rev Neurother 15(1):7–17. https://doi.org/10.1586/14737175.2015.995639
Delbono O, Gopalakrishnan M, Renganathan M, Monteggia LM, Messi ML, Sullivan JP (1997) Activation of the recombinant human alpha 7 nicotinic acetylcholine receptor significantly raises intracellular free calcium. J Pharmacol Exp Ther 280(1):428–438
Dineley KT, Pandya AA, Yakel JL (2015) Nicotinic ACh receptors as therapeutic targets in CNS disorders. Trends Pharmacol Sci 36(2):96–108. https://doi.org/10.1016/j.tips.2014.12.002
Dinklo T, Shaban H, Thuring JW, Lavreysen H, Stevens KE, Zheng L, Mackie C, Grantham C, Vandenberk I, Meulders G, Peeters L, Verachtert H, De Prins E, Lesage AS (2011) Characterization of 2-[[4-fluoro-3-(trifluoromethyl)phenyl]amino]-4-(4-pyridinyl)-5-thiazolemethanol (JNJ-1930942), a novel positive allosteric modulator of the {alpha}7 nicotinic acetylcholine receptor. J Pharmacol Exp Ther 336(2):560–574. https://doi.org/10.1124/jpet.110.173245
Eskildsen J, Redrobe JP, Sams AG, Dekermendjian K, Laursen M, Boll JB, Papke RL, Bundgaard C, Frederiksen K, Bastlund JF (2014) Discovery and optimization of Lu AF58801, a novel, selective and brain penetrant positive allosteric modulator of alpha-7 nicotinic acetylcholine receptors: attenuation of subchronic phencyclidine (PCP)-induced

cognitive deficits in rats following oral administration. Bioorg Med Chem Lett 24(1):288–293. https://doi.org/10.1016/j.bmcl.2013.11.022

Fabian-Fine R, Skehel P, Errington ML, Davies HA, Sher E, Stewart MG, Fine A (2001) Ultrastructural distribution of the alpha7 nicotinic acetylcholine receptor subunit in rat hippocampus. J Neurosci 21(20):7993–8003

Faghih R, Gopalakrishnan SM, Gronlien JH, Malysz J, Briggs CA, Wetterstrand C, Ween H, Curtis MP, Sarris KA, Gfesser GA, El-Kouhen R, Robb HM, Radek RJ, Marsh KC, Bunnelle WH, Gopalakrishnan M (2009) Discovery of 4-(5-(4-chlorophenyl)-2-methyl-3-propionyl-1H-pyrrol-1-yl)benzenesulfonamide (A-867744) as a novel positive allosteric modulator of the alpha7 nicotinic acetylcholine receptor. J Med Chem 52(10):3377–3384. https://doi.org/10.1021/jm9003818

Florian H, Meier A, Gauthier S, Lipschitz S, Lin Y, Tang Q, Othman AA, Robieson WZ, Gault LM (2016) Efficacy and safety of ABT-126 in subjects with mild-to-moderate Alzheimer's disease on stable doses of Acetylcholinesterase inhibitors: a randomized, double-blind, placebo-controlled study. J Alzheimers Dis 51(4):1237–1247. https://doi.org/10.3233/JAD-150978

Freedman R (2014) alpha7-nicotinic acetylcholine receptor agonists for cognitive enhancement in schizophrenia. Annu Rev Med 65:245–261. https://doi.org/10.1146/annurev-med-092112-142937

Freedman R, Olincy A, Ross RG, Waldo MC, Stevens KE, Adler LE, Leonard S (2003) The genetics of sensory gating deficits in schizophrenia. Curr Psychiatry Rep 5(2):155–161. https://doi.org/10.1007/s11920-003-0032-2

Freedman R, Olincy A, Buchanan RW, Harris JG, Gold JM, Johnson L, Allensworth D, Guzman-Bonilla A, Clement B, Ball MP, Kutnick J, Pender V, Martin LF, Stevens KE, Wagner BD, Zerbe GO, Soti F, Kem WR (2008) Initial phase 2 trial of a nicotinic agonist in schizophrenia. Am J Psychiatry 165(8):1040–1047. https://doi.org/10.1176/appi.ajp.2008.07071135

Garai S, Raja KS, Papke RL, Deschamps JR, Damaj MI, Thakur GA (2018) B-973, a novel alpha7 nAChR ago-PAM: racemic and asymmetric synthesis, electrophysiological studies, and in vivo evaluation. ACS Med Chem Lett 9(11):1144–1148. https://doi.org/10.1021/acsmedchemlett.8b00407

Gault LM, Ritchie CW, Robieson WZ, Pritchett Y, Othman AA, Lenz RA (2015) A phase 2 randomized, controlled trial of the alpha7 agonist ABT-126 in mild-to-moderate Alzheimer's dementia. Alzheimers Dement (N Y) 1(1):81–90. https://doi.org/10.1016/j.trci.2015.06.001

Gault LM, Lenz RA, Ritchie CW, Meier A, Othman AA, Tang Q, Berry S, Pritchett Y, Robieson WZ (2016) ABT-126 monotherapy in mild-to-moderate Alzheimer's dementia: randomized double-blind, placebo and active controlled adaptive trial and open-label extension. Alzheimers Res Ther 8(1):44. https://doi.org/10.1186/s13195-016-0210-1

Gee KW, Olincy A, Kanner R, Johnson L, Hogenkamp D, Harris J, Tran M, Edmonds SA, Sauer W, Yoshimura R, Johnstone T, Freedman R (2017) First in human trial of a type I positive allosteric modulator of alpha7-nicotinic acetylcholine receptors: pharmacokinetics, safety, and evidence for neurocognitive effect of AVL-3288. J Psychopharmacol 31 (4):434–441. https://doi.org/10.1177/0269881117691590

Gill JK, Savolainen M, Young GT, Zwart R, Sher E, Millar NS (2011) Agonist activation of alpha7 nicotinic acetylcholine receptors via an allosteric transmembrane site. Proc Natl Acad Sci U S A 108(14):5867–5872. https://doi.org/10.1073/pnas.1017975108

Giniatullin R, Nistri A, Yakel JL (2005) Desensitization of nicotinic ACh receptors: shaping cholinergic signaling. Trends Neurosci 28(7):371–378. https://doi.org/10.1016/j.tins.2005.04.009

Gotti C, Clementi F (2004) Neuronal nicotinic receptors: from structure to pathology. Prog Neurobiol 74(6):363–396. https://doi.org/10.1016/j.pneurobio.2004.09.006

Gronlien JH, Hakerud M, Ween H, Thorin-Hagene K, Briggs CA, Gopalakrishnan M, Malysz J (2007) Distinct profiles of alpha7 nAChR positive allosteric modulation revealed by structurally diverse chemotypes. Mol Pharmacol 72(3):715–724. https://doi.org/10.1124/mol.107.035410

Haig G, Wang D, Othman AA, Zhao J (2016a) The alpha7 nicotinic agonist ABT-126 in the treatment of cognitive impairment associated with schizophrenia in nonsmokers: results from a randomized controlled phase 2b study. Neuropsychopharmacology 41(12):2893–2902. https://doi.org/10.1038/npp.2016.101

Haig GM, Bain EE, Robieson WZ, Baker JD, Othman AA (2016b) A randomized trial to assess the efficacy and safety of ABT-126, a selective alpha7 nicotinic acetylcholine receptor agonist, in the treatment of cognitive impairment in schizophrenia. Am J Psychiatry 173(8):827–835. https://doi.org/10.1176/appi.ajp.2015.15010093

Hajos M, Hurst RS, Hoffmann WE, Krause M, Wall TM, Higdon NR, Groppi VE (2005) The selective alpha7 nicotinic acetylcholine receptor agonist PNU-282987 [N-[(3R)-1-Azabicyclo [2.2.2]oct-3-yl]-4-chlorobenzamide hydrochloride] enhances GABAergic synaptic activity in brain slices and restores auditory gating deficits in anesthetized rats. J Pharmacol Exp Ther 312 (3):1213–1222. https://doi.org/10.1124/jpet.104.076968

Han ZY, Le Novere N, Zoli M, Hill JA Jr, Champtiaux N, Changeux JP (2000) Localization of nAChR subunit mRNAs in the brain of Macaca mulatta. Eur J Neurosci 12(10):3664–3674. https://doi.org/10.1046/j.1460-9568.2000.00262.x

Han ZY, Zoli M, Cardona A, Bourgeois JP, Changeux JP, Le Novere N (2003) Localization of [3H] nicotine, [3H]cytisine, [3H]epibatidine, and [125I]alpha-bungarotoxin binding sites in the brain of Macaca mulatta. J Comp Neurol 461(1):49–60. https://doi.org/10.1002/cne.10659

Harvey AJ, Avery TD, Schaeffer L, Joseph C, Huff BC, Singh R, Morice C, Giethlen B, Grishin AA, Coles CJ, Kolesik P, Wagner S, Andriambeloson E, Huyard B, Poiraud E, Paul D, O'Connor SM (2019) Discovery of BNC375, a potent, selective, and orally available type I positive allosteric modulator of alpha7 nAChRs. ACS Med Chem Lett 10(5):754–760. https://doi.org/10.1021/acsmedchemlett.9b00001

Hauser TA, Kucinski A, Jordan KG, Gatto GJ, Wersinger SR, Hesse RA, Stachowiak EK, Stachowiak MK, Papke RL, Lippiello PM, Bencherif M (2009) TC-5619: an alpha7 neuronal nicotinic receptor-selective agonist that demonstrates efficacy in animal models of the positive and negative symptoms and cognitive dysfunction of schizophrenia. Biochem Pharmacol 78 (7):803–812. https://doi.org/10.1016/j.bcp.2009.05.030

Hendrix M, Boess F-G, Erb C, LFlessner T, van Kampen M, Luithle J, Methfessel C, Wiese WB (2010) U.S. Patent 7,732,477

Huang M, Felix AR, Flood DG, Bhuvaneswaran C, Hilt D, Koenig G, Meltzer HY (2014a) The novel alpha7 nicotinic acetylcholine receptor agonist EVP-6124 enhances dopamine, acetylcholine, and glutamate efflux in rat cortex and nucleus accumbens. Psychopharmacology 231 (23):4541–4551. https://doi.org/10.1007/s00213-014-3596-0

Huang M, Felix AR, Kwon S, Lowe D, Wallace T, Santarelli L, Meltzer HY (2014b) The alpha-7 nicotinic receptor partial agonist/5-HT3 antagonist RG3487 enhances cortical and hippocampal dopamine and acetylcholine release. Psychopharmacology 231(10):2199–2210. https://doi.org/10.1007/s00213-013-3373-5

Hudzik TJ, Basso AM, Lynch JJ 3rd, Bracken WM, Mohler EG, Kohlhaas KL, Xu H, Haig G, Gault L (2017) Preclinical abuse liability assessment of ABT-126, an agonist at the alpha7 nicotinic acetylcholine receptor (nAChR). Pharmacol Biochem Behav 158:22–31. https://doi.org/10.1016/j.pbb.2017.05.010

Hunter BE, de Fiebre CM, Papke RL, Kem WR, Meyer EM (1994) A novel nicotinic agonist facilitates induction of long-term potentiation in the rat hippocampus. Neurosci Lett 168 (1–2):130–134. https://doi.org/10.1016/0304-3940(94)90433-2

Hurst RS, Hajos M, Raggenbass M, Wall TM, Higdon NR, Lawson JA, Rutherford-Root KL, Berkenpas MB, Hoffmann WE, Piotrowski DW, Groppi VE, Allaman G, Ogier R, Bertrand S, Bertrand D, Arneric SP (2005) A novel positive allosteric modulator of the alpha7 neuronal nicotinic acetylcholine receptor: in vitro and in vivo characterization. J Neurosci 25 (17):4396–4405. https://doi.org/10.1523/JNEUROSCI.5269-04.2005

Hurst R, Rollema H, Bertrand D (2013) Nicotinic acetylcholine receptors: from basic science to therapeutics. Pharmacol Ther 137(1):22–54. https://doi.org/10.1016/j.pharmthera.2012.08.012

Jones IW, Wonnacott S (2004) Precise localization of alpha7 nicotinic acetylcholine receptors on glutamatergic axon terminals in the rat ventral tegmental area. J Neurosci 24(50):11244–11252. https://doi.org/10.1523/JNEUROSCI.3009-04.2004

Keefe RS, Meltzer HA, Dgetluck N, Gawryl M, Koenig G, Moebius HJ, Lombardo I, Hilt DC (2015) Randomized, double-blind, placebo-controlled study of Encenicline, an alpha7 nicotinic acetylcholine receptor agonist, as a treatment for cognitive impairment in schizophrenia. Neuropsychopharmacology 40(13):3053–3060. https://doi.org/10.1038/npp.2015.176

Kem WR, Olincy A, Johnson L, Harris J, Wagner BD, Buchanan RW, Christians U, Freedman R (2018) Pharmacokinetic limitations on effects of an Alpha7-nicotinic receptor agonist in schizophrenia: randomized trial with an extended-release formulation. Neuropsychopharmacology 43(3):583–589. https://doi.org/10.1038/npp.2017.182

King D, Iwuagwu C, Cook J, McDonald IM, Mate R, Zusi FC, Hill MD, Fang H, Zhao R, Wang B, Easton AE, Miller R, Post-Munson D, Knox RJ, Gallagher L, Westphal R, Molski T, Fan J, Clarke W, Benitex Y, Lentz KA, Denton R, Morgan D, Zaczek R, Lodge NJ, Bristow LJ, Macor JE, Olson RE (2017) BMS-933043, a selective alpha7 nAChR partial agonist for the treatment of cognitive deficits associated with schizophrenia. ACS Med Chem Lett 8(3):366–371. https://doi.org/10.1021/acsmedchemlett.7b00032

Kitagawa H, Takenouchi T, Azuma R, Wesnes KA, Kramer WG, Clody DE, Burnett AL (2003) Safety, pharmacokinetics, and effects on cognitive function of multiple doses of GTS-21 in healthy, male volunteers. Neuropsychopharmacology 28(3):542–551. https://doi.org/10.1038/sj.npp.1300028

Krause RM, Buisson B, Bertrand S, Corringer PJ, Galzi JL, Changeux JP, Bertrand D (1998) Ivermectin: a positive allosteric effector of the alpha7 neuronal nicotinic acetylcholine receptor. Mol Pharmacol 53(2):283–294. https://doi.org/10.1124/mol.53.2.283

Lieberman JA, Dunbar G, Segreti AC, Girgis RR, Seoane F, Beaver JS, Duan N, Hosford DA (2013) A randomized exploratory trial of an alpha-7 nicotinic receptor agonist (TC-5619) for cognitive enhancement in schizophrenia. Neuropsychopharmacology 38(6):968–975. https://doi.org/10.1038/npp.2012.259

Liu H, Stresser DM, Michmerhuizen MJ, Li X, Othman AA, Reed AD, Schrimpf MR, Sydor J, Lee AJ (2018) Metabolism and disposition of a novel selective alpha7 neuronal acetylcholine receptor agonist ABT-126 in humans: characterization of the major roles for Flavin-containing Monooxygenases and UDP-Glucuronosyl Transferase 1A4 and 2B10 in catalysis. Drug Metab Dispos 46(4):429–439. https://doi.org/10.1124/dmd.117.077511

Livingstone PD, Srinivasan J, Kew JN, Dawson LA, Gotti C, Moretti M, Shoaib M, Wonnacott S (2009) alpha7 and non-alpha7 nicotinic acetylcholine receptors modulate dopamine release in vitro and in vivo in the rat prefrontal cortex. Eur J Neurosci 29(3):539–550. https://doi.org/10.1111/j.1460-9568.2009.06613.x

Malysz J, Gronlien JH, Anderson DJ, Hakerud M, Thorin-Hagene K, Ween H, Wetterstrand C, Briggs CA, Faghih R, Bunnelle WH, Gopalakrishnan M (2009) In vitro pharmacological characterization of a novel allosteric modulator of alpha 7 neuronal acetylcholine receptor, 4-(5-(4-chlorophenyl)-2-methyl-3-propionyl-1H-pyrrol-1-yl)benzenesulfonamide (A-867744), exhibiting unique pharmacological profile. J Pharmacol Exp Ther 330(1):257–267. https://doi.org/10.1124/jpet.109.151886

Malysz J, Anderson DJ, Gronlien JH, Ji J, Bunnelle WH, Hakerud M, Thorin-Hagene K, Ween H, Helfrich R, Hu M, Gubbins E, Gopalakrishnan S, Puttfarcken PS, Briggs CA, Li J, Meyer MD, Dyhring T, Ahring PK, Nielsen EO, Peters D, Timmermann DB, Gopalakrishnan M (2010) In vitro pharmacological characterization of a novel selective alpha7 neuronal nicotinic acetylcholine receptor agonist ABT-107. J Pharmacol Exp Ther 334(3):863–874. https://doi.org/10.1124/jpet.110.167072

Mazurov AA, Kombo DC, Hauser TA, Miao L, Dull G, Genus JF, Fedorov NB, Benson L, Sidach S, Xiao Y, Hammond PS, James JW, Miller CH, Yohannes D (2012) Discovery of (2S,3R)-N-[2-(pyridin-3-ylmethyl)-1-azabicyclo[2.2.2]oct-3-yl]benzo[b]furan-2-car boxamide

(TC-5619), a selective alpha7 nicotinic acetylcholine receptor agonist, for the treatment of cognitive disorders. J Med Chem 55(22):9793–9809. https://doi.org/10.1021/jm301048a

McGehee DS, Role LW (1995) Physiological diversity of nicotinic acetylcholine receptors expressed by vertebrate neurons. Annu Rev Physiol 57:521–546. https://doi.org/10.1146/annurev.ph.57.030195.002513

Mehta D, Jackson R, Paul G, Shi J, Sabbagh M (2017) Why do trials for Alzheimer's disease drugs keep failing? A discontinued drug perspective for 2010-2015. Expert Opin Investig Drugs 26 (6):735–739. https://doi.org/10.1080/13543784.2017.1323868

Meyer EM, Tay ET, Papke RL, Meyers C, Huang GL, de Fiebre CM (1997) 3-[2,4-Dimethoxybenzylidene]anabaseine (DMXB) selectively activates rat alpha7 receptors and improves memory-related behaviors in a mecamylamine-sensitive manner. Brain Res 768 (1–2):49–56. https://doi.org/10.1016/s0006-8993(97)00536-2

Mitrofanis J, Guillery RW (1993) New views of the thalamic reticular nucleus in the adult and the developing brain. Trends Neurosci 16(6):240–245. https://doi.org/10.1016/0166-2236(93)90163-g

Murray TA, Bertrand D, Papke RL, George AA, Pantoja R, Srinivasan R, Liu Q, Wu J, Whiteaker P, Lester HA, Lukas RJ (2012) alpha7beta2 nicotinic acetylcholine receptors assemble, function, and are activated primarily via their alpha7-alpha7 interfaces. Mol Pharmacol 81 (2):175–188. https://doi.org/10.1124/mol.111.074088

Ng HJ, Whittemore ER, Tran MB, Hogenkamp DJ, Broide RS, Johnstone TB, Zheng L, Stevens KE, Gee KW (2007) Nootropic alpha7 nicotinic receptor allosteric modulator derived from GABAA receptor modulators. Proc Natl Acad Sci U S A 104(19):8059–8064. https://doi.org/10.1073/pnas.0701321104

Nikiforuk A, Popik P (2014) The effects of acute and repeated administration of ketamine on attentional performance in the five-choice serial reaction time task in rats. Eur Neuropsychopharmacol 24(8):1381–1393. https://doi.org/10.1016/j.euroneuro.2014.04.007

Nikiforuk A, Kos T, Holuj M, Potasiewicz A, Popik P (2016) Positive allosteric modulators of alpha 7 nicotinic acetylcholine receptors reverse ketamine-induced schizophrenia-like deficits in rats. Neuropharmacology 101:389–400. https://doi.org/10.1016/j.neuropharm.2015.07.034

Olincy A, Harris JG, Johnson LL, Pender V, Kongs S, Allensworth D, Ellis J, Zerbe GO, Leonard S, Stevens KE, Stevens JO, Martin L, Adler LE, Soti F, Kem WR, Freedman R (2006) Proof-of-concept trial of an alpha7 nicotinic agonist in schizophrenia. Arch Gen Psychiatry 63 (6):630–638. https://doi.org/10.1001/archpsyc.63.6.630

O'Neill HC, Rieger K, Kem WR, Stevens KE (2003) DMXB, an alpha7 nicotinic agonist, normalizes auditory gating in isolation-reared rats. Psychopharmacology 169(3–4):332–339. https://doi.org/10.1007/s00213-003-1482-2

Orr-Urtreger A, Broide RS, Kasten MR, Dang H, Dani JA, Beaudet AL, Patrick JW (2000) Mice homozygous for the L250T mutation in the alpha7 nicotinic acetylcholine receptor show increased neuronal apoptosis and die within 1 day of birth. J Neurochem 74(5):2154–2166. https://doi.org/10.1046/j.1471-4159.2000.0742154.x

Papke RL (2014) Merging old and new perspectives on nicotinic acetylcholine receptors. Biochem Pharmacol 89(1):1–11. https://doi.org/10.1016/j.bcp.2014.01.029

Papke RL, Trocme-Thibierge C, Guendisch D, Al Rubaiy SA, Bloom SA (2011) Electrophysiological perspectives on the therapeutic use of nicotinic acetylcholine receptor partial agonists. J Pharmacol Exp Ther 337(2):367–379. https://doi.org/10.1124/jpet.110.177485

Phillips E, Mack R, Macor J, Semus S (2000) U.S. Patent 6,110,914. 6,110,914

Post-Munson DJ, Pieschl RL, Molski TF, Graef JD, Hendricson AW, Knox RJ, McDonald IM, Olson RE, Macor JE, Weed MR, Bristow LJ, Kiss L, Ahlijanian MK, Herrington J (2017) B-973, a novel piperazine positive allosteric modulator of the alpha7 nicotinic acetylcholine receptor. Eur J Pharmacol 799:16–25. https://doi.org/10.1016/j.ejphar.2017.01.037

Preskorn SH, Gawryl M, Dgetluck N, Palfreyman M, Bauer LO, Hilt DC (2014) Normalizing effects of EVP-6124, an alpha-7 nicotinic partial agonist, on event-related potentials and cognition: a proof of concept, randomized trial in patients with schizophrenia. J Psychiatr Pract 20(1):12–24. https://doi.org/10.1097/01.pra.0000442935.15833.c5

Prickaerts J, van Goethem NP, Chesworth R, Shapiro G, Boess FG, Methfessel C, Reneerkens OA, Flood DG, Hilt D, Gawryl M, Bertrand S, Bertrand D, Konig G (2012) EVP-6124, a novel and selective alpha7 nicotinic acetylcholine receptor partial agonist, improves memory performance by potentiating the acetylcholine response of alpha7 nicotinic acetylcholine receptors. Neuropharmacology 62(2):1099–1110. https://doi.org/10.1016/j.neuropharm.2011.10.024

Reid CA, Fabian-Fine R, Fine A (2001) Postsynaptic calcium transients evoked by activation of individual hippocampal mossy fiber synapses. J Neurosci 21(7):2206–2214

Role LW, Berg DK (1996) Nicotinic receptors in the development and modulation of CNS synapses. Neuron 16(6):1077–1085. https://doi.org/10.1016/s0896-6273(00)80134-8

Rousseau SJ, Jones IW, Pullar IA, Wonnacott S (2005) Presynaptic alpha7 and non-alpha7 nicotinic acetylcholine receptors modulate [3H]d-aspartate release from rat frontal cortex in vitro. Neuropharmacology 49(1):59–72. https://doi.org/10.1016/j.neuropharm.2005.01.030

Sabbagh MN (2009) Drug development for Alzheimer's disease: where are we now and where are we headed? Am J Geriatr Pharmacother 7(3):167–185. https://doi.org/10.1016/j.amjopharm.2009.06.003

Sahdeo S, Wallace T, Hirakawa R, Knoflach F, Bertrand D, Maag H, Misner D, Tombaugh GC, Santarelli L, Brameld K, Milla ME, Button DC (2014) Characterization of RO5126946, a novel alpha7 nicotinic acetylcholine receptor-positive allosteric modulator. J Pharmacol Exp Ther 350 (2):455–468. https://doi.org/10.1124/jpet.113.210963

Schrimpf MR, Nersesian DL, Sippy KB, Ji J, Li T, Scanio M, Shi L, Lee C-H, Bunnelle WH, Zhang GGZ, Brackemeyer PJ, Chen S, Henry RF (2012) U.S. Pattent 8,314,119

Seguela P, Wadiche J, Dineley-Miller K, Dani JA, Patrick JW (1993) Molecular cloning, functional properties, and distribution of rat brain alpha 7: a nicotinic cation channel highly permeable to calcium. J Neurosci 13(2):596–604

Simosky JK, Stevens KE, Kem WR, Freedman R (2001) Intragastric DMXB-A, an alpha7 nicotinic agonist, improves deficient sensory inhibition in DBA/2 mice. Biol Psychiatry 50(7):493–500. https://doi.org/10.1016/s0006-3223(01)01093-9

Sinha N, Karche NP, Verma MK, Walunj SS, Nigade PB, Jana G, Kurhade SP, Hajare AK, Tilekar AR, Jadhav GR, Thube BR, Shaikh JS, Balgude S, Singh LB, Mahimane V, Adurkar SK, Hatnapure G, Raje F, Bhosale Y, Bhanage D, Sachchidanand S, Dixit R, Gupta R, Bokare AM, Dandekar M, Bharne A, Chatterjee M, Desai S, Koul S, Modi D, Mehta M, Patil V, Singh M, Gundu J, Goel RN, Shah C, Sharma S, Bakhle D, Kamboj RK, Palle VP (2019) Discovery of novel, potent, brain-permeable and orally efficacious positive allosteric modulator of alpha 7 nicotinic acetylcholine receptor [4-(5-(4-chlorophenyl)-4-methyl-2-propionylthiophen-3-yl) benzenesulfonamide], structure activity relationship and preclinical characterization. J Med Chem. https://doi.org/10.1021/acs.jmedchem.9b01569

Sinkus ML, Lee MJ, Gault J, Logel J, Short M, Freedman R, Christian SL, Lyon J, Leonard S (2009) A 2-base pair deletion polymorphism in the partial duplication of the alpha7 nicotinic acetylcholine gene (CHRFAM7A) on chromosome 15q14 is associated with schizophrenia. Brain Res 1291:1–11. https://doi.org/10.1016/j.brainres.2009.07.041

Stoner EJ, Christesen A, Sheikh AY, Wang XC, Pal A, Murphey DB, Ji J, Law D, Zou D (2010) U.S. Patent 7,655,657

Sydserff S, Sutton EJ, Song D, Quirk MC, Maciag C, Li C, Jonak G, Gurley D, Gordon JC, Christian EP, Doherty JJ, Hudzik T, Johnson E, Mrzljak L, Piser T, Smagin GN, Wang Y, Widzowski D, Smith JS (2009) Selective alpha7 nicotinic receptor activation by AZD0328 enhances cortical dopamine release and improves learning and attentional processes. Biochem Pharmacol 78(7):880–888. https://doi.org/10.1016/j.bcp.2009.07.005

Timmermann DB, Gronlien JH, Kohlhaas KL, Nielsen EO, Dam E, Jorgensen TD, Ahring PK, Peters D, Holst D, Christensen JK, Malysz J, Briggs CA, Gopalakrishnan M, Olsen GM (2007) An allosteric modulator of the alpha7 nicotinic acetylcholine receptor possessing cognition-enhancing properties in vivo. J Pharmacol Exp Ther 323(1):294–307. https://doi.org/10.1124/jpet.107.120436

Umbricht D, Keefe RS, Murray S, Lowe DA, Porter R, Garibaldi G, Santarelli L (2014) A randomized, placebo-controlled study investigating the nicotinic alpha7 agonist, RG3487, for cognitive deficits in schizophrenia. Neuropsychopharmacology 39(7):1568–1577. https://doi.org/10.1038/npp.2014.17

Wallace TL, Bertrand D (2013) Alpha7 neuronal nicotinic receptors as a drug target in schizophrenia. Expert Opin Ther Targets 17(2):139–155. https://doi.org/10.1517/14728222.2013.736498

Wallace TL, Porter RH (2011) Targeting the nicotinic alpha7 acetylcholine receptor to enhance cognition in disease. Biochem Pharmacol 82(8):891–903. https://doi.org/10.1016/j.bcp.2011.06.034

Wallace TL, Chiu G, Dao H, Lowe DA, Porter R, Santarelli L (2009) R3487/MEM 3454, a novel nicotinic alpha 7 receptor partial agonist, improves attention and working memory performance in cynomolgus macques. Biochem Pharmacol 78(7):912

Wallace TL, Callahan PM, Tehim A, Bertrand D, Tombaugh G, Wang S, Xie W, Rowe WB, Ong V, Graham E, Terry AV Jr, Rodefer JS, Herbert B, Murray M, Porter R, Santarelli L, Lowe DA (2011) RG3487, a novel nicotinic alpha7 receptor partial agonist, improves cognition and sensorimotor gating in rodents. J Pharmacol Exp Ther 336(1):242–253. https://doi.org/10.1124/jpet.110.171892

Walling D, Marder SR, Kane J, Fleischhacker WW, Keefe RS, Hosford DA, Dvergsten C, Segreti AC, Beaver JS, Toler SM, Jett JE, Dunbar GC (2016) Phase 2 trial of an Alpha-7 nicotinic receptor agonist (TC-5619) in negative and cognitive symptoms of schizophrenia. Schizophr Bull 42(2):335–343. https://doi.org/10.1093/schbul/sbv072

Weed MR, Polino J, Signor L, Bookbinder M, Keavy D, Benitex Y, Morgan DG, King D, Macor JE, Zaczek R, Olson R, Bristow LJ (2017) Nicotinic alpha 7 receptor agonists EVP-6124 and BMS-933043, attenuate scopolamine-induced deficits in visuo-spatial paired associates learning. PLoS One 12(12):e0187609. https://doi.org/10.1371/journal.pone.0187609

Wheeler JW, Olubajo O, Storm CB, Duffield RM (1981) Anabaseine: venom alkaloid of aphaenogaster ants. Science 211(4486):1051–1052. https://doi.org/10.1126/science.211.4486.1051

Whiteaker P, Davies AR, Marks MJ, Blagbrough IS, Potter BV, Wolstenholme AJ, Collins AC, Wonnacott S (1999) An autoradiographic study of the distribution of binding sites for the novel alpha7-selective nicotinic radioligand [3H]-methyllycaconitine in the mouse brain. Eur J Neurosci 11(8):2689–2696. https://doi.org/10.1046/j.1460-9568.1999.00685.x

Whitehouse PJ, Price DL, Struble RG, Clark AW, Coyle JT, Delon MR (1982) Alzheimer's disease and senile dementia: loss of neurons in the basal forebrain. Science 215(4537):1237–1239. https://doi.org/10.1126/science.7058341

Williams DK, Wang J, Papke RL (2011) Investigation of the molecular mechanism of the alpha7 nicotinic acetylcholine receptor positive allosteric modulator PNU-120596 provides evidence for two distinct desensitized states. Mol Pharmacol 80(6):1013–1032. https://doi.org/10.1124/mol.111.074302

Winterer G, Gallinat J, Brinkmeyer J, Musso F, Kornhuber J, Thuerauf N, Rujescu D, Favis R, Sun Y, Franc MA, Ouwerkerk-Mahadevan S, Janssens L, Timmers M, Streffer JR (2013) Allosteric alpha-7 nicotinic receptor modulation and P50 sensory gating in schizophrenia: a proof-of-mechanism study. Neuropharmacology 64:197–204. https://doi.org/10.1016/j.neuropharm.2012.06.040

Xie W, Herbert B, Nguyen T, Gauss C, Tehim A (2008) U.S. Patent 7,429,664, 30 Sep 2008

Zhu PJ, Stewart RR, McIntosh JM, Weight FF (2005) Activation of nicotinic acetylcholine receptors increases the frequency of spontaneous GABAergic IPSCs in rat basolateral amygdala neurons. J Neurophysiol 94(5):3081–3091. https://doi.org/10.1152/jn.00974.2004

Zoltewicz JA, Prokai-Tatrai K, Bloom LB, Kem WR (1993) Long Range Transmission of Polar Effects in Cholinergic 3-Arylideneanabaseines. Conformations Calculated by Molecular Modelling. Heterocycles 35(1):171–180

Zwart R, Vijverberg HP (2000) Potentiation and inhibition of neuronal alpha4beta4 nicotinic acetylcholine receptors by choline. Eur J Pharmacol 393(1–3):209–214. https://doi.org/10.1016/s0014-2999(00)00002-9

Zwart R, De Filippi G, Broad LM, McPhie GI, Pearson KH, Baldwinson T, Sher E (2002) 5-Hydroxyindole potentiates human alpha 7 nicotinic receptor-mediated responses and enhances acetylcholine-induced glutamate release in cerebellar slices. Neuropharmacology 43(3):374–384. https://doi.org/10.1016/s0028-3908(02)00094-1

The manufacturer's authorised representative in the EU is Springer Nature Customer Service Centre GmbH, Europaplatz 3, 69115 Heidelberg, Germany. If you have any concerns regarding our products, please contact ProductSafety@springernature.com

Printed and bound by CPI Group (UK) Ltd, Croydon, CR0 4YY
22/09/2026
02221060-0007